Climate Crisis and the Democratic Prospect

Climate Crisis and the Democratic Prospect

Participatory Governance in Sustainable Communities

Frank Fischer

OXFORD
UNIVERSITY PRESS

OXFORD
UNIVERSITY PRESS
Great Clarendon Street, Oxford, OX2 6DP,
United Kingdom

Oxford University Press is a department of the University of Oxford.
It furthers the University's objective of excellence in research, scholarship,
and education by publishing worldwide. Oxford is a registered trade mark of
Oxford University Press in the UK and in certain other countries

First Edition published in 2017
Impression: 1

Published in the United States of America by Oxford University Press
198 Madison Avenue, New York, NY 10016, United States of America

British Library Cataloguing in Publication Data
Data available

Library of Congress Control Number: 2016959793

ISBN 978–0–19–959491–7 (hbk)
978–0–19–959492–4 (pbk)

Printed and bound by
CPI Group (UK) Ltd, Croydon, CR0 4YY

For my grandfather, Frank L. Fischer

Preface and Acknowledgments

Can contemporary democratic governments tackle climate crisis? Some say that democracy has to be a central part of a strategy to deal with climate change. Others say that it is not up to the challenge in the time frame available—that it will require a stronger hand, even a form of eco-authoritarianism. A question that does not lend itself to easy answers, this is the issue we seek to sort out and assess in these pages. While we come down on the side of an environmentally oriented democracy, establishing and sustaining its practices will not take place under the existing arrangements of a capitalist-dominated democratic state and its politics, described as the politics of unsustainability. Democratic governance during climate crisis, we argue here, will have to invent a new way forward.

The situation we find ourselves in—"the start of a global climate emergency"—presses for serious attention (Christoff 2016: 781). While we carry on with our regular activities, in particular those of uncontrolled consumerism, climate change and its worrisome impacts are regularly reported to be getting worse and faster than expected. We are, in short, running out of time to make the kinds of changes needed to avert a very serious climate crisis, even potential catastrophe. Even if talk of catastrophe turns out to be exaggerated, climate change can still result in serious upheavals leading to various states of emergency. Without doubt, measures will need to be introduced to deal with pressing emergencies—heat, flooding, hunger, migration, civil violence, and more. Still, under such circumstances, it is far from certain that contemporary political systems, including democratic political systems, will be able to adequately cope with these pressures.

We are thus approaching a stage of climate change in which the democratic prospects for the future look increasingly troublesome. Given the failures of governments to rise to the challenge so far, Lester Brown (2003), founder of the Worldwatch Institute, has suggested that we need a "Plan B." As a political-theoretical exercise devoted to the question of "what if?," this book can be seen as part of the search for such an alternative plan. But the search in this volume has a particular focus on the role of democracy.

While it is commonplace in mainstream environmental political theory to posit democracy as an essential component of a strategy to deal with climate change, others say that the failure of democratic politics to act decisively is itself the problem. While we side with the democratic theorists, we also believe that there is nothing all that obvious about the uses of democratic practices to resolve the problem ahead, even perhaps about democratic prospects generally, given the short amount of time remaining before we confront climate crisis. Not only is time short, but democracies already face "democratic deficits" and contemporary authoritarian trends grow ever stronger around the world. In short, democratic governments are already having trouble stepping up to the climate crisis challenge and there is little reason to believe that democracy will do any better if or when we start to face permanent states of emergency, leading perhaps to the so-called "life-boat" scenario.

Under such alarming circumstances the door will open wider for the eco-authoritarians. And they will have less and less trouble finding support from political and economic elites and worried publics. Casting their arguments in terms of emergency, perhaps even "survivalism," they will in all likelihood push for crash technology programs to be governed by technocratic decision-makers. Closely related would be calls for a strong police and military role to quell increasing civil unrest in the face of dire circumstances. This work seeks to address the issue by turning the question around. Instead of taking democracy to be the inevitable solution, it asks if and how democratic values and practices might survive the crisis. In so far as the conditions will be less and less conducive to democracy, much of contemporary environmental theory is speaking to a reality that is unlikely to exist.

This inquiry, then, seeks to move beyond dominant theoretical discussions focused on the democratic greening of the state or on deliberative environmental democracy—theoretical discussions addressed to realities that will not be available—and turns toward the search for an appropriate political-ecological strategy capable of preserving a measure of democratic governance. We need, in this view, to think strategically about what kinds of theory and practices might help those who will almost surely confront the social and political dislocations caused by dramatic climate disruptions.

After an examination of competing arguments for both environmental democracy and eco-authoritarianism, as well as the disappointing state of current democratic politics, the exploration identifies an alternative path in participatory environmental governance, a growing participatory relocalization movement, and a new environmentalism of everyday life (Schlosberg and Coles 2015). We find here dynamic and vibrant socio-ecological movements that not only speak directly to the crisis ahead, but are already well established and thriving on the ground. Although these movements are at present under the radar of contemporary political and social theory, and written off by many as irrelevant in an age of global environmentalism, local participatory environmental governance and the eco-local movement—in particular, intentional communities such as ecovillages, eco-communes, eco-neighborhoods, and local transition initiatives—already practice much of what academics generally preach, including deliberative participatory democracy in one form or another. This turn also leads to a "rediscovery" of a significant body of earlier environmental, social, and political theory that speaks directly to the ecological climate crisis ahead, especially when it comes to participatory governance and environmental democracy. The present work makes a case for bringing these long-neglected, even ignored theorists back into the discussion. Most importantly for the central question in this effort, these movements and their theorists offer insightful ideas about how to maintain authentic democratic values and practices. Not least important is the recognition that authentic democracy has to be anchored to local community life rather than the bureaucratic state. There are, of course, no guarantees. But this is, we argue, where democracy can make what may be its last best stand against an unfriendly future.

Turning to the local does not mean walking away from the struggle for global climate agreements. Nor does it mean ignoring the related need for eco-technological innovation. Rather it is to recognize that such political and technical efforts to respond to climate change are not working fast enough—and that even if they were, their effectiveness will still depend on a vibrant local citizenry, and local communities have always been a foundation of authentic democratic governance. The argument here is thus more pragmatic than idealistic. The relocalization movement and its eco-communes already provide important practical experiences and lessons that can be useful—perhaps in some cases even essential—in a time of serious ecological crisis. Their lessons can be passed along not only to people who flee uninhabitable cities, but also to those who have chosen to restructure city neighborhoods in the form of urban ecovillages or eco-neighborhoods.

With the help of these ideas and projects, the task is to shift the discourse of environmental political theory in ways that better prepare future generations for the hard times ahead. It is a call for an environmental political theory that is more strategically oriented to the contemporary political-ecological context and the looming circumstances of climate crisis. It should be a theory that can be of assistance to those who will face the climate crisis in its true magnitude in real terms. While we cannot deliver that theory in this book, we attempt to provoke a discussion in directions that can help to develop it.

Finally, many people have contributed to this work in various ways. I would thus like to acknowledge the fellow scholars and activists who have helped me along the way. Special thanks go to Alan Mandell, Doug Torgerson, Tim Luke, Piyapong Boossabong, and Hubertus Buchstein, for commenting on numerous chapters of the book. Thanks also go to colleagues and friends for commenting on particular chapters—Pheroze Wadia, Raul Lejano, Susan

Fainstein, Edgar Göll, Igofür Blühdorn, Holger Strassheim, Gabriela Kütting, Dieter Plehwe, Hemant Ojha, Hubert Heinelt, Robert Hoppe, Jens Newig, John Grin, Alex Demirovic, Selen Ercan, Ricardo Mondonca, Denis Smith, Tim Forsyth, Sonja Thiegles, and Leonardo Secchi. Further, I owe a special debt to Karin Litfin, Iris Kunze, Gabi Bott, Christian Strünke, Ulrika Schimmel, Lena Ferreira, Tiasa Matteo, Lois Arkin, Lara Morrison, May East, Maya Norton, and Rosana Boullosa for helping me understand the ecovillage and eco-neighborhood movements. My thanks go as well to Stefanie Burkhart, Piyapong Boossabong, and Dean Smith from whom I learned a great deal about the Transition Town movement. I am grateful to John Dryzek, Bob Jessop, Herbert Gottweis, Thomas Saretzki, Maarten Hajer, Robin Eckersley, Miranda Schreuers, Reiner Grundmann, Jan-Peter Voss, Navdeep Mathur, Martin Bierbaum, Shunsaku Komatsuzaki, Robert Schneider, and Lance Bennett, who were kind enough to discuss particular issues with me. I am indebted to Hemant Ojha for generously inviting me to Nepal, without which the Nepal chapter would not have been possible. I also thank Peter Feindt and James Meadowcroft who helpfully permitted me to explore some of these ideas at a European Consortium for Political Research workshop on "Greening the Leviathan?," as did Karin Bäckstrand and Eva Lövbrand at a conference on "Global Environmental Governance" in Linköping, Sweden. I also appreciatively benefitted from discussions of various chapters at conferences on "Politische und Epistemische Autorität" at the Wissenschaftszentrum, Berlin and on "Demokratische Transformation" at the Rosa-Luxemburg-Stiftung, Berlin, as well as seminars at New York University, the University of Tokyo, the University of Vienna, and the Free University of Berlin. My appreciation is also extended to the students and faculty members of the Technical University of Darmstadt and the University of Kassel in Germany, Mahasarakham University in Thailand, the University of Nagoya in Japan, the University of Strasbourg in France, and Rutgers University in the U.S. for comments and suggestions. And last but not least, thanks go to Dominic Byatt and Olivia Wells of Oxford University Press for steady support along the way, and to Sally Evans-Darby and Andrew Hawkey for their very helpful editorial assistance.

Frank Fischer
Humboldt Universität zu Berlin

November 2016

Contents

Introduction

The global warming climate crisis is the challenge of this century. Although one would not always know this from the daily newspapers or political campaigns speeches in the U.S. and elsewhere, climate change threatens to throw the world as we know it into social and political turmoil. While the public focus, for understandable reasons, has been mainly on the ecological consequences, the coming crisis has implications for all aspects of modern life, including future modes of governance (Dryzek, Norgaard, and Schlosberg 2011).[1] When environmental academics and activists discuss the topic of governance, they often maintain that democracy has to be an important part of climate change policy. More than a few even see a vigorous participatory democracy involving an active citizenry as an essential part of the solution. Some formulate this in terms of the concept of "environmental" or "ecological democracy,"[2] which refers to the idea or principle that citizens should have equal rights to meaningfully participate in environment decisions that affect their lives.[3] Others support environmental democracy with the concept of "ecological" or "environmental citizenship."[4] Indeed, this emphasis on environmental or ecological democracy and participatory environmental politics is for many taken as an article of faith.

In this Introduction, we explore these issues by first taking note of the potentially devastating consequences of the climate crisis not just for the global ecological system, but also for social and political systems. We then outline the failures of the dominant environmental approaches to adequately confront these very large challenges. Against this background, we take up the contention, central to much of environmental political theory, that more democracy is a necessary part of the solution.[5]

Without questioning the importance of democracy, we turn the issue around and ask instead to what degree democracy is likely to withstand the social and political turmoil that the ecological consequences of the climate crisis will bring about. Under less turbulent conditions democratic participation is surely an important component for sustainability. But these conditions in the future are unlikely to prevail during crisis. Thus, the ecological consequences of genuine climate crisis will just as likely—perhaps even more likely—bring to the fore calls for alternative forms of governance, including calls to replace democracy with expert-oriented modes of eco-technocratic authoritarianism. Indeed, many political and scientific leaders draw on a military analogy to argue that democracy has to be put on the shelf until ecological survival is secured (see Chapter 3). Democracy, as these leaders point out, does not always ensure that the right decisions will be made. In view of the seriousness of the situation ahead, they call for a form of environmental "guardianship" (Dahl 1989).

In search of an alternative theory and practice to preserve a democratic mode of governance in the face of threatening political-ecological disruptions, we turn to a reconsideration of the local level—where the consequences will take a heavy toll—and advocate the need for social and political experimentation. Toward this end, we take up consideration of the all-too-neglected relocalization movement that is rapidly growing around the world and draw out its connections to an earlier body of environmental thinking and practices, especially ecological self-help and local environmental democracy, that speaks to the critical challenge ahead. It is at the local level, we suggest, that the future of democratic governance not only has its best chances to play an essential role in the struggle for environmental sustainability, but even to take a firm stance against the enemies and opponents of democratic governance.

Environmental Democracy in Political Perspective

The tensions between calls for environmental democracy and eco-authoritarianism are scarcely new to environmental political thought. Where one stresses the central role of the citizen in environmental decision-making, the other calls for political and scientific elites. Both orientations have been present from the outset of modern environmentalism. The call for ecological citizenship and environmental democracy, however, has generally succeeded in marginalizing the voices of eco-authoritarianism. One reason for this is that eco-authoritarianism has been closely related to theories of "survivalism," which others—reformists in particular—have portrayed as too pessimistic. Just as participatory environmental politics was advanced to counter the technocratic eco-authoritarian perspective, ecological citizenship and environmental democracy have also been put forward as essential foundations of a sustainable way of life.

Today we confront a different situation. In view of the limited amount of time left to mitigate planetary warming, pessimism is making a comeback. As scientists offer additional evidence that shows the accelerating pace of the crisis, coupled with growing criticisms of democracy's failure to come to grips with these realities, more and more people are losing faith in the hope of avoiding tragedy. Although the argument here runs against much conventional thinking on the matter, namely a relatively complacent hope that we will somehow avert tragedy, we find that the time has come to take these worries seriously. Even if the solutions of the survivalists are unpalatable—even wrong-headed—their portrayal of the future might not be as inaccurate as we would prefer to think. With regard to the democratic prospect, one could even argue that the future of democratic governance itself will depend on our ability to meet the ecological challenge by imagining ways to construct and practice new forms of democracy.

While we take no issue here with the call for democracy put forward by environmental democratic theorists—indeed very much to the contrary—we point to a pressing need to give more attention to the relationship of their theories to the realities of both political power and the limited time frame now available for realizing such a challenging societal eco-transformation. In light of a crisis ahead, the effort in this work seeks to make a contribution to environmental political theory and environmental democracy in particular by stimulating a much-needed reexamination of this theoretical orientation and its assumptions. In reality, it is argued, the question is much more complicated than current discussions often suggest. Although the case for environmental democracy is morally and politically commendable, it is difficult to find people who believe that such democratic practices are anywhere near taking hold. The discussion here thus starts from the concern that when we consider the democratic prospect in light of the realities of climate crisis, it seems more likely that democratic values and practices will be under heavy stress, even in some quarters under outright attack. Indeed, these values and practices could be in serious jeopardy further down the road.

The problem may be even further compounded by the fact that many more people may well argue that democracy—at least in the form we have come to know it—has been itself part of the problem. In some circles the view is bound to give rise to noisy eco-authoritarian voices (taken up in Chapter 3). For them, democratic governments have failed to act in a timely fashion to forestall dire consequences, a situation that necessitates a stronger role for knowledgeable ecological experts, centralized forms of resources planning, and restrictions on certain economic and political freedoms. In many places around the globe, including the U.S. and the U.K., this critique will also lead to calls for a stronger role of the military—even to military rule in some places—to ensure social and political stability during increasing states of emergency. At the same time, accompanying these calls for action will be demagogic politicians invoking the politics of fear, with significant numbers of anxious people willing to follow them.

Even where such forces are successful, however, this will not give rise to some sort of ecologically minded post-political consensus that can resolve the problem—that is, a setting aside of politics to deal with the problem in technical terms. There will, in short, be no sudden end to the struggles of environmental and other social movements. Indeed, as the economic and social circumstances of the majority of the people of the world take a turn for the worse, many voices will get louder and some of them will be accompanied by increasing levels of violence. In response, one can imagine at the same time stronger cries from other quarters for tougher measures to maintain social order, perhaps for some at all costs.

For social and environmental movements the call will be first and foremost for environmental or climate justice; activists will still speak of democracy but it will take the backseat. While to say this is not to ignore a relationship between democracy and social justice, the inequalities and hardships that will become part of daily life will increasingly focus attention on human injustices, a point captured by the political slogan of a prominent social movement in Thailand: "give us democracy that we can eat." And as in Thailand, currently governed by a military regime, police and military will play a central role in the response to the demands of such movements.

This possible future suggests that we should set aside the improbable—some would surely say naïve—idea that environmental democracy will save the day and begin to think more about strategies to preserve some measure of democratic values and governance practices, even perhaps in anticipation of a future point in time when a new generation might be able to bring them back. As the suggestion of social and political demise will strike many as unlikely, we need to consider the question of political democracy against the proportions of the crisis heading our way, faster and faster. It is important to appreciate, as Naomi Klein (2014) has put it, this will "change everything." That is, whereas much of environmental theory and politics today proceeds on the basis of contemporary social and political assumptions, these will most likely no longer be operative in this different and more precarious world.

If this sounds pessimistic, there are plenty of reasons to rationally justify this pessimism. Before we focus more specifically on environmental democracy, we do well to set the topic in the context of the ecological crisis more generally. While we can clearly perceive a growing awareness of the threat, and increasing moves to deal with it, these efforts remain small against the immense proportions of the challenge. A main message, we argue here, can be stated simply: We are running short on time (Orr 2009). Although there is still probably a chance that the worst of the consequences might be mitigated, we can only ignore the trouble ahead at great risk. We need therefore to think more strategically about ways to adapt to these risky consequences, both ecologically and politically.

Climate Crisis: Why Are We Waiting?

The distinguished economist Nicolas Stern (2015) titled his recent book *Why Are We Waiting?* Indeed, there is today no question more important in environmental discourse. As we head for temperatures that will wreak a worrisome degree of havoc on modern advanced societies, and cause devastation for many developing and underdeveloped countries, the long list of consequences resulting from the failure to take dramatic action can boggle the mind. These include rising sea levels and receding shorelines; extreme weather that floods towns and cities; agricultural drought causing hunger and famine; unbearable temperatures that will make many places unlivable; mass climate migrations by desperate people seeking to escape widespread misery; migrating diseases giving rise to new and unknown public health hazards; and most important for the present discussion, an increase of social, civil, and political disorder resulting from these desperate human catastrophes, often violent in nature (taken up in detail in Chapter 1). Given that these alterations are seen to be accelerating at disturbing—and often unprecedented—rates measured against ecological changes over the past 10,000 years, it is difficult to overestimate the effects these socio-ecological changes will have on human societies (Barnett and Adger 2007: 640).

To be sure, we do not know for certain how disastrous the consequences will be, or how fast these changes will take place. Even if they have not yet fully registered in the consciousness of the average citizen, scientists tell us that climate change is already here, with many of its consequences already visible in a growing number of countries around the globe. The best estimates indicate that conditions will range from very uncomfortable to disastrous in different parts of the world.

As noted above, there is a strong possibility that future conditions in many countries will lead to increasing states of emergency to deal with the resulting social and political conflicts, many of them severe and potentially destabilizing. We can hope for the best, but everything we know so far suggests it is time to take a serious look at the political dimensions of our potential ecological future and consider how we might best respond. As Homer-Dixon (2006: 291) has put it, "we can't know exactly what breakdown will look like, and we don't know when it will happen, but we can start figuring out now how we'll respond." This will require, as he says, a "vigorous, wide-ranging, yet disciplined conversation among ourselves." And it is to the political dimensions of this conversation that this book hopes to contribute. How will democratic governance fare, and how might we prepare ourselves to deal with its further marginalization, if not decline?

One would think that the environmental facts before us would lead to concerted action. But so far this has not happened. It is not that nothing is being done. Little of it, however, rises to the challenge of the climate crisis we will be facing. A good example is the global climate conferences. Rather than taking decisive steps, world leaders have mainly procrastinated, despite one global environmental conference after another. Some will say that the recent Paris COP 21 conference changes the story. They can point to the lead role of the U.S. and China, including the signing of the Paris agreement. This, however, is more a hope than a foregone conclusion. Although nearly all of the 196 nations on Earth agreed to reduce their use of carbon-based fuels, the agreement is in fact weak both ecologically and politically.

The Limits of Global Climate Governance: Paris COP 21

The initial optimism accompanying the Paris agreement can partly be understood in terms of the long twenty-year history to arrive at it. Against this background, it was a political achievement. But after the clapping and handshaking, it was already clear that the agreement is very disappointing in terms of ecological solutions, if not a failure. Lacking concrete actions, many experts write it off as empty words (Taz 2015: 3; Milman 2015). The renowned climate researcher, James Hanson (Milman 2015), went so far as to describe it as a "half-baked" deceit and "a fraud." As he argued, "it's just worthless words. There is no action, just promises." This, to be sure, is harsh language, but there are plenty of others who share this view.

The agreement can be judged as a precondition for climate action, but it is in no way a sufficient response to the coming turbulence (Rees 2016). After a detailed assessment of the outcomes, largely seeking to emphasize the positive, Christoff (2016: 781) concludes that the agreement is only a "promissory note," whose "value remains unclear." For one thing, the agreement is mainly voluntary; there are no legal or binding enforcement mechanisms. Missing also are robust commitments on the part of the developed nations to assist the poorer countries, financially and economically. There is little in past experience to suggest that all of the pledges in the agreement will be fulfilled. The conference organizers themselves point to the need to ratchet up regularly the commitments to decarbonize. As it stands, climate scientists say that even if all of the pledges were fulfilled, the agreement would only bring us halfway to the goal of diverting serious consequences.

There is also the worry that this agreement will contribute to a form of symbolic politics. People will now say that finally something is being done, without really paying attention to the follow-through or the lack of it. This will make it easy for countries to backslide. One sign of this concern is that after the agreement was achieved, newspapers around the world returned to other topics, devoting dwindling attention to the global ecological challenge ahead.[6] Another indication of the weakness of the accord is the limited attention it has gotten in the 2016 presidential candidate debates in the U.S. (Krugman 2016). Given that the politics of the United States remains crucial to any future transition to a sustainable world, the fact that President Trump and the Republicans in the U.S. Congress mainly remain climate skeptics or outright deniers casts doubt on the future of the agreement.

Why then is this agreement so weak? For one thing, the effort to get nearly 200 nations to sign onto such a global agreement required substantial watering down of the pledged commitments. On a deeper level, all of these nations represent economies dependent on fossil fuels. The task of turning these economies around in the time available is huge and appears unlikely to many observers. Another reason why such a transformation is unlikely has to do with the fact that these economies are dominated by powerful carbon-based energy sectors with strong economic interests in maintaining the status quo (Mitchell 2013). Through massive lobbying activities, large financial campaign contributions, and well-funded think tanks that advocate their cause, these industries have managed to delay a transition from carbon-based fuels to renewable sources of energy.

There is yet a fundamental ideological dimension to the limited response. In addition, thanks to these same powerful economic leaders, the nations of the world and their leaders are imbued with a deep-seated belief in the primacy of the role of the market system, reflected in the dominant ideology of neoliberalism. Focused single-mindedly on individual freedoms—those of entrepreneurs in particular—and a belief in a need to limit the role of government, this creed impedes government environmental regulations, public subsidies for the innovation of green technologies, and the role of government guidance and planning of resource uses. Within such a context, the solution then becomes voluntary action, which is in fact the foundation of the Paris agreement. This belief system—"let the market do it"—serves as ideological cover for industries that prefer to maintain the existing arrangements in order to maintain company profit margins. To be sure, the market will do things when it is profitable; but transitioning away from the carbon economy is generally not profitable for the fossil fuel industry in the short run.

Oreskes and Conway (2014) also suggest that the empirically oriented positivist conception of science has contributed to the slow pace of change. Modern science, as the product of a particularly narrow empirical conception of the scientific method, focuses on ever smaller parts of the climate problem and is thus unable to speak about the problem as a whole. As such, the larger picture of the crisis headed in our direction tends to get lost. The idea that scientists should focus only narrowly on the empirical validity of their findings has led to a comparatively muted research community that accepts the notion that the larger climate questions in need of assessment are not only beyond their competences, but their responsibilities as well. Some scientists, however, take exception to the idea.[7]

Underlying this, as well, is the positivist principle of value-neutrality; that is, the claim that scientists should avoid statements about the normative implications of their findings for society more generally. Many climate scientists, in fact, go out of their way to avoid overstepping these boundaries and spend a great deal of time worrying about being attacked by the climate deniers and their think tanks, often funded by the fossil fuel industry itself.

It is this economic, political, and scientific configuration rather than technical know-how per se that impedes our ability to address this ecological crisis head on. To be fair, it is not that these forces have blocked all climate-related action; it is more a matter of the way they have watered down the actions that have been taken. Indeed, it is just this watered-down reformist approach that defines much in the Paris agreement. Anything but new, this reformist approach, often known as ecological modernization, has come to replace the earlier push for "sustainable development," advanced by the U.N. Brundland Report (World Commission on Environment and Development) in 1987.

To offset the more dramatic, even radical, societal changes called for by the concept of sustainable development, business and government leaders have put forth ecological modernization (now sometimes referred to as "ecological modernism") as a weaker alternative. As a programmatic expression of the neo-liberal approach to the environmental problems (Bernstein 2002), ecological modernization seeks to *reform*—rather than *transform*—the existing political-economic system into a sustainable form of green capitalism (Parr 2012; 2009). Toward this end, ecological modernization translates the otherwise radical climate-oriented call for societal change into business-friendly reform-oriented tasks, emphasizing the use of markets, cost-benefit analysis, green marketing, and the search for the technological fix (examined in Chapter 2). Indeed, ecological modernization is the perspective of most political and economic leaders today and has been adopted by many academics.

For the most part, ecological modernization is an effort to skirt around or set aside the socio-ecological challenge of sustainable development, namely to restructure the fundamental economic and societal processes responsible for the problem (taken up in Chapter 3). As the dominant environmental ideology in the corridors of power, ecological modernization is advanced as an effort to undercut the more transformative alternatives put forth by progressive environmentalists, including calls for public participation and environmental democracy (examined in Chapter 4). Such reformists often explicitly argue against attaching the ecological challenge to other social and political issues, in particular as they relate to matters of environmental justice. Advanced primarily as part of an elite agenda, it was not surprising that a major thrust to promote eco-modern technological solutions in Paris was led by Bill Gates.

With regard to technical solutions, many of the alternative technologies we need are by and large available. Rapid technological advances with renewable energy have been made, despite all of the efforts along the way to discredit them. While it is true that solar and wind energy still account for only a small percentage of energy usage, the story would be much different if renewable sources received anything near the kinds of subsidies that coal, gas, and nuclear energy have received, thanks to the fossil fuel lobby (Elliott 2016). One source of hope is a recent global agreement to help eliminate coolants that contribute to global warming, said to have a potentially greater impact than the Paris agreement (Davenport 2016b). Another is the growing interest of large pension funds in the United States and Europe in disinvesting in the fossil fuel industry and turning to greener investments in renewable energy (Corkery 2016).

However, even if the political-economic commitment was stronger, retooling the large economic systems of the world with clean energy poses a gigantic challenge. While there is movement to reduce the carbon emissions, a process that will perhaps even accelerate because of post-Paris initiatives, such reductions will not happen fast enough. The introduction of these new technologies, in short, will occur too slowly to avoid the crisis altogether. While these efforts can surely help to mitigate some of its worst effects, there is at this point little likelihood that many of the consequences of atmospheric warming can be avoided.

Equally problematic, ecological modernization and technological innovation do not attack the roots of the problems causing global warming. The underlying commitment to maintaining and expanding economic growth, even if a greener form of growth, is unlikely to bring the planet back into ecological balance. But if the strategy somehow could, it seems even less likely to adequately confront the problem of time. At best, one could argue, it appears to be a strategy to buy time. We can thus proceed on the assumption that we will not succeeded in dodging the crisis.

Environment Reform and the Social Sciences

Mainstream social science has both implicitly and explicitly been influenced by the same positivist influences that Oreskes and Conway (2014) single out for the natural sciences. One aspect of this influence is the still dominant epistemological understanding that facts and values can be separated (Fischer 1995). As such, mainstream social science tends to be removed from the dominant value-laden political controversies of the day by limiting itself to what is otherwise an overly narrow empirical conception of social and political inquiry.

Although the earlier environmental movement included a strong value-oriented call for societal transformation, what Porritt (1986) called a "radical overhaul," from the 1980s onward such an agenda of transformation was pushed aside by a more techno-managerial effort to achieve sustainability through public regulatory policies, market incentives, and subsidies. In the social sciences implicit acceptance of ecological modernization is, for example, easily recognized by the dominant empirical methodological emphases in environmental policy analysis on cost-benefit analysis, risk, and technology assessment. While one needs to acknowledge that the project has helped to mitigate—at least to some degree—many of the earlier environmental problems, such as automobile pollution, the waste reduction, and the ozone hole, climate change poses a different order of challenge. With the later arrival of global warming, the nations of the world find themselves confronted with a much more difficult order of crisis that the reform-based techno-managerial regulatory approach can address in only limited ways.

When these crisis conditions are acknowledged as potential concerns in contemporary social science, the proportions of the problem—especially the accumulation of interrelated environmental problems and their unknown, nonlinear synergistic effects—are generally lost to reform-oriented discussions that take for granted existing economic and socio-political arrangements. Leading writers in both political and academic spheres typically attempt to propose solutions that are essentially based on an extension of today's realities. How can the existing systems of capitalism, state bureaucracies, and policy agendas better respond to environmental problems, and so on? Often these discussions focus on finding ways to better "frame" the problem that might lead to consensus for concerted action (Koehane 2015). If we could just find the right way to frame the problem, so the argument goes, the public and its politicians would suddenly recognize it and spring into action. In these discussions, however, there is far too little recognition of or attention paid to the fact that current conditions will no longer exist. There is far too little projection forward to the kinds of conditions that will prevail under the looming crises. Dealing with the new circumstances will require very different modes of thinking—that is, "thinking outside the box."

One of the most critical elements missing from environmental political and policy analysis is the issue of time. Absent from most political analysis generally is the crucial role time plays in decision-making, a topic extremely important for the issue at hand (Strassheim and Ulbricht 2015). Although the element of time is not new to politics, it has scarcely received adequate attention in either policy studies or political theory (Pollitt 2008). This is especially problematic for the topic here as deadlines and short time frames militate against the more deliberative, time-consuming practices of democratic decision-making (Zahariadis 2015: 129). The relatively limited amount of time available before we might expect climate emergencies to become commonplace does not appear to bode well for the future of democratic practices. While the role of time as a theoretical issue is beyond the scope of this work, we nonetheless invoke it along the way (in the everyday common-sense understanding of time).

Democratic Theory and the Environment: The Question of Political Relevance

Within the social sciences our concern here is mainly with the current state of democratic theory and its relevance to the climate confrontations ahead. For the most part, much of this theory has today neglected the eco-political realities portended by climate change and thus given insufficient attention to these contemporary real-world developments. By and large, the theoretical emphasis is on broad concepts such as freedom, democracy, and justice, particularly as discussed by the classic theorists. With regard to democracy, much of the discussion is dominated by the normative prescriptions of deliberative democracy, a theoretical orientation generally quite removed from the realities of political struggle (discussed in Chapter 4). Theorists, as Isaac (1995) has argued, have been far too reluctant to analyze contemporary political events. Referring to what he calls "the strange silence of political theory," he asks if we could imagine classical political philosophers such as Locke, Paine, or Marx ignoring the burning issues of their times. Largely missing to date are reflections on how contemporary democratic struggles already need a different kind of democratic theory, a point that will surely be even more relevant to future political thinkers confronting the climate crisis. What we require, from this view, is a theory that is rooted more concretely in political-ecological events. It would be a worldly political theory that can flexibly alter its explanations and prescriptions in the context of actual political strivings. In this regard, as Schwartzmantel (2012) argues, political theorists should develop conceptual frameworks that help to enlighten us about the political conflicts and struggles that are taking place and, in the process, identify particular problems, explore barriers blocking particular political movements, examine the implications of particular events, and put forward substantial ideas and analyses that speak to the political turmoil of the times. As he puts it, the question then is: "Whether from practice to theory, or from theory to practice, how might (or how should) democratic theory be developed in the light of current events throughout the world?" (2012: n.p.).

As democratic governance will have to defend against unfriendly or even hostile forces, its supporters would benefit from theoretical guidance for the struggle. Of particular importance might be the insights gained from insurgent democratic movements that have confronted authoritarian forces (see Chapter 11). Toward this end, we need further investigation into how the theory and practices of insurgent democracy could be extended to climate change (Brecher 2016). We also need to better understand how the theoretical issues raised can be related to an effective agenda of political contestation. This suggests, as Schwartzmantel observes, a need for better understandings of the relationship of popular insurgent democracy to established institutions such as representative bodies and political parties that rely on professional politicians.

Moving beyond interesting intellectual exercises, too often lacking relevance for present-day challenges, a relevant political theory will have to find opportunities in this new world that capture the minds and imaginations of contemporary democratic activists with practical interests in taking on and resolving the pressing political issues of the day (Schwartzmantel 2012). As de Tocqueville (2000: 7) wrote in the introduction to *Democracy in America*, "A new political science is needed for a world itself quite new."

It is striking at present to notice how little effort there is to engage this challenge. It is thus important to take up this concern. The history of the social sciences shows that during times of social and political turmoil, social scientists often come under pressure from the world outside of the academy to engage the burning issues of the time. During such periods students and critics typically accuse an academically preoccupied professoriate of lacking social or political relevance. Insofar as the pressing social and political questions—hunger, migration, and civil violence, among others—brought forward by climate change fail to receive sufficient attention, it is quite likely that strident voices concerned with social relevance will again return.

In this work we seek to address the question of relevance by returning to an earlier but largely neglected tradition in social and political thought, namely attention to the "human scale" (the focus of Chapter 10). This human dimension at the local level has always been a critical component of environmental theory, especially prominent at the outset of the environmental movement. Eckersley (1987: 19–22) captured the thrust of the earlier radical localism in a call for a transformational "ecopraxis" built around structures "that foster the development of self-help, community responsibility and free activity." Such "person-in-community" structures, she argued, need to be "consistent with the ecotopian ideals of a loose federation of regions and communes." A perspective largely neglected in the age of globalization, we argue that in the face of potentially dire socio-ecological consequences ahead, it is time to return to a reconsideration of such structures and practices.

Return to the Local: Participatory Environmental Governance

With the foregoing discussion as context, the chapters turn to the topic at hand, namely the future of democratic environmental politics and the ideal of environmental democracy. Given the limited amount of time available for getting global warming under manageable control, coupled with the unlikelihood that democratic governance will suddenly bloom around the world, it is time to begin rethinking eco-political strategies for confronting the ecological hardships that many people will confront in the not-all-that-distant future and the democratic malaise that will accompany them. Insofar as the ecological systems of the world will radically change, it is seen here as time for radical solutions.

In search of an alternative political-ecological orientation that would better meet the challenge, a sharp move away from much of conventional environmental thought in the social and political sciences is required. In a turn to the local we need not—cannot—abandon the global level, despite the failure of global environmental agreements. But, as protesters in Paris pointed out, we can no longer rely on global agreements or national strategies. Without losing sight of the fact that powerful transnational forces are responsible for much of the crisis, we need to shift our orientation as much or more to the sorts of things that people will need to do in their efforts to adapt to the consequences of the climate crisis, or what Schlosberg and Coles (2015) refer to as a "new environmentalism of everyday life." We need, as such, to look more carefully at the things around us, things closer to us, and the possibilities they might offer.

As de Young and Princen (2012) point out, there is no one single model that will bring about a more sustainable transition. Many efforts have something to contribute and the time has come to approach these various projects experimentally, collecting the better elements and discarding those that seem less workable. Given the combination of uncertainty and danger that comes with climate crisis, we can best take the advice of the renowned ecologist C. S. Holling (2004), who has argued that "the only way to approach such a period, in which uncertainty is very large and one cannot predict what the future holds, is not to predict but to experiment and act inventively and exuberantly via diverse adventures in living."

Since environmental problems always have consequences—and often causes—at the local level of society, a good place for this experimentation is the local community. Indeed, such experiments are already underway, even if largely under the radar of mainstream political and social science. Of particular importance is the relocalization movement, explored in the second half of the book, where we look as well at the transition town and ecovillage movements.

There are two basic reasons for turning to the local level. One is that the local level has always been seen as essential for an effective sustainability strategy for reducing the human footprint. Citizens, in short, have to be engaged in the effort. Although much rhetoric has been devoted to this point, it has largely been neglected or marginalized in much of mainstream environmental practice. Relocalization and eco-localism generally seeks to carry through on this essential commitment.

The second is related to politics and democratic politics in particular. Local face-to-face relations have long been recognized as the foundation of authentic democratic self-governance, a point lost to large-scale mass democracy (Chapter 10). Toward this end, the vibrant and often imaginative relocation movement offers a venue not only for preserving democratic values but, even more specifically, for advancing the participatory deliberative practices of environmental democracy.

To better grasp the ways in which the movement can supply a home for the deliberative politics of democratic environmental governance, we first consider the conditions that would be required for the realization of deliberative democracy more generally. Without denigrating the importance of these efforts, one recognizes rather quickly that deliberative democracy remains too theoretical and abstract when it comes to the crisis at hand (examined in Chapter 4). It also offers no strategy for governance.

Further, it is important to face the fact that we do not have flowering democracies around the world. Indeed, countries classified as democratic are widely seen to suffer from a "democratic deficit," with others moving away from democracy altogether. Many writers in fact speak of a "crisis of democracy," while others speak of "post-democracy" and "simulative democracy" (discussed in Chapter 5). It is thus reasonable to assume that the time between now and the emergencies of the climate crisis is not going to be long enough to turn this situation around. Indeed, there is plenty of reason to believe that the political winds will blow in a different direction. Arguments in the environmental literature that democracy is not up to the challenge date back to the beginning of modern environmentalism and they have returned. Today, as then, democracy's critics often foresee the need for a form of eco-authoritarianism, as citizens, they believe, will not freely and knowledgeably support the necessary ecological changes.

Over the past several decades these eco-authoritarian views have largely been written off as misbegotten. But they do raise questions that need attention. A primary theme throughout this literature, then as well as today, is a call for a greater role by the experts, including forms of expertocracy or technocratic governance (see Chapter 2). Indeed, it is not an argument without support, especially in more powerful circles, at least in its more benevolent forms. And it is likely to gain more supporters as the consequences of the crisis become more visible.

As Dobson (2003: 7) has pointed out, much of the work in environmental democracy is designed to counter such eco-authoritarian arguments. But it is very unlikely that deliberative or discursive environmental democracy is going to win the day when the going gets rough. We thus argue that instead of focusing on parliaments, bureaucracies, societal deliberative systems, or global politics, deliberative theorists should refocus more on the local level. Toward this end, we suggest that a turn to a variant of deliberative politics referred to here as participatory governance offers a more contextually grounded way to advance the principles and values of deliberative environmental democracy. Not only does it relate more specifically to the larger tasks of governing, it captures and embraces important participatory democratic experiments. Toward this end, participatory governance is identified with concrete illustrations such as participatory budgeting in Brazil, people's planning in India, community forestry in Nepal (taken up in Chapter 7), or the participatory local governance practices of Khon Kean municipality in Thailand (Fischer and Boossabong 2016). While these projects incorporate insights from deliberative democracy, they also move out of the realm of theory into the practical world of political struggles. Not only do these efforts enrich deliberative theory, they are more relevant to the democratic challenges ahead, particularly as they pertain both to the theory and the potential practices of environmental democracy.

Indeed, these local endeavors pertain to the larger and growing relocalization movement. Despite the fact that this movement has many insights to offer the social sciences, it has largely escaped notice. What we discover upon closer investigation is a thriving movement laboring to make good on the green call to "think globally but act locally." Of particular importance, in this regard, are the transition town, ecovillage, and eco-neighborhood movements spreading around the world (topics discussed in Chapters 8 and 9).

There are three main reasons for shifting attention to these efforts. One has to do with the fact they involve active citizens across the globe who want to practice what they preach; the people in these movements—ecological citizens—are all dedicated to reducing their ecological footprint. Second, such efforts may well become more important as the social logic of the crisis changes. As Arendt documented (1965: 260), during periods of severe duress or crisis throughout history, large numbers of people have found it necessary to flee the cities (with a country like Greece serving as a contemporary illustration). If this proves to be the case during the climate crisis, these local projects can provide important forms of learning—or what Hopkins refers to as the "Great Reskilling"—that can speak to the realities such climate migrants will confront, growing their own food being a primary example (discussed in Chapter 10). As Daly (1977: 170) has argued, it is quite "unrealistic" to expect people to choose "simplicity and frugality except under ecological stress"—in other words, in the face of a new and different social logic ushered in by climate stress.

Third, and most important for this discussion, relocalization can provide the ground for genuine participatory democracy. Not only has participatory democracy always been associated with small groups, but locally based decision processes are widely recognized as facilitating a larger degree of citizen participation. Greens have always seen a shift to decentralization as a way to both foster greater environmental consciousness and reduce the ecological impact (Connelly and Smith 2003: 327–32). Decentralized grassroots democracy has indeed long been considered a requirement for achieving sustainability.

Once we explore this shift in focus, we discover something of a lost world of environmental political thought that speaks directly to this return to localization. The earlier environmentalism of writers such as Bookchin (1982), Sale (1980), and Bahro (1987) theoretically worked out political foundations for such a move (to which we turn in Chapter 10). Their ideas, however, were too radical for the times. Moreover, as environmental politics turned its attention to the global level, theories of decentralized localism—radical or otherwise—tended to be written off as irrelevant to the new world of globalization. The environmental problem could, in this view, be best dealt with from the top down, in practice if not theory.

Now, however, we see the limitations of this global approach. While we cannot suddenly ignore the global level, we need a much more diverse and multidimensional approach that includes a stronger recognition of the essential role of the local. But this has to be more than a recognition of the place of the local in the global—that is, as an instrumental lower-level adjunct to the larger global system. It requires an understanding and acceptance of the local on its own terms. Whereas we can empirically find the practical basis of this understanding in eco-localism and the relocalization movement, we can recover its theoretical expression in the works of the earlier environmental theorists. We conclude the book by relating these otherwise radical ideas to a number of the broader mainstream themes in contemporary political and democratic theory. Toward this end, we suggest several conceptual bridges that can connect contemporary political theory to the challenges of future climate-induced socio-ecological developments. In addition to participatory environmental governance, these are (as we shall see in Chapter 11) the theory of associative democracy, the politics of insurgent democratic struggles, and the practices of democratization of policy expertise.

It is not that relocalization and the political writings of Sale, Bookchin, and Bahro provide *the* answer to the democratic prospect per se. Rather the argument here is that they provide useful starting points for beginning to rethink political theory for a time that lies ahead. In addition to ideas that can be connected to concerns of political theory more generally, they suggest ways that further practical experimentation might take place.

Before closing this introductory discussion, one final observation seems in order. It is more than a bit unusual for an author to write a book based on events that he or she hopes will not take place, or propositions that may prove to be false. One can indeed hope that the climate change somehow fails to materialize as a major crisis for life as we know it. But, as we have argued, it seems increasingly important to confront this risk. To neglect the potential implications of climate crisis for democratic governance would be a mistake. What is more, even if the concern turns out to be overblown, the path proposed here would nonetheless be a significant step for renewing democratic prospects. Independent of the climate crisis and its potential ramifications throughout society, recovering and honoring participation at the local level can only help to revitalize democratic governance in an age demarked by various forms of post-democratic politics.

Notes

1. Different writers use different terms for the climate challenge. Some speak of global warming; some refer to climate disruption, climate destabilization, and climate collapse, among others. All of these descriptions fit the central concern of this book, but we prefer principally to employ "climate crisis," as this term not only captures the changes to the climate but also points to the nature of the political-ecological dynamic underlying the problem. See the discussion of crisis in Chapter 1.
2. A variety of terms have emerged to conceptualize the relationship between environmental or ecological systems and democracy. Most have generally spoken of environmental democracy, but others employ ecological democracy, eco-democracy, and green democracy. Many writers use these terms interchangeably, but others have sought to draw out distinctions among them. In particular, Dobson (1990) has argued that the emphasis on ecology, or what he calls "ecologism," captures the need for a deeper understanding of green politics than the concept of environment. Part of this deeper understanding is often seen to connote a recognition of a role for non-human as well as human participation in ecological systems, a theoretical orientation yet to be fully developed and established. However, others employing environmental democracy also sometimes consider the nonhuman dimension (Mason 1999). In any case, the difference between these two concepts is not obviously captured through a semantic conceptual change. As both Torgerson (1999) and Luke (2009) argue, the differences among these conceptualizations remain at best matters for theoretical debate, as ecologism is itself not without problems. What we can agree upon here is that achieving sustainable social systems requires radical changes in both our social and political ways of life, especially as they pertain to

our interactions with the natural world, changes that move far beyond techno-managerial environmental reforms. But we tend in this work to stay with the more traditional terminology and refer to the call for more fundamental changes as "radical environmentalism." Environmentalism has always focused on the interaction of humankind with the natural environment or the ecological system; it is as such concerned with the social and political dimensions of the environmental crisis. Environmental democracy, as a radical practice of environmentalism, is taken here to refer to a political process designed to restore and maintain the life-sustaining capacities of ecological systems.

3. Writers use somewhat different terms to define environmental democracy, but all of the definitions involve the belief that citizens affected by environmental concerns should have equal rights in participating in the environmental policy decision processes (Worker and Ratté 2014). Hazen captures this in these words: it reflects the "recognition that environmental issues must be addressed by all those affected by their outcome, not just by governments and industrial sectors. It captures the principle of equal rights for all those in the environment debate—including the public, community groups, advocates, industrial leaders, workers, governments, academics and health care professionals. For those whose daily lives reflect the quality of their environment, participation in environmental decision-making is as important as in education, health care, finance and government" (Hazen 1997).
4. Environmental or ecological citizenship refers to the idea that each citizen is a component of the larger ecological system and that, as such, a healthy and stable environment requires each citizen to act responsibly toward the ecosystem as a whole, both local and global.
5. For a good picture of the field of environmental political theory, see Gabrielson, Hall, Meyer, and Schlosberg (2016). Others use the term "green political theory" to refer to the same field of inquiry.
6. For an example of such retreat, see Innis (2016) on the financial cuts of the Australian Climate Center and its redirection to focus on the commercial benefits of its research.
7. "Scientists are trained to be objective," Hansen told *The Guardian*. But, he says, "I don't think we should be prevented from talking about the implications of science." He explained that his former employer, NASA, "appointed a media overseer who vetted what he said to the press. They held practice press conferences where any suggestion that fossil fuels be reduced was considered political and unscientific, and therefore should not be uttered" (quoted in Milman 2015).

Part I

Climate Change, Crisis and the Future of Democracy: Setting the Stage

1

Ecological Crisis and Climate Change: From States of Emergency to "Fortress World"?

This first chapter looks at the possible futures associated with climate change, in particular the ecological crisis it will bring for many people on the planet. It does this with a special focus on the political challenges that will accompany this crisis, especially as they relate to democratic politics. It asks how states will manage, and in some cases even survive, in the face of a very serious or catastrophic social-ecological crisis. The ecological consequences of climate change are now widely discussed and the outlines of some of them are already visible. Indeed, climate change has already begun. The long list of consequences includes rising oceans leading to receding shorelines; flooded towns and cities; agricultural drought causing hunger and famine; unbearable temperatures making many places unlivable; mass migrations by desperate people seeking to escape widespread misery; the ecological destruction of life-sustaining oceans and the death of coral reefs; migrating diseases giving rise to new and unknown public health problems; and most important for the present discussion, an increase of social, civil, and political disorder, often violent in nature. Given that these changes are seen to be accelerating at disturbing—and often unprecedented—rates of change measured against ecological realities over the past 10,000 years, it is difficult to overestimate the effects of these socio-ecological alterations on human societies (Barnett and Adger 2007: 640).

To be sure, we do not know the degree to which all of this will happen or how fast. But best estimates indicate that conditions will range from very uncomfortable to disastrous in different parts of the world. We can in fact already see such disasters in a growing number of countries around the globe. In many places, these future conditions will surely lead to increasing states of emergency that will usher in difficult political conflicts, many of them severe and destabilizing (Kunstler 2006). The responses will differ, but in many countries strong political leaders supported by anxious or desperate publics will turn to authoritarian forms of rule, perhaps something like Lasswell's (1941) "Garrison State." A phenomenon that can already be identified in various parts of the world, it is likely to spread to others.

Much of the world goes on with little apparent worry about climate change—that is, as if the future can still be understood, to one degree or another, in terms of today's social and political realities. An important exception to this are the militaries around the world, the United States Pentagon in particular (Webb 2007). Indeed, the one place where these crisis potentials are receiving very serious attention is in the military-sponsored research field of environmental security. While more authoritarian forms of governance need not be inevitable, in the face of the currently low levels of preparation for these coming environmental conditions around the world, coupled with the weak performance of liberal democratic regimes, it does not seem wrong to worry that various forms of eco-authoritarianism will be on the rise. As careful consideration of the political implications of these extreme conditions are relatively secondary, often even marginal, in contemporary discussions of environmental political theorists and most social scientists generally, whether empirically or normatively oriented, the idea of beginning to think more seriously about them certainly cannot be misplaced. While we don't know exactly what the crisis will look like, or just when it will occur, it is not too early to begin thinking about how we should respond (Homer-Dixon 2006: 291).

Environmentalism: The Politics of Crisis

Crisis has always been part of environmental politics. Over the past fifty years, as the planet has steadily confronted an increasing degree of ecological degradation, the focus has been on one crisis after another, with the primary question turning on the degree of crisis. Given that the concept of "crisis" is central to the discussion at hand, it is important to be clear on the nature of the challenge posed by a daunting ecological crisis.

Defining crisis, however, can be as difficult as it is important. It is a concept with a long history but no simple definition (Starn 1971). In general, crisis refers to a situation in which a complex, highly uncertain, and often unexpected system starts to function or operate poorly and the causes of the malfunctioning are not well known. It is a situation in which a decision or course of action to stop further decline is necessary. As in the case of environmental crisis, the causes are often subject to multiple interpretations, a point basic to the political struggle over environmental policy, climate change policy in particular. The causes can come from sources external to the system, such as the changing atmosphere, or they can arise from the inability of the internal institutions and decision-mechanisms to effectively develop a strategy to mitigate the problem. Because the causes—external, internal, or both—can be numerous and unknown, crisis does not lend itself to rationally structured problem-solving. A crisis typically means that the established routines of a system—e.g., societal system, economy, family, or polity in this case—have stopped working normally and thus create a threat to the goals and priorities of the institutional arrangements. Crisis, as such, often portends a turning point in the face of a dark and foreboding time; if a crisis cannot be managed and/or the old system defies change within a specific period of time, the result is disaster, even potential catastrophe (Jessop 2015). It is in this sense that we refer to climate crisis, resulting from both external and internal causes.

Such a conception of crisis, as imposing a dark future, has been basic to environmental politics from the outset (Gail 2016). With regard to contemporary environmentalism, crisis is usually traced back to Rachel Carson's (1962) book, *Silent Spring*. Long associated with the origins of the modern environmental movement, her book widely publicized the presence of toxins in our food supply and their serious implications for human health. Since then, environmental problems—both concerns about pollution and resource scarcities—have grown in number and size. They range from the early stages of environmentalism—focused on air and water pollution—to more catastrophic disasters like the Chernobyl nuclear tragedy followed later by the explosion at Fukushima, major oil spills such as the case of Exxon Valdez in Alaska and BP's Deepwater Horizon in the Gulf Coast, increasing deforestation, the rapid loss of biodiversity, and now climate change, just to name the most obvious. These crises, especially occurring together, are seen by many to threaten our very way of life in numerous ways.

The underlying origins of our ecological problem are clear enough. They are the consequence of industrial stress on nature's capacity to supply for and carry up to 9 billion people (the projected population by 2050) on a scale far beyond anything previously experienced in human history. As such, these realities raise the question as to whether life on Earth can sustain itself as we know it in the face of such assaults, especially when the implications of global warming are added to the picture. In short, the growing number of environmental challenges has become progressively more dangerous. Altogether, they clearly constitute a very serious ecological crisis.

Understanding the complexities of environmental degradation and finding policy strategies to deal with them are primary challenges of the twenty-first century, many say *the* primary challenge (Gore 1992; Catton 1982; Stern 2006; 2015). Although recognition of environmental problems did not come without struggle, they have clearly made it onto the political agendas of countries around the globe. Moreover, progress has been achieved on many fronts; the air is cleaner in lots of places, rivers have been cleaned up, greater care is taken with toxic wastes, and more. But even here, to use a standard metaphor, the "glass is only half full." Further, the most threatening of the environmental problems—climate change—presents an unprecedented challenge much more worrisome in its consequences than the earlier environmental problems. Many ask if environmental rescue is still possible. Some say that it is no longer entirely possible.

The question of how severe the consequences will be remains open. Toward this end, a number of ecologists and environmental writers have sought to scope out various scenarios to estimate these consequences. These range from regular disruptions to the worst of all possible consequences. Many economic and political leaders have focused on the possibility of reforming the current political-economic system, often described as a process of "ecological modernization." Others have advanced more fundamental alternative paradigms such as "sustainable development" and "degrowth," which in their strongest versions call for a retreat from capitalism as we know it and the adoption of a more modest way of life organized around a different set of values, emphasizing greater respect for nature through an ecological balance between production and needs (Kallis 2015). But none of these perspectives, at this point, offer a great deal of hope for turning the situation around in the time frame now available. They can surely help mitigate some of the effects, but there is no longer much chance that they will be able to stop altogether an irreversible ecological decline and the social turmoil that will accompany it. We are, in short, simply running out of time to forestall very problematic, even dire, consequences. Indeed, the predictions related to climate change appear regularly to get more and more pessimistic.

Climate Change: Crisis Par Excellence

Climate change, emerging in the 1990s as a major issue for the political agenda, requires special attention for all of the reasons noted above. It raises environmental crisis to a new and unprecedented scale. Not only is it closely related to most of the other environmental problems, it seriously exacerbates many of them—such as air and water pollution, loss of biodiversity, deforestation, soil erosion, and more. Indeed, it has rapidly become nothing less than the greatest challenge of our time. As Klein (2014) has put it, "this changes everything."

Climate change needs little introduction, as it is widely discussed in the public media, as well as in countless books and at environmental conferences. It is, however, a topic that generates a good deal of controversy (Hulme 2009). Although a small number deny the human impact on the changing climate, the vast majority of the scientific community—at least 97 percent of scientists—support the view that humankind has been significantly contributing to global warming (Oreskes and Conway 2010). Given that this work is based on the speculative question of "what if?," we do not need to address this debate. But we also assume that it is better to side with the dominant scientific consensus on the grounds that doing nothing and being wrong about the sources of the changing weather would most likely have disastrous consequences.

Although climate change is a very complex phenomenon, far too complex for the discussion here, its basic contours can be easily summarized (Archer and Rahmstorf 2010). The main question is not whether the climate is changing, as this is an established fact (Cullen 2016). As is well known, climate change is the product of an alteration in the flow of energy coming from the sun. Human life depends on the sun's warm rays passing through the atmosphere to heat the Earth's surface, but it also depends on a certain percentage of that warmth being reflected back into space to keep the Earth's temperature in balance. The heat that is reflected upward can get blocked from leaving the Earth's atmosphere by the growing concentration of carbon dioxide, methane, and nitrous oxide that forms a barrier in the upper atmosphere. All of these greenhouse gases ascend to form a layer above the Earth that traps the heat from exiting, creating the "greenhouse effect." Not only is this layer scarcely new, its presence is necessary to regulate the Earth's temperature. Indeed it has long held the Earth's average temperature at 15°C, warm enough to sustain life on the planet. In the absence of these gases the average temperature would drop to about −18°C, a temperature too cold for most forms of life. Today, however, as the increase of greenhouse gases builds, less heat escapes and the planet threatens to warm up to dangerous levels. Atmospheric carbon dioxide now is 30 percent higher than it was in pre-industrial times and is higher than it has been for at least 420,000 years (Archer and Rahmstorf 2010).

The scientific evidence shows that the Earth continues to warm significantly. Some of the warming is the result of historical alterations in climate and is thus unavoidable. But scientists nearly unanimously believe that human activities such as the burning of fossil fuels—coal, oil, gas—and the clearing of forests are significant causes of this additional warming, particularly as they are generated by various industrial, transportation, and agricultural activities. Drawing on various sources, including sophisticated computer projections and measures taken from the frozen Arctic tundra, the predictions are quite worrisome. An increasing number of studies now find that the Earth might heat up by 6°C (9°F) around the end of this century, a temperature that would have devastating consequences for life on Earth, with some areas being more adversely affected than others. More recent studies have worried about the possibility of 10°F, considered unlikely but possible if nothing is done to mitigate the increases (Gillis 2014).

Essentially, this steadily increasing change in the history of the world's temperature is dated back to the Industrial Revolution. It is the result of the dramatic increase in economic and industrial production that the Industrial Revolution unleashed in the mid-nineteenth century and which is still moving forward today at even faster speeds. This development accelerated rapidly after World War II in the advanced industrial countries, followed by the emergence of newly developing countries adopting similar industrial practices—India, China, and Brazil being leading examples. Since then energy consumption, especially coal for electricity, has expanded dramatically. Today it is said that our common future rests in many ways in the hands of these developing countries and their accelerating pace of economic development, further complicated by high levels of population growth. But not only in their hands; a lot will now depend as well on the Trump Administration and its threats to withdraw from the effort to reduce greenhouse gases.

Further, it is not just the rise in carbon dioxide over the past fifty some years that we have to worry about. Also problematic is methane, which is around thirty times more dangerous than CO_2 as a greenhouse gas (Krupp 2014). Released from animals, the melting permafrost in the Arctic regions, and the process of fracking to obtain oil and gas from shale rock underground, methane has been entering the atmosphere at increasing rates. This is thanks in significant part to the raising of animals to meet a growing demand for meat consumption around the world.

Nineteen of the twenty hottest years on record have occurred since 1980, the top ten since 1990, with 2015 and 2016 successively recorded as the warmest years in the 135 years of temperature recording (Mathiesen 2015). We are already starting to see the effects of these increased temperatures in various ways—glaciers are melting (Filkins 2016), plants and animals are being forced from their habitat (Lai et al. 2014), and the number of severe storms, heavy flooding, and serious droughts is increasing (MET Office 2014). A recent Intergovernmental Panel on Climate Change (IPCC) report, in fact, announced that climate change is now here; the only question is how much worse it will get (Gillis 2014).

If current rates of climate change continue, we can expect catastrophic consequences. These would include the increasing frequency of extreme weather, including hurricanes, cyclones, and tropical storms, all of which will become more and more intense. Global seas are predicted to rise by up to 6 meters or more, destroying coastal regions around the world, affecting many millions of coastal dwellers. The Arctic Ocean is predicted to be ice free in the summer by the year 2050. The world's water management, agriculture, and forestry sectors will require additional annual costs in the tens of billions of dollars. Many of the world's coral reefs are expected to be destroyed by 2050, devastating the underwater ecology upon which many essential forms of sea life depend, as well as about 275 million humans.[1]

Hundreds of millions of additional people will be at risk from malaria as it moves northward to warmer climates and billions will suffer from water shortages. Heat waves will occur more often and become more severe, causing an increase in deaths, especially among the elderly. Water shortages and devastating wildfires will become more common, as already witnessed in places such as California. More than a million species worldwide could be driven to extinction by 2050, as part of what is described as the beginning of the "sixth mass distinction" in the history of the planet.

Even if we were tomorrow to undertake dramatic efforts to mitigate the increase of greenhouse gases, the Earth would still experience global warming, though most likely at more manageable levels. Some of this, as noted, is the result of historical climate shifts. But the added warming today due to the greenhouse effect is mainly the result of emissions from earlier centuries with the effects of current emissions yet to manifest themselves.

It would be an understatement to say that this picture does not look good. Despite efforts since 1972 to take action to ameliorate the coming crisis, not nearly enough has been done at the international level. Since then international conference after conference has mainly ended with promises to keep working on the attempt to come up with some sort of meaningful global agreement to reduce greenhouse gases. But it is widely agreed that efforts to date have been disappointing failures, Paris COP 21 being at best only a partial exception (Death 2013). Neither have they been able to establish mandatory emissions targets or the kind of crash technological research program needed to develop decarboned energy alternatives. In some ways there have been more meaningful actions at national, regional, and local levels than in the global realm. Here too, though, there is nothing that addresses the size and scope of climate change. Indeed, one of the main responses at all levels has been to shift from the attempt to mitigate carbon emissions to strategies to adapt to the consequences. In other words, serious consequences seem to have more or less come to be accepted and the goal is now how to protect against the worst forms of damage.

Also closely related to this ecological crisis is the problem of scarcity. Increasingly, we are running out of the natural resources basic to the functioning of contemporary industrial systems and climate change exacerbates the loss. In his book, *Peak Everything: Waking up to the Century of Declines*, Heinberg (2007) presents a very disturbing picture of the ecological limits we are running up against. Beyond the widely discussed questions concerning the long-run decline of available oil, the list includes grain production, fresh water, wild fish harvests, arable land, and the extraction of particular minerals and metals, such as uranium production.

These ominous changes, even if they should be only partly accurate, portend a very different world ahead.

Particularly worrisome from this perspective is the energy supply. The Australian ecologist Holmgren (2009) has written about what he calls "energy descent" and its implications for modern civilization. He argues that the fast-converging crises of peak oil and climate change will lead to a future far different from our own past. A future of less energy, as he explains, requires living more locally with less economic and social complexity. In his worst-case scenario, as energy descent and climate change intensify, technologically advanced societies geared to high-energy lifestyles will attempt to cling to their existing ways of life through aggressive measures, including a turn to authoritarian leaders.

It is important to note that while ecologists have long pointed to peak oil as a primary problem leading to decline and catastrophe, the situation seems to have changed. Indeed, it now appears that the assessment for oil has been premature. With the discovery and development of shale oil, many now even speak of a new energy revolution (Morse 2014). Despite the technical difficulties and costs of pumping oil from more traditional sources, suddenly, thanks to fracking, tar sands, and other new oil discoveries, the world is awash in oil, reflected today in low prices. In various ways this makes matters even worse, as the burning of this oil only pushes us further toward genuine climate crisis. For one thing, the process of fracking associated with shale oil can, beyond polluting ground water, releases methane which significantly contributes to the greenhouse effect. For another, major oil companies are now aggressively searching for oil in Arctic regions. As the ice cover melts because of global warming, this opens up the possibility of exploring new sources that were heretofore unreachable. Some energy experts believe that there is *enough* new oil there to keep modern industry going for a long time. But, in this case, there is considerable irony: the oil companies can now drill into those reaches because global warming has opened up the area, only to pump up more oil that will end up further heating up the planet (Krauss et al. 2005; Lempinen 2013). It is now estimated that to stay under a temperature increase of 2°C, 80 percent of the known oil reserves have to stay in the ground, including the new sources of oil and gas made available by the melting of the Arctic regions (Mckibben 2012). The chances of slowing the warming thus seem slim.

From a political perspective, the challenge is enormous. The degree of difficulty involved in turning the current situation around is not easy to estimate, but the political struggle toward that end has been likened to that of the abolition movement in the United States, the very long and arduous social and economic battle to end slavery, which took more than a century (Hayes 2014).

Worst-Case Scenarios

These dimensions of climate change have led to any number of dire warnings about what lies ahead. Many of them, if not most, have been advanced by a number of scientists deeply worried about what they see in an extensive body of climate change research. But social scientists have joined in as well. Indeed, Urry (2011) has surveyed the field of what he calls the "new catastrophism" and counted more than twenty books over recent years, a list that continues to grow. All point to very worrisome times ahead, some even speaking of a new "Dark Age" (Gail 2016; Berman 2000). Various predictions seem rather farfetched—for example, Rees (2003) gives only a one in two chance of the human species surviving the twenty-first century. In any case, they all offer plenty of reason to worry about climate change and its severe implications for future societies.

A typical starting point for such analysis is the work of Diamond (2011), who has demonstrated that environmental degradation throughout human history—mainly in earlier times, but not only—has brought about the collapse of societies and civilizations. Diamond lists a number of factors typically responsible for the disappearance of such societies. All are basically the kinds of things that are today associated with climate change—from deforestation and overpopulation to water shortages and agricultural problems, to name just a few. From such a perspective, we can see how ephemeral our period is against the weight of human and planetary history, just how brief and fragile the twentieth-century period of wealth and abundance has been (Kolbert 2014). And now, with soaring levels of carbon dioxide and growing shortages of affordable energy sources, many knowledgeable experts believe that the current epoch, now referred to as the "anthroprocene," will come to an end and be replaced with another (Stromberg 2013). What that new future will look like is the open question.

Also basic to these writings, and in some ways the most theoretically interesting, is the way these collapsed societal systems contained the seeds of their own destruction. Frequently these works draw on Marxist-like analyses, which sees capitalist societies to contain the internal contradictions that can bring them to an end (Foster 2015; Demirovic 1994). Where Marx saw a capitalist system emerging in the seventeenth and eighteenth centuries that was constructed around internal contradictions based on productive relationships that could only be managed for so long, the catastrophe theorists—or "epochalists"—see industrial capitalism using up—wasting and exploiting—the scarce resources that made the system possible. Although there is still plenty of coal and oil, they are principal sources of air pollution and global warming. Without these sources, high-energy economies as currently structured like those in U.S. and Europe, but also China and Japan, cannot continue to function.

Such internal logics, however, are not so easy when it comes to the complexity of nature, which does not lend itself to linear forms of thinking. There is, in short, no way to predict what will happen by extrapolating from where we are, or even have been over the past 300 years. As Urry writes (2011: 41), "there are no simple unchanging stable states, or states to which there is equilibrium-establishing movement," making predictions difficult and unreliable if not impossible. Policy changes will be unable to restore balance to a "normal state," as policymakers often contend. In fact, as Urry continues, actions can "often generate the opposite or almost the opposite from what is intended," a result of the unexpected, interrelated, and multiple consequences associated with complex societal systems.

We thus appear to confront a cataclysmic future driven by the interrelated impacts of global warming. What we get are unpredictable interactions at a time of declining resources necessary for confronting the consequences. Many see these shortages and their "tipping points" leading to resources wars in various parts of the world (Klare 2002). Once the various interacting systems are under stress, or further stress as many of them already are, the shocks are likely to be intense and multiple. The response of many countries and regions of the world will surely be to fortify themselves against these shocks. A big issue will be the large numbers of people attempting to flee their unlivable cities and homelands, setting off waves "climate refugees." Some believe that this will mostly lead the wealthy north to evolve as a fortified enclave designed to keep out the desperate poor from the south. Outside of these fortified regions will be "wild zones," where the multitudes—tribal, ethnic, or religious—will be left to fend on their own (Urry 2011). Impoverished areas will often be wracked by civil disorders created by peoples seeking to survive. We have already witnessed the beginnings of this as thousands attempt to reach European shores, especially from Africa.

The question then is, what would the future in the second half of this century or early into the next one look like? We focus here on the worst-case scenario, as it seems more and more likely that many in the next generations might well confront it, to one degree or another. Raskin and his fellow researchers (2002) describe this future as a descent into some form of "fortress world," with barbarism sweeping up many parts of the world. This scenario may strike many as too harsh and we can only hope that it is. But even if future generations in parts of the world are spared from the worst scenario, many of these economic, social, and political features will still have emerged, if only to a lesser degree. A large part of the world, the developing countries in particular, will most likely confront the full force of the crisis. We thus turn in the next section to the broad outlines of the future as drawn from a number of important scenarios. Scenarios, of course, are not science; rather they represent best guesses on the part of thinkers who have devoted a considerable amount of time to looking at the facts and trying to project their implications into the future. Despite their limitations, we ignore them at our own peril.

Peering Into the Future: Fortress World?

For the purposes of the discussion here we draw primarily on the work of Raskin and his colleagues (2002), with additional support from the writings of Dyer (2010), Holmgren (2009), Welzer (2012), Homer-Dixon (2006), and Oreskes and Conway (2014). The scenarios of the Raskin group are especially useful as they set the fortress world scenario in a larger discussion of historical transitions, moving from the Stone Age through Early Civilization to the Modern Era. Not only does this help to put the nature of the transition in perspective, it helps us understand the challenges it poses.

Each historical transition is unique, as Raskin and colleagues point out. The particulars shaping an emerging transformation in time and space will contingently forge or open particular pathways and not others. Such historical transitions, as they explain, are complex socio-political junctures which involve alterations in both the cultural bases of societies and the underlying relation of humans to the natural ecological system. These interactions play themselves out at critical thresholds. The gradual but steady "processes of change working across multiple dimensions—technology, consciousness and institutions—reinforce and amplify the changes underway" (Raskin et al. 2002: 3).

At later points along the way, the structures and processes of a socio-ecological order stabilize in a revised system with new conditions and dynamics that guide and govern further change. As a result of each transition, we can witness critical, interrelated changes in the forms of social and political organization, the nature of the productive economic system, and the capabilities to communicate.

We can restrict ourselves in this discussion to the current transition from the Modern Era to what Raskin and colleagues refer to as the "Planetary Phase." The Modern Era is defined by the nation-state, the Industrial Revolution, the capitalist economy, and the rapid spread of communication through printing, whereas the new emerging period features globalization of economic systems, a communications revolution driven by the internet, and the expansion of global governance institutions. While the transition has taken off, it is not yet clear how it will continue to evolve. Such a transition, as Raskin and colleagues explain, is typically not a total break with the past; various ideas and practices get carried over from one era to the next. As the new social system is intellectually refracted by economic, philosophical, and political thinkers, an extensive range of ideologies and worldviews engage the question of transformational change.

In the case of climate change, these range from the technological optimists to social engineers, market enthusiasts, pessimistic survivalists, and eco-anarchists. All of these thinkers attempt to grasp the forces—uncertain and often contradictory—that will usher in the transformation. As Raskin and his group explain, they struggle to comprehend the uncertainties, risks, and indeterminacies lodged deep in the social and natural fabrics of the old and new societal realities. These can involve discontinuous shifts between political and economic arrangements; courses of development confronting conflicting futures at critical junctures; systems absorbing and integrating external phenomena until they clash with fundamental beliefs and values. At some point, societal systems transform into one of various possible alternatives of societal configurations. In the process, even small disturbances can have a big influence at critical intersections.

There are two potential outcomes of this catastrophe scenario. The first outcome is called "breakdown." This results from the failures of powerful institutions—international organizations such as the United Nations, alliances among nations, transnational corporations, and powerful militaries—to take necessary precautionary actions. As the global crisis starts to get out of control, many of the more prosperous nations will also experience hardships as their technologies and infrastructures begin to fail. Fearing being engulfed by rampant social disorder, affluent minorities will turn to concerted military actions. A global emergency would probably be declared by the United Nations and its agencies. Powerful international actors—military, corporate, governmental—would most likely establish alliances for their own protection. Peace-keeping units would replace whatever would be left of global aid. A campaign of overwhelming force and "rough justice" are seen to sweep through hot spots of conflict and discontent. This would give rise to something of a "planetary apartheid." As Raskin and colleagues aver, the separate realms of the well-off and the poor would be asymmetrically divided by authoritarian institutional and legal frameworks. Police states would be expected to hold the impoverished many outside the fortress. Vast populations mired in abject poverty would be denied basic rights. In the view of these theorists, a new type of militant activist—"educated, excluded and angry"—will stoke the flames of misery and protest. The greatest concern, in Homer-Dixon's (2006: 259) words, is that these groups "will achieve technological leverage when they start using weapons of mass destruction, including chemical, biological, radiological and nuclear weapons," which many specialists believe will only be a matter of time.

As ecological degradation takes its toll, these agents will be increasingly unable to offer anything more than fragmentary piecemeal reactions, which eventually give rise to conflicting struggles among them—possibly even rivalries that block efforts to establish a workable order. When these institutions and their leaders start to grasp fully the dangers of the breakdown, many of them will separately regroup in an effort to protect their own economic and political interests. This might evolve rapidly to prevent the erosion of their economic resources and governing systems, including any alliances they have been able to forge in the process. The powerful wealthy nation-states are foreseen to pull back in an effort to fortify their enclaves and the ability to maintain them, although this will happen as well in parts of the less developed countries.

The ability of global political elites to manage such a retreat to "fortress world" will depend on their institutional and political capabilities to control the dispossessed beyond the walls. This will involve putting down militant uprisings of the poor and oppressed populations that surround them, both from without and within, including waves of terrorism against their citadel. An important part of the strategy to rally support for this effort, as the Raskin study notes, will be the "politics of fear."

If they fail to establish a workable fortress, we then can expect what Holmgren (2009) calls the "life-boat" scenario. In this scenario, we find a similar description. The adverse consequences of the inability to shape a workable way around the crisis converge to initiate a steady decline and eventual collapse of the various forms of social and economic organization.[2] The breakdown will give rise to increasingly violent conflicts in various areas of the world—ranging from local wars to perhaps even the use of nuclear weapons. Mass hunger resulting from waves of famine will be a common feature. Thousands of people will die from heat exhaustion, as foreseen today in places like India where temperatures now reach the mid-40s Celsius (Di Liberto 2015). Disease of all sorts—from malnutrition and contaminated water to malaria—will inevitably spread on a large scale. Holmgren suggests that it could be on a scale as large or larger than the "Black Death" in Europe during the Middle Ages, reducing the world population in a matter of decades to half its current size.

Like Raskin et al., Holmgren sees the institutional capacities of national and international regimes diminishing as ecological conditions pushed by the multiple stresses of climate change and declining resources amplify the decline (Nadesan 2016). The Southern Hemisphere will be hardest hit, as warming will dramatically reduce arable croplands, with Africa suffering the most. But northern countries will also have to struggle, as the situation in places like California starts to make clear. Countries like the United States and Russia will no longer have enough foodstuffs to export, especially as development aid for agriculture (Dyer 2010). There will, in short, be no way to feed 9 million people or more, leading to a shrinkage of the world's population.

Vulnerable people from the affected regions will migrate northward to affluent regions in a desperate search of food, employing any means necessary (Forsyth 2015).

As poverty increases, further widening the gap between rich and poor, vast and spreading health crises resulting from malnutrition and the lack of drinkable water will emerge as sources conflict. In regions with shared water basins, conflicts over scarce water resources will be a regular occurrence, often violent in nature. Violence will in fact become common (Welzer 2012; Barnett and Adger 2007). Media images of the affluence in northern parts of the world will only fan the restiveness of the excluded billions. In this atmosphere of deepening unrest, disputes and conflicts of various sorts will feed on established nationalist, religious, and ethnic tensions, already witnessed in parts of Europe and North America. Homer-Dixon (2006: 255) writes that "in poor countries, where environmental, population, and economic stresses are already severe and social capacity to manage them remains low, we'll probably see a steady increase in outbreaks of civil violence—including riots, insurgency, ethnic cleansing, and terrorism." Shortages of food and extensive hunger can precipitate guerrilla war in the countryside. "Such events," he continues, "could culminate in larger social earthquakes, as multiple shocks topple regimes in zones of geopolitical importance, like the Middle East, South Asia, and East Asia." Should this turbulence go unchecked, "world order could disintegrate in stages—from the poorest countries at the periphery to the richest countries at its core, much as happened in the western Roman empire."

As poor countries start to fragment under the pressure of civil collapses, spreading crime will fill the vacuum. Anarchic conditions will provide openings for criminal activities of all types, including powerful global syndicates quite capable of fielding armed groups to do battle with international policing organizations (Raskin et al. 2002; Dyer 2010). Terrorism will be a central part of this picture with frequent suicide bombings at public gatherings and on the symbols of globalization. In some cases this will involve biological weapons and perhaps even nuclear arms. In the case of nuclear conflict, Pakistan and India have often been thought to be the most likely candidates. But other countries like Iran and North Korea will pose serious worries (Dyer 2010).

Dyer (2010) and Welzer (2012) are particularly vivid when it comes to offering a detailed look at the violent conflicts resulting from climate crisis. Dyer's book on climate wars provides an eye-opening portrait based on extensive interviews with a wide range of scientists, government leaders, business leaders, and military experts. Peering into different corners of the globe, Dyer plausibly suggests how various conflicts will emerge and play out. He offers, for instance, the case of Pakistan and the need to protect its vast irrigated agricultural lands on rivers that have their origins in India. Assuming that India would consider its own interests first, its long-time rival would be quite vulnerable. This is already part of the world where water wars are a reality, especially when we add Bangladesh to the equation. Also, Pakistan and India, as just noted, both possess nuclear weapons.

Water has in fact been described as the "next oil." It is now seen as the substance over which wars will be fought later in this century. Two recent Secretary Generals of the United Nations, Kofi Annan and Ban Ki-Moon (Carius et al. 2004; Wall 2008), have warned that water scarcities are likely to lead to violent struggles. As Ki-Moon put it in 2007 at the Security Council, water shortages can transform "peaceful competition into violence" with both droughts and floods triggering "massive human migrations, polarizing societies and weakening the ability of countries to resolve conflicts peacefully." Experts say that four major rivers—the Nile, Tigris-Euphrates, Jordan, and Indus—will in particular spark "high levels of tensions along with periods of outbreaks of violence" (Klare 2002). A river like the Nile is controlled by eight other countries in addition to Egypt. This poses an ongoing danger of violent conflict in a troubled part of the world.

Or consider the possibility of conflict between China and Russia. China, Dyer (2010) explains, will have trouble growing enough wheat, as the monsoons in those parts of the country are already failing to supply enough water. In this case, military experts speculate that China might try at some point to reclaim parts of Siberia that were appropriated by Russian Czars. As Siberia will become more suitable for agricultural production after temperature increases of 5–6°C, coupled with rich deposits of oil, minerals, and gas, the option might become attractive. Given the thinness of the population in the area, it would be easy for China to swiftly move into that region. As historical observers have noted, there are few examples in human history of starving populations that did not at some point attempt to take what they can get. For this reason, one expert argues that the chance of a nuclear conflict between these two superpowers could increase significantly.

In Europe, as we have already mentioned, the primary issues will be migrations from the south and the east. By 2040, in various scenarios, the idea of free movement of citizens within the European Union would most likely be discontinued in northern countries—the U.K., France, Germany, the Netherlands, and the Scandinavian countries in particular—nations that will be confronting millions of migrants fleeing from the intolerable temperatures and agricultural drought plaguing southern European peoples. This might well give rise to a speedy reorganization of the EU into a two-level-like structure, with the northern countries providing whatever emergency aid they can, in particular food supplies, but no longer permitting any citizens of the southern countries to move northward.

In the case of Africa, after more countries become failed states, it will be increasingly difficult for Europe to hold back migrating African refugees.

The northern states would most likely place soldiers along the Turkish border and send naval ships into the Mediterranean to keep migrants from the Middle East and Africa from entering the EU countries, a policy already witnessed on a lesser scale. Given that many refugees will in any case try to make it across the borders, it has been estimated that up to a quarter of the peoples from southern Italy, Greece, Bulgaria, and Romania could live as illegal immigrants in various European countries. In moves that recall Rome bringing its legions back home, in Dyer's view, northern Europeans would withdraw their military forces from more distant boundaries of Europe and concentrate on securing their own borders. This could lead to regular military border engagements, increasingly with armed refugees. The northern countries would probably try as long as possible to supply enough food to southern European states that find themselves collapsing under the pressures of desperate populations. But this will not continue indefinitely, especially as the northern countries begin themselves to experience drought in their own agricultural fields. These tensions, as we already see today, would give rise to citizens in the north blaming the illegal refugees for their own plight, giving rise to nationalistic tendencies designed to keep them out.

Then there is the issue of the Arctic region. The thawing now taking place makes available new sources of oil, gas, and other rare minerals. When the Russians managed to plant their flag at the bottom of the North Pole in 2007, more than 2.5 miles below the thick layer of ice, it set off protests among other nations, appealing to the United Nations Convention on the Law of the Sea. It has been estimated by U.S. Geological Survey that around 15 percent of the planet's undiscovered oil and some 30 percent of its natural gases can be found at the bottom of the Arctic, a large percentage for a rather small portion of the globe's surface. Since then, countries like Canada, the United States, Russia, and China have entered a new sovereignty race for these untapped hydrocarbons. Already during the Cold War, the Russian and U.S. military forces taunted each other in the region with military outposts, nuclear submarines, and spy planes. Now with the opening of the Arctic, offering new shipping routes from the Atlantic to Asia, military rather than diplomatic activities, particularly on the part of Russia and Canada, threaten to be the primary mode of dispute resolution. The powerful Arctic countries are thus "fast approaching diplomatic gridlock that could eventually lead to the sort of armed brinkmanship that plagues other territories" (Borgerson 2008).

Environmental Security and the Military

While the disarray associated with the "life-boat" world is seldom the primary focus of most environmental scholars, it is the subject of intense research in other domains, in particular the U.S. Defense Department (Webb 2007). For some time now the Pentagon has had a special office of environmental security devoted to the study of the social and political disorganization that severe climate change will bring. Indeed, analysts there have developed sophisticated scenarios and readiness plans to deal with such crises when they come about, as have other military organizations. According to a recent Pentagon report (Davenport 2014), climate change represents an immediate threat to the national security of the United States and other parts of the world. As Davenport summarized the document in the *New York Times*, the immediate threat involves "increased risk from terrorism, infections from diseases, global poverty and food shortages" with a "rising demand for military disaster responses as extreme weather creates more global humanitarian crisis." In other studies the Pentagon has also warned of energy shortages, drought, and coastal flooding that can have big impacts on cities within the next twenty years (Schwartz and Randall 2003).

But this is not just a concern of the United States military. NATO's website states that the organization addresses "security challenges emanating from the environment" including "extreme weather conditions, depletion of natural resources, pollution and so on—factors that can ultimately lead to disasters, regional tensions and violence."[3] NATO leaders describe environmental and resource constraints, in particular the consequences of climate change, to be basic considerations shaping the future of European security.[4] They call for the need to prioritize the integration of environmental collapse into their more general plans for European security.

Environmental security, as a field of investigation, has emerged as a special branch of both environmental studies and security studies more generally (Homer-Dixon 1991; Myers 1989; Dalby 2002). Although thought and inquiry about the relationship between security and the environment have been around for a long time, environmental security in its contemporary interdisciplinary form can roughly be dated to the early 1980s. Many environmental scholars have avoided this topic because it is widely seen as an acceptance of the coming crisis and in no way an appropriate solution; the discussion thus has been more about sustainable development than environmental security. In recent years, however, a growing number—including politicians and environmental activists—have argued that the environmental crisis should in fact be linked to matters of security, as security gets the attention of both politicians and the public. Whereas environmental protection is underfunded, the military generally gets, more or less, what it asks for in the name of security.

Environmental security, as such, is concerned with the human impact of environmental problems on nations, communities, and citizens. The U.N.'s Millennium Project somewhat more specifically defines environmental security as an emphasis on the ecological viability of life-support systems as they relate to both ecologically caused conflicts and the prevention of military destruction to the environment.[5]

Examining the ability of nations and peoples to deal with environmental changes, in particular resource shortages and the risks they pose, the field draws heavily on international relations theory and development studies. Researchers investigate the political conflicts, social tensions, and civil disorders that result from global and regional changes, such as the impact of climate change on food production, water shortages, and fisheries. The scholars in the field, a small but growing group, pay special attention to conflicts in the developing and underdeveloped worlds, especially as they tend to be vulnerable to state failures and political destabilization.

As a contemporary research field, environmental security owes much to the work of the noted environmental political scientist, Thomas Homer-Dixon (2006), who has served as a military consultant. Arguing that environmental scarcities will have dramatic implications, leading to civil unrest and violence in urban areas, particularly in developing countries, he maintains that social and political conflicts in Chiapas, Mexico and unrest in numerous Asian and African nations are already in part a result of resource scarcities. While the effects of such scarcities can be indirect, and work in tandem with other social and economic stresses, the potential for violence, he maintains, is not to be underestimated, as roughly 50 percent of the population of the planet depends on local sources of renewable food and energy for daily survival. In the coming decades, chronic scarcities will affect a large part of the world's population with unprecedented levels of hardship at an accelerating pace.

The purpose here is not to examine the field of environmental security per se, but rather to point to it as evidence of—and attention to—the issues raised in the previous sections of this chapter. Even if such research is not the major focus in the field of environmental studies, there are serious and influential people who devote large amounts of time to anticipating the states of emergency ahead, including the possibility of fortress world. For example, the CAN Corporation, a think tank serving the U.S. Navy, concludes that "climate change can act as a multiplier for instability in some of the most volatile regions of the world and thus presents significant national security challenges for the United States" (Busby 2007). To take another illustration, the Association of Climate Change Officers published a report on the implications of environmental sustainability for security that calls for a shift from sustainability to adaptability and resilience. The document asserts that sustainability does not spell out positive goals for protection. Whereas sustainability calls for fundamental societal changes, adaptation for security simply means the ability of the established institutions to "bounce back" from major disturbances (Association of Climate Change Officers 2012).

Finally, it is important to note, as Luke (1997; 2011: 96–109) has pointed out, that this environmental security discourse is mainly technocratic and managerial. Oriented to prediction and control, it is thus well suited for engineering and planning, particularly in the service of the military. And as critics are quick to point out—correctly—the military has no positive record when it comes to dealing with environmental destruction. Although we seldom think of war and the preparation for it as an environmental issue, it is one of the most severe causes of ecological damage (Austin and Bruch 2000; Thomas 1995).

Environmental Survivalism and Eco-Authoritarian Politics

Associated with the worst-case scenarios is an environmental discourse referred to as "survivalism" (Dryzek 2005: 26–50). Survivalism is oriented to the belief that we already—and will increasingly—overshoot the Earth's carrying capacity (Catton 1982). Although the theory has its early origins in the writings of nineteenth-century economist Thomas Malthus, who saw continued population growth leading to socio-economic decline, its most prominent contemporary representative has been the population biologist Garrett Hardin (1968), particularly as spelled out in his profoundly influential article on the "tragedy of the commons." But one has to also cite the well-known works of political scientist William Ophuls (1977), the economist Robert Heilbroner (1974), and socio-ecological philosopher Rudolf Bahro (1987), among others. Survivalists disagree on various points of emphasis, but they all believe that current rates of economic growth, abuse of the collective commons, depletion of unrenewable resources, ecological destruction, and population increases are out of control and leading to an unprecedented—even apocalyptic in the view of some—ecological crisis.

Pointing to historical examples, coupled with present-day illustrations that are already observable, survivalism sends a strong wake-up call to those who see the possibility of circumventing ecological crisis with reform-oriented economic and social innovations. More or less, the survivalism discourse—in one form of another—has been a fundamental strand of *radical* green environmentalism from the 1960s onward (Lewis 1992: 9–10). Based on early influential arguments about the "limits of growth," green survivalists have nearly always focused on a return to a simpler, less materialist mode of human existence, often emphasized by communal ways of life and political decentralization. By and large, they were later pushed aside by the reform orientation that was to dominate environmentalism (although in more recent years the "limits debate" has returned with the "degrowth" movement).

The survivalism argument, as Dryzek (2005) points out, can run up against problems when applied to particular areas of the world, especially impoverished regions and localities, where humanitarian aid and at times trade relationships can sometimes help a country stay afloat even after it has reached its carrying capabilities, various countries in Africa and Latin America offering illustrations. It is also the case that prosperous cities such as London and New York can manage to live beyond their carrying capacities by continuing to exploit distant resources. New York City, for example, has no place for the massive quantities of waste that it generates; it thus deals with the problem by transporting its waste to other states like West Virginia and Ohio. But none of this in and of itself renders the survivalist perspective false. Its traction takes hold when we speak of carrying capacity in terms of the global ecological system as a whole. At this level, on balance, human activity on the planet has clearly pushed the global system to its limits with implications for all countries, especially when it comes to the planetary climate.

Underlying the basic survivalist narrative, then, is the assumption of shortages, both of resources and capacities. As Dryzek (2005: 38) explains the view in his detailed explication of the discourse, "human demands on the carrying capacity of ecosystems threaten to explode out of control, and draconian actions need to be taken in order to curb these demands." The resources upon which human life depends are being depleted, in particular non-renewal resources such as farm cropland, fuels such as oil and gas, and metallic ores. On top of that, the diminishing ability of the ecosystem to produce renewable resources such as oil, wood, and fish and the Earth's capacity to assimilate pollution and waste is under extreme stress. Driven by a combination of population growth, excessive material production, and overconsumption, the long-term possibility of life as we know it is put in question. Toward this end, many survivalists appeal to metaphors such as "Spaceship Earth," which requires that the life-support systems be maintained so that those on board do not perish. In general, the message is gloomy, often pointing to doom. The famous physicist Stephen Hawking told the BBC that human beings are now in danger of being eliminated in the next 100 years, adding that we have created our own demise. Our main hope, he suggested, is the establishment of space colonies.[6]

Most of the survivalists take a more or less Hobbesian view of human nature, at least of human nature as it will express itself under the stresses of extreme climate change. They thus emphasize conflicts, rivalries, and the need for hierarchical control systems to deal with them. There are, though, different views on the nature of these controls. For some survivalists the controls will depend on strong political leaders and for others scientific experts should be in charge. Several survivalists argue that the controls need not be hierarchical and centralized, but rather only require coordinated forms of action on the part of elite groups. Only a few such as Brown (2003) include a requirement for extensive citizen action.

While most of these writers focus on the ecology of the crisis itself, some have drawn out its political implications, many of which are quite harsh. The most famous example is Hardin (1968), who envisions a few "life-boat" nations surrounded by a world of nations sinking in abject misery. Challenging liberal capitalist regimes, he argues that "freedom in the commons will bring ruin to all," going so far as to include the freedom to reproduce the human species. The solution to this problem of limits, he writes, is "mutual coercion mutually agreed upon." That is, people need to willfully surrender themselves to strong forms of authority. Other survivalists are less harsh, but like Hardin they are also skeptical about the ability of the existing systems to come to grips with the crisis, not only their ability to slow its advance but to deal with the challenges once it arrives full-blown. While most see the need for a strong hand at the stern, others such as Daly (1977) have worked out the economics of a "steady state economy" which would permit a controlled form of growth at the level of replacement.

Survivalism has supplied a vision of an environmental horizon which, in any case, serves as a serious warning of what could happen. Much of the contemporary academic world has sought to ignore or discredit this vision. Challenges have come as well from Marxists and ecofeminists. But most of these critiques have focused on issues of social justice or social construction without managing to undercut the basic argument about limits on its own terms.

Some argue that we should dismiss survivalism's message, as doom is a nonstarter. If we are doomed, why bother? Most writers, understandably, feel the need to keep up the struggle for reform and thus show a degree of optimism. Without a measure of hope, there is little use or need to carry on with the effort. And then just maybe the survivalists are wrong: Who can know for sure? That would be the best of outcomes.

To hope for the best at this point, however, is not much of a strategy. Given the relatively limited nature of the environmental response since the early 1960s, especially at the global level, there is every reason to worry about the future. With some fifty years of modern environmentalism behind us, there have been few actions that have adequately risen to the challenges ahead—about this there is little disagreement. There are also few efforts that hold out real promise. Progress with developing alternative sources of power at competitive prices has, to be sure, proceeded much faster than expected. But while such technological advances hold out the possibility of slowing the processes of warming, they are not proceeding fast enough to allow us to sidestep the coming crisis. Even if we stopped emitting greenhouse gases today, the globe would still warm because of the gases already in the atmosphere.

Although the solutions advanced by survivalists do not sit well with the world as it is structured today, this view may well change under the conditions to which the discourse points. For this reason, it is time to take survivalism more seriously and speculate on how we might best politically orient ourselves to the drastic circumstances it portends. In this regard, we can take the message seriously without necessarily accepting the draconian, authoritarian prescriptions that many of its exponents have discussed, if not advanced. Indeed, that is the focus of this work.

It is clear that an eco-dictatorship cannot solve the ecological problem. But what it can do, problematically, is strategically defend and preserve to some degree various parts of the world against others, as the "fortress world" scenario would suggest. At the same time, however, we need to interrogate democracy as well. Whether democratic systems—not only as we know them today but most likely will have them in the future—can solve the ecological problem is anything but certain. The question is a serious one when we place it in the time frame ahead, say the next thirty to fifty years. Few honestly believe that contemporary democratic systems are on the verge of blooming into authentic forms of participatory democratic governance, and the chances of that happening during a future state of emergency is difficult to imagine. Indeed, the political trend appears to be heading more in the direction of oligarchy, if not plutocracy. While democracy can and should remain the ideal, environmental scholars need to begin thinking more about what to do in its absence—and how to do it in ways that might still keep democratic values alive in one way or another.

Conclusion

We examined in this chapter the environmental crisis both generally and specifically as it pertains to climate change, seen to be the major crisis confronting the latter half of this century. The focus was on the political implications of the states of emergency to which it will undoubtedly give rise. Toward this end, we outlined the worst-case scenarios, fortress world and the life-boat society, and considered the kinds of social and political disorders that would be associated with them. While all countries will be affected, some will be hit much harder than others. Parts of the world will have to grapple with hardships less severe than those portrayed by the fortress world scenario, while others will experience the full brunt of the catastrophe, particularly those living in the developing world. In some places clear signs of these future conditions are already present. Numerous writers have suggested that circumstances will give rise to strong authoritarian leaders. Many parts of the world are already governed by such leaders and many more will turn to them in the name of security, including parts of the world that are generally considered democratic, to one degree or another. The situation can be analogized to that of a time of war, when democracy is put on the shelf in an all-out effort to deal with a major threat to survival. The focus on environmental security is seen to already involve preparations for such contingencies.

Democracy, it is argued, will have difficulties standing up to these authoritarian forces. Surrounded by "wild zones," many political leaders will gain support by advancing demagogic appeals to fear and the need for strong leadership. Such leadership will most likely not present itself as anti-democratic per se, but will rather wrap itself up in technocratic appeals to technological innovations, assisted by military forces. This suggests the need to strategically rethink democratic theory in an effort to save democratic values for a later time after the crisis has played itself out, as in the case of war. But nobody knows how long this will take; surely it will last much longer than the World Wars. At that later point in time, many might view democracy as a thing of the past. Indeed, it might be seen as the very form of government that was unable to respond in the face of a coming crisis.

Before taking up this question directly, it is important to gain a broader understanding of the "apolitical"—mainly anti-democratic—techno-managerial mentality that will move to the front. In Chapter 2 we thus examine the rise and progress of techno-environmentalism. Not only is it already a dominant force in environmental politics, its representatives are very busy searching for a technological "fix" that will sidestep social and political solutions to the crisis, including the democratic alternative.

Notes

1. https://deepgreenresistance.org/en/why-resist/ecological-collapse.
2. A study by the technology assessment bureau of the German Parliament in 2011 predicts that within a short period of time after a widespread total electricity blackout, unpredictable chaos could break out, as foodstuff would perish, security would break down, communication would collapse, institutions would cease to function, and transportation would come to a halt, to name just some of the more obvious consequences. "Was bei einem Blackout geschieht," www.tab-beim-bundestag.de.
3. http://www.nato.int/cps/en/natolive/topics_49216.htm.
4. http://climateandsecurity.org/2014/09/05/nato-summit-declares-climate-change-will-shape-future-security-environment.
5. http://www.aepi.army.mil/reports/docs/2011/June-2011.pdf.
6. http://www.aol.com/article/2016/01/19/stephen-hawking-humans-may-not-survive-another-100-years/21299325/?icid=maing-grid7|main5|dl1|sec1_lnk3%26pLid%3D-1267646288_htmlws-main-bb.

2

Democracy at Risk: From Citizen Activism to Techno-Environmentalism

Can democratic government stand up to the social and political stresses that will accompany environmental crisis, climate change in particular? This question is as difficult as it is important. As we saw in Chapter 1, it has led various prominent environmental writers to call for greater reliance on experts. Not only have they called on experts for technical solutions, but also for a greater role in the policy decision processes. These writers argue that the physical and social stresses in many places in the world will in time necessitate states of emergency, both to take quick action to deal with hazards and to confront the resultant civil unrest on the part of desperate populations. This will pose the threat of more authoritarian forms of rule, as central governments will step in with strong measures. In some places, it will no doubt mean military rule. In all ways, the search for technological solutions will intensify, which will move engineers and other technologists more to the front of the advisory processes, even the decision-making process itself in many cases. Indeed, some writers have argued that this is *the* solution and that it is time to look it in the face.

It is a view embraced by many powerful politicians in the developed world. The answer to the environmental problem is for them to be found in such technologies as carbon capture and storage, biofuels, nuclear power, and even geoengineering of the atmosphere. President George W. Bush expressed such enthusiasm for technological solutions in these words: "We've identified a problem, let's go solve it together. We will harness the power of technology. There is a way forward that will enable us to grow our economies and protect the environment, and that's called technology. We'll meet our energy needs. We'll be good stewards of this environment."[1]

In this chapter we seek to sort out the issues and questions posed by this appeal to technical expertise. Insofar as a call for technocratic expertise is not new to environmental politics, we begin by looking at the evolution of techno-managerial expertise in modern environmental politics and policy practices. This is followed by an examination of the technocratic mode of reason, in particular its emphasis on the depoliticization of the policymaking and political and epistemological considerations that are often invoked to support it. The discussion then moves from the realm of theory to examine specific contemporary technocratic arguments calling for the ascent of technical experts. The chapter concludes with a discussion of the nature of technocratic power, both as a result of the expert's ability to frame policy problems and a deeper form of discursive power that more fundamentally shapes the underlying political structures and policy processes of modern society.

From Concerned Citizens to Technical Experts: Techno-Managerial Environmentalism

Techno-managerial practices are not difficult to identify in the realm of contemporary environmental protection. In fact, we can identify them in various phases in the development of environmental expertise generally. From the very outset of environmentalism, to be sure, technical information has played an important role. Science, for example, discovered the harm to bird eggs exposed to pesticides and the presence of radioactive elements in reindeer meat. Rachel Carson (1962) marshalled such information in her seminal book, *Silent Spring*. But it was citizens rather than scientists who initiated the environmental movement. At that time, many of the pressing environmental problems required few scientific instruments: laypersons simply detected dirty air and polluted water with the eye and nose. Indeed, in the initial period science and technology were often vilified. Technologies such as power plants and automobiles were seen to be dangerous sources of pollution. In fact, remnants of such hostility to the forces of technology are still today to be found among many environmentalists, radical environmentalists in particular. Often founded on the argument that "small is beautiful," such denunciations have been accompanied by calls for "appropriate technologies" (Schumacher 1973; Fischer 2000).

The 1970s began to see significant changes in environmentalism. Although the environmental movement started in the 1960s with street protests by concerned citizens, the focus in the subsequent decades involved a shift from citizens to an emphasis on science and technical information accompanied by more technocratic approaches to policymaking. Part of this had to do with the reconfiguration of environmental problems in terms of risk. While citizens were organizing to protect themselves from dangerous technologies, the shift to a discussion of the levels of risk posed by such technologies pushed technical questions about ecological destruction to the center of environmental politics. As the discussion turned to the empirical identification of "acceptable risks," laypersons were put at a disadvantage, especially when it came to environmental deliberation and argumentation.

In this second phase of environmentalism scientists and engineers themselves increasingly took center stage, including "movement scientists" working with environmental organizations. During this period, many new environmental problems required more than sensory perceptions.

Complex biological and physical sciences began to define the environmental challenge, with the detection of invisible and infinitesimal sources of risk necessitating precise measurements. It took, for instance, extensive scientific investigation and technical deliberation to determine the health hazards of toxic wastes and the causes of acid rain. Indeed, many of the problems to be discovered by science during the 1970s and 1980s posed threats even more worrisome to health and well-being than concerns about polluted water and air. Moreover, invisible problems such as the ozone hole and climate change emerged to increasingly threaten the very future of economic and social life as we know it. As a consequence, a technocratic environmentalism moved prominently to the fore, both in the corridors of power and in parts of the environmental movement (Fischer 2000). By the mid-1970s and beyond, the technical languages of environmental management dominated much of environmental policy politics, a phenomenon that writers such as Swyngedouw (2007) and Mouffe (2005) have referred to as a form of "post-politics."

Transferred from the public arenas of political protest to the administrative realm of policy expertise, the emphasis shifted to difficult technical issues in search of solutions. As environmental policymaking became increasingly mediated by the languages of environmental impact assessment, cost-benefit analysis, technology assessment, and risk-benefit analysis, environmental politics featured encounters among industry-oriented experts and environmental counter-experts arguing over the validity of competing assessments. In short, there were still debates about the environment, but they were now more and more between different groups of techno-environmental experts.

Basic to this development was the rise of a form of "regulatory science." The environmental policymaking process has since then been increasingly governed by what Jasanoff (1990) has called the "science policy paradigm." In this approach, regulatory decision-making involves a mix of interrelated scientific, administrative, and legal elements. Combined, these components have been shaped into a practice that has had a profound influence on both the making of agency regulatory procedures and the assessment of findings, particularly in the face of technical uncertainties. By bringing scientific experts together with administrative decision-makers, the paradigm serves to harness scientific advice to the goals and requirements of a system of governance (Fischer 2000). In the process, a small industry of regulatory science experts has emerged to serve the arrangement. Described as a profession designed to ensure safe and effective products, they are now commonly found giving advice to decision-makers in government, industry, and academia (www.regulatoryscience.org).

Fundamental to this regulatory science is an emphasis on the interrelated methodologies of risk assessment and cost-benefit analysis, or risk-benefit analysis when integrated. Risk assessment is employed to evaluate risks resulting from both hazardous technologies and toxic health threats. In the latter case, for example, the goal is to accurately predict the health implications of a hazard before or after it exists and to establish valid safety standards to protect the exposed population. After the hazards are identified, there is an assessment of human exposure, an evaluation of the adverse effects, and a characterization of the overall risk posed by the substance, typically expressed as a statistical probability of the frequency of specific health problems occurring in the population as a whole (Fischer 2000). The second phase of a risk-benefit analysis is to submit the risk assessment to a cost-benefit analysis. Here the goal is an explicit comparison of the benefits derived from a hazardous activity and the risks involved in that activity (Sunstein 2004). The objects of investigation are defined in terms of specific levels of risks rather than monetary values. The method thus involves calculating the benefits of a project (adjusted against regular costs, such as plant construction and maintenance costs), comparing the ratio of the risks to the benefits, and multiplying the resulting figure by the total number of people affected.[2]

The concern here is not that one should be uninterested in costs and benefits per se. Rather it has to do with the way this data is organized and used in environmental decision-making processes. Part of the problem is that the approach is often presented as the *essence* of rational decision-making, neglecting the equally important social, political, and moral considerations that are also in play. Moreover, this conception of rationality pre-structures the decision-making process in ways that favor business interests (Noble 1987). Business can deal with costs and benefits; they are basic to the business way of thinking. But the environment is about more. Environmental protection is not something that can just be understood and measured in terms of divisible costs and benefits. It is also a matter for the larger public interest. Among other reasons, we save the environment for future generations and the costs and benefits relevant to that goal cannot easily be figured into the equation, if at all.

With these developments in mind, we can at this juncture examine technocratic expertise more specifically as a mode of thought and a type of reason. We begin with its basic assumption about the need to depoliticize policy decision-making.

Technocratic Expertise as Apolitical Politics

The concept of technocracy is a subset of the more general topic of expertise in modern society. For this reason, it is useful to begin by noting that most things in modern society are dependent on expertise in one way or another (Giddens 1990). We live in societies surrounded by and bound together with expert systems that are largely independent of space and time. We thus generally have little alternative but to rely on judgments made by faceless authorities, whether medical doctors, aeronautical engineers, food inspectors, or environmental regulators. Trust in experts therefore takes on special significance (Fischer 2009).

Such reliance on experts is central to the ideology and practices of technocratic governance. Essentially technocracy is a belief system that not only urges us to trust the work of these experts, but also, given their central importance to modern life, to turn over the tasks of social guidance to them (Fischer 1990). An idea that has its origins in the Age of Enlightenment and the writings of St. Simon and August Comte, considered to be the founders of modern sociology, technocratic thought has risen and fallen throughout modern history, but never altogether disappeared. In many ways, it lives on today through technocratic assumptions often embedded in practices of decision-making and government.

Technocratic reason and its assumptions are can be thought of as a mentality. In ideal type, this mentality generally focuses on two separate but interrelated concerns. One concerns scientists and engineers who call for technical innovation and the use of new ecologically relevant technologies to solve problems such as climate change. Although most of these scientists are driven by practical or pragmatic considerations, the general orientation is typically supported by the belief that technology is the key to social and material progress, even the driving force behind the development of modern civilization more generally. As the President of the British Royal Academy of Engineering has put it, history is a linear "race to advance technology," which "will determine the future of the human race" (Broers 2005). It is, moreover, a belief enthusiastically advanced today by numerous technological futurists who envision something of a "brave new technological world" on the horizon (Kurzweil 2005; Vinge 1993).

The second concern involves the application of the methodological techniques of the administrative sciences to plan and implement the use of these technologies. It is to such decision-making methods that the word "technocratic" is usually applied. These writers and practitioners, as the heirs of the scientific managerial approach to decision-making, rely on the instrumental methods of "administrative rationalism": strategic planning, quantitative analysis, cost- and risk-benefit analysis, formal modeling, computer simulations, systems analysis, and evidence-based policymaking (Dryzek 2005). Indeed, given the technically oriented nature of environmental policy, the practices of technocratic environmental expertise have long been a form of "evidence-based policymaking" (officially adopted as best policy practice in the U.K. and elsewhere in Europe).

The historical coherence that defines technocratic thought is an underlying dislike of—even animosity toward—politics, particularly democratic politics, coupled with an unswerving commitment to scientific decision-making. As Stone (2002) puts it, the common mission of this "rationality project" is to rescue public policy from the irrationalities and indignities of politics, hoping to conduct it instead with rational, analytical, and scientific methods. Hostile to politicians and unsympathetic to democracy, dedicated technocrats argue that scientific techniques can treat problems in "apolitical" terms and replace political decision-making with superior modes of technical analysis.[3]

Thus for the technocrat the solution is to replace the "irrational" decision processes of democratic politics (such as group competition, bargaining, negotiation, and compromise) with "rational" empirical/analytical methodologies of scientific decision-making, or what has been called "methodological decision-making" (Fischer 1990). Nothing is more irrational to technocratically minded theorists than the disjointed, incremental forms of decision-making typically described as "muddling through," in significant part the result of a political commitment to democratic bargaining and compromise. Whereas many democratic political theorists have long celebrated these features as the marks of a well-functioning and politically legitimate government, technocratic thinkers see them as an exercise in irrationality—a mode of governance which produces ineffective policies that generally compound the very problems they attempt to solve. Such processes, in this view, should have little or no place in a modern complex technological society. In the high-tech information society they are taken to be the road to folly (Fischer 1990). Climate change policy is taken to be a case in point.

To put it succinctly, technocratic thinkers see politics—especially democratic politics—as the problem rather than a solution. In the Western context, this view is reflected in a belief that the contemporary malaise—budgetary crises, extensive poverty, educational decline, energy shortages, environmental pollution, and not least climate change—is largely attributable to the way decision-making processes are organized in democratically structured political systems. In the exacting world of technology, according to technocratic thought, it is absurd to ask the politician, let alone the person on the street, to decide complex policy issues concerning genetically modified foods or nanotechnology. Political issues, for such thinkers, need to be redefined and dealt with in scientific and technical terms (Fischer 1980; 1990).

Vivid contemporary examples of this underlying belief in technocracy have been witnessed in Europe after the 2008 onslaught of the ongoing financial crisis, particularly as it pertains to the plight of southern European states such as Greece, Italy, Spain, and Portugal. Given the failures of the political systems of these countries to bring their expenditures under control, as well as the inability to restart the processes of economic growth, numerous writers and other public officials called for a turn to technocrats as a way to sidestep the corrupt and incompetent practices of governing political elites. The most celebrated case has been that of Mario Monti in Italy, who replaced Berlusconi as the prime minister of the country. An economist by profession, he was widely praised for his ability to apply professional reason to the country's problems. Numerous influential publications such as the *Wall Street Journal* and *The Economist* readily supported such measures. The implication of this support of technical expertise was clear: democratic politics took a backseat, to the degree that it was a concern at all.

Many accept the merits of such arguments, even if they find this awkward to admit. But just as significantly, one does not need to reflect only on the Euro crisis. Even more generally, Müller (2016: 83) describes the European Union itself as technocratic. In an essay about the "democratic deficit in the EU" he writes that "for many observers the accounting language reveals that even EU democracy can only be talked about in technocratic terms."

Western democratic regimes are, one has to concede, rampant with contradictory if not irrational politics. Many environmental policies that are decided politically do not work very well, work at cross-purposes, or don't work at all. Democracy, as such, has its troubles; with this it is difficult to take issue. What is far less clear, however, is whether the turn to technocratic strategies offers a better alternative.

Technocratic experts are not in political control, surely to the dismay of some, but their information is a key resource in the governance of contemporary society. Access to such technical knowledge and skill, moreover, helps to sustain the power of top-level political and economic elites. And, conversely, it is the lack of such access that hinders the possibility of an active and meaningful political engagement in the decision processes for a large segment of the public. Thus, the allegiance of these "technostructures" to the top elites—liberal or conservative—is one of the critically important features of the modern-day political-economic power structure (Galbraith 1967).

Although one can write off technocratic theory as an ideology of the experts, there are aspects of the technocratic argument that are difficult to push aside when confronting the environmental challenge. There is, in fact, a fundamental tension between science and democracy that many democratic theorists tend to ignore or downplay. As this can be at the expense of fully grasping the problem of democratic environmental governance, we turn to an examination of this underlying tension, if not conflict.

Ecological Reason and Democratic Decision-Making: Clash of Rationalities

Representing two different modes of inquiry, science and democracy are in numerous ways incompatible, at least on the surface of the matter. Whereas democracy follows the logic of open, inclusive discussion, the science of ecology—like science in general—privileges the possession of expert knowledge, rejecting ordinary opinion. Where one defers decision-making until all interests have been heard and then seeks to find a compromise between them, the other accepts only those with the proper methodological training and chooses on the merits of the analysis (Fischer 2009; Diesing 1962).

Environmentalists, from their perspective, have long seen science and technology to be a big part of the problem. One of the primary reasons has to do with the close relationship of science and technology to industrial development. Although science and technology clearly made the industrial revolution possible, and thus the high standard of living for many in the West, technologies based on coal and oil have later been overextended, leading to many forms of environmental degradation and the health problems associated with them, both for humans and the animal world.

This "Faustian bargain" further manifests itself today in the fact that we need the scientific techniques of an "epistemic community" to identify problems that have resulted from the steady advance of modern technologies (Haas 1992; Bernstein 2002: 122–77). Many of today's problems, as already noted, require a high-powered microscope to even detect them, such as invisible cancer-causing particles in the air that we breathe. We become, in short, dependent on the very forces—some say culprits—which created the problem.

On a deeper theoretical level, the techno-scientific worldview is seen to have its origins in the scientific domination of nature (Horkheimer and Adorno 2002). As the modern-day legacy of the eighteenth-century Enlightenment, science and technology, long since the privileged modes of reason, have led to an instrumental rationalization of the world, both natural and social. Instrumental rationality, geared to functional tasks, has increasingly subjugated traditional social structures to a form of technical-managerial control. From the organization of labor and modes of communication to economy and government, scientific prediction and technical control rather than political and social deliberation have come to shape the dominant ideologies intended to guide society. Environmental destruction can, in this sense, be interpreted as nature's revenge for this technical oppression.

Many environmentalists have argued that further reliance on science and technology, at least the kinds of science and technology we have had in the past, will only lead to further ecological disaster. But whether, as some argue, more democracy is the answer is not altogether straightforward. Indeed, many have maintained that democracy and science are ultimately incompatible. There are two issues here in particular, one political and the other epistemological. The political worry is that the demands for new technologies to deal with the consequences of environmental crisis will simply overshadow and push democracy and social justice to the side. This would especially be the case if or when the consequences reach emergency proportions. Simply "solving" the problem will take precedence in the minds of many. Essentially the worrisome problem that we examined in the first chapter, it is fundamentally a political consideration.

In this regard, Devall and Sessions (1985) have argued that modern science typically employs narrow mechanistic definitions and assumptions about knowledge and facts. While such definitions work in a field like physics, this misses the *perspectival* character of politics, in particular democratic politics. Given that the ecological system is always relational by nature, traditional science is seen to be blind to the underlying dynamics of the environmental challenge. In this view, science is poorly suited to deal with multiple socio-ecological realities in a democratic society. Habermas (1970: 63) more or less echoes this argument in his early work. Lamenting the instrumentalization of modern life, he has argued that we must confront the "pessimistic assertion that technology excludes democracy." Although he does not see this as a necessary conclusion, Habermas recognizes that events could go in this direction.

The second concern is more deeply epistemological in nature. Democracy and science are seen to simply follow different logics, which at times—at least on their own terms—are irreconcilable. Whereas democracy is about compromise among competing interests and values, science seeks the best possible answer (or solution), independent of interests. While one seeks to preserve the capacities of the political decision-making structure by discussing issues until all major opposition disappears, scientific rationality focuses more on things as they are and the effectiveness of interventions designed to improve or alter them (Diesing 1962).

With regard to the environmental issue, Goodin (1992) has put forth a strong argument in support of this incompatibility, pointing to a sharp distinction between the procedural nature of democracy and the substantive requirements of environmentalism. As he has put it, "To advocate democracy is to advocate procedures, to advocate environmentalism is to advocate substantive outcomes: What guarantee can we have that the former procedures will yield the latter sorts of outcome?" (Goodin 1992: 168).

Goodin argues that the tension between the two is lodged in a green theory of value that environmentalists see to take precedence over political institutions. In this view, there is no guarantee that democracy, deliberative or otherwise, will ensure environmental norms and values. A green theory of political agency, it is argued, cannot be derived from a green theory of value. The main goal of green politics should from this perspective be the promotion of core green values and the protection of the environment, as a commitment to a particular form of democracy cannot be derived from these core values; it is therefore a secondary consideration. About this Goodin is not altogether wrong, although it is not the end of the debate.

We can also detect this tension in a much-cited conceptualization of "ecological rationality" put forward by Dryzek (1987). He defines ecological rationality as "the capacity of ecosystems consistently and effectively to provide the good of *human life* support." From this view of ecological rationality, "what one is interested in is the capacity of human systems and natural systems in combination to cope with human-induced problems" (1987: 36). Such a conceptualization poses problems for ecological democracy as it only refers to the support of human life with no reference to the values of democracy and social justice. Although Dryzek has long been an advocate of environmental democracy, the definition conveys the functional nature of ecology and thus the tension with democratic deliberation. If the ecological system has its own functional imperatives, then they must be recognized and dealt with on their own terms, leaving little to discuss. Humans simply need to adapt. But later, we shall see, that adaptation itself is not so simple; not so surprisingly, it too involves political choices.

Torgerson sees this orientation within green political thought to demand, even guarantee, a "latent authoritarian tendency." It does this by seeking to put the green principle beyond dispute (Torgerson 1999: 126). Or as Smith (2003: 67) explains it, "the contingency and uncertainty inherent in decision making within democratic institutions becomes unacceptable to more fundamentalist greens." And just here we find a convergence with both the technocratic mindset and authoritarian calls for an environmental elite to guide economic and social development.

For technocrats, arguments about the perspectival nature of democracy overlook a more important point, namely its apparent inability to respond to the crisis at hand. In their view, democracy, at least as currently practiced, is itself an ideology. This is captured in the writings of Shearman and Smith (2007), who unapologetically put forward a full-blown case for enlightened and benevolent rule by experts.

The Case for Technocracy: Benevolent Guardians or Eco-Dictators?

Many technical people, as we have noted, tend to lose patience with democracy. And this frustration is not always difficult to understand. In view of the seriousness of these environmental problems, it would often not be unfair to describe many—or most, some would say—current political-economic efforts on the part of contemporary democratic systems to deal with problems such as global warming as little more than limited symbolic gestures, especially given the pressing constraints of time. Frustrated by political disappointments, many see technological innovation as the best hope of sustainably addressing the problems ahead. Not only do they call for new technologies, particularly for a transition to a low-carbon future—technological innovation is seen for more than a few as a way to bypass the failures of democratic politics.

There is no starker statement of this argument to look beyond democracy than that put forward by the famous British scientist, James Lovelock (1979), founder of the Gaia theory that holds that the Earth is a huge, self-regulating organism. Humankind, in his view, is not smart enough to come to grips with climate change. As he has put it, "I don't think we are yet evolved enough to the point where we're clever enough to handle a situation as complex as climate change. The inertia of humans is so huge that you can't really do anything meaningful." For this reason, he argues that government by the people will have to be put on hold. In his words, "Even the best democracies agree that when a major war approaches, democracies must be put on hold for the time being. I have a feeling that climate change may be an issue as severe as war" (quoted in Hickman 2010).

This can sound draconian, but some see it as an enlightened intervention in the name of saving us from ourselves. From the technocratic environmental perspective, our existing liberal government or capitalist democracies (subsequently referred to here as "liberal capitalist democracy") have in fact not proven to be up to the task of saving us from the coming crisis. Given the failures of political and economic leadership under this system of governance, so it is argued, a more authoritarian polity will be required to introduce stronger environmental controls (Hardin 1968; Ophuls 1977; Ophuls and Boyan 1992; Heilbroner 1974), referred to by its critics as "eco-authoritarianism" (Humphrey 2007). In no way overlooking the implications, Hardin (1968: 145) argued that under the circumstances "injustice is preferable to total ruin." Heilbroner (1974: 130) put it this way: "candor compels me to suggest that the passage through the gauntlet ahead may be possible only under governments capable of rallying obedience far more effectively than would be possible in a democratic state. If the issue is survival, such governments may be unavoidable, even necessary." Ophuls

(1977: 15) wrote that "democracy as we know it cannot conceivably survive" and called for a "Priesthood of Technologists." Bahro (1987) similarly argued for the introduction of an "Ecological Council of knowledgeable leaders" (a "Salvation Council") with overriding power to protect the environment, at least until a sustainable state is reached. For many of the critics Ophul's priesthood and Bahro's ecological council are in effect forms of eco-dictatorship.

This early discussion of authoritarianism in the 1970s seemed at first short-lived. As Taylor (1996: 87) has pointed out, it faded because the crisis seemed exaggerated and because it was self-destructive for environmentalists to advance arguments against democracy, members of Earth First! in the U.S. being perhaps the exception. But the argument never altogether disappeared. In more recent years, for example, Westra (1993: 128) impugns "the sacred cow of democracy." In her view, liberal democracy is in significant part responsible for the ecological crisis, rather than part of the solution. It is unlikely, she argues, that the citizens of such democracies will "freely embrace the choices that would severely curtail their usual freedoms and rights . . . even in the interests of long term health and self-preservation" (Westra 1998: 198). She thus calls for a global top-down regulatory expert commission that should have a direct role in benevolently ensuring the ecological future of life on the planet. Somewhat similarly, Barber (2013) argues that the existing forms of liberal democracy are the "enemy" of sustainability; democracy as constituted will fail to rise to the challenge. Jamieson (2014) argues that global warming confronts us with the greatest collective action problem that humankind has ever confronted, but our evolution has not supplied us with the ability to deal with it. He asks if "democracy is up to the challenge of climate change" (2014: 44). Thus far we have yet to develop governance processes well enough designed to help us solve such problems. It is not clear, he concludes, that democracy will be able to deal with the crisis of global warming. Burnell (2012) also sees the future emphasis on democracy as potentially dysfunctional for the protection of the environment, and climate change in particular. For him, more important than a blanket call for more democracy is an evaluation of the political capacities of states, developing nations in particular, to respond to ecological destruction and its social and political consequences. And not least important, Giddens (2009) makes a plea for removing climate change policy from politics and placing it in the hands of a more expert-oriented centralized planning process geared to achieving specific ecological goals.

Other writers, following the line of "survivalist" thinking, have asserted that it is better to choose the kind of authority that we prefer to live under ourselves. As such, an enlightened expertocracy is seen to offer the best choice (Shearman and Smith 2007). In this view, experts will not only most wisely introduce the kinds of difficult technological choices that are needed, they will supply the kinds of informed central guidance that will be called for.

This technocratic mentality is shared by many, although those who share the view are often unwilling to openly admit to it as such. Many or most of those who hold it today either don't present their beliefs as technocratic, mainly because of its negative implications for democratic society, or they don't recognize or accept their views as technocratic. Others have addressed the issues head on, calling for the enlightened rule of scientific technocrats. In this latter case, the example par excellence is a book by Shearman and Smith (2007). Boldly stating the case, they present a stark modern-day argument for technocracy; they offer an updated version of the technocracy argument as applied to climate change. Although this argument is unlikely to prevail politically, at least before we confront the full force of the climate change crisis, it is instructive to examine its direct attack on democracy.

Democracy, argue Shearman and Smith (2007: 1), is the "Holy Grail of Western Culture." It is, in their view, taken to be synonymous with both freedom and prosperity. The emphasis placed on personal liberty, as they see it, does not take seriously enough the dangerous threats to "the continuation of human life and civilization." Referring here to the coming global climate crisis, they seek to present the reader with a discussion of a "problem of such magnitude that issues of personal liberty pale into insignificance." Indeed, they offer an argument against democracy, at least as contemporarily understood, contending that it portents to propagate an unprecedented form of "environmental tyranny." In fact, they go so far as to argue that "the fruits of liberal democracy may prove to be more bitter than even the gulags of the Soviet system, as horrible as the gulags were" (2007: 2). This is strong language, even a serious exaggeration. But it does convey their unwillingness to accept contemporary versions of democracy.

It is not that authoritarian systems have been good on environmental protection. The environmental record of ecological destruction under Eastern European socialism was often far worse than that encountered in the West. But this fact, they argue, does not in and of itself establish the case for democracy, especially when it comes to environmental survival. Moreover, they contend that there are other forms of authoritarian government and one of them, based on expertise, offers us the best chance to deal with the coming crisis. Returning to Plato's call for rule by "philosopher kings," they see the need to reject liberal democracy and substitute it with a benevolent system of technocratic government.

The starting point for their analysis is thus to question "the true record of democracy in addressing and preventing major issues besetting humanity today, such as war, equity, and especially environmental damage" (2007: 3). Citizens living under liberal democratic-capitalist systems "are cushioned from these happenings and their consequences" by the ideologies of consumerism and the pursuit of unsustainable Western lifestyles. The result is a form of ideological and physical manipulation that ultimately usurps the very system that advances it. It is a system based on power rather than truth; it is a regime that keeps its citizens in a state of ignorance. This edifice of political deceit, in their words, "is surrounded by a shell of spin and bureaucracies and scientists chosen for their compliance," while at the same time thoughtful, informed citizens "accept their impotence to influence events, and the professional image of politicians rank at the lowest along with used car salesmen."

Are these democratic systems, they ask, "able to grasp and remedy the emerging ecological crisis facing the entire human race"? Not only are the institutions questioned, but also the citizens they represent do not always appear to be up to the task of intelligently judging these issues. Indeed, the level of knowledge about climate change and other environmental issues in the general public is often disturbingly low. From this perspective, the reformist approach only serves to perpetuate the very system responsible for the problem. Although Shearman and Smith concede some merits to democracy, they argue that the system as a whole is unsatisfactory and unsustainable in the long run. In their view, we will have to trade the liberty offered by liberal democratic government for a system of governance which recognizes survival to be the paramount concern. Here they echo writers such as Hardin and Lovelock.

To support their point, they appeal to a medical analogy. The Earth's environmental systems, in short, require intensive care. The Earth is suffering from "multiorgan failure." Under such conditions we cannot predict the outcome in advance: "ecological and medical science cannot tell us whenever the human body or the ecological system has reached the point of irretrievable collapse." But we can draw lessons from patients in intensive care treatment. It is here that we encounter the technocratic argument—the case for expertocracy. They put it this way (2007: 7):[4]

> The patient's resuscitation is in the hands of a leader, the expert doctor in intensive care, and a team of nurses and scientists, which combines leadership with expert knowledge, decision making, speed, dedication, and compassion. The leader does not explore the public opinion polls to see what can be tolerated or is popular. He or she does not act to preserve their position at the next election and is not influenced by corporatism or the perceived state of the economy. There is one collective, unsullied goal, to recognize the emergency, to make a skilled diagnosis based upon scientific assessment and to restore health before the situation becomes irreversible.

Seen from the perspective of a "sick Earth" in need of intensive care, the experiences drawn from medicine, as they see it, suggest the need for a strategy based on more substantive criteria than those of liberal institutions and the electoral strategies to which they give rise. The problem is found in the meshing of liberal capitalism and liberal democracy and the critique is harsh. Liberal capitalism and the modern corporation have "become part of the genetic material of democracy" which is seen to be directing the system (2007: 15). For this reason, in their view, large-scale environmental destruction, both ongoing and impending, will not be ameliorated by liberal political systems. The argument that we have to rely on democratic reforms, election after election, decade after decade, is seen as a course for certain disaster. When major heart surgery is required, nobody wants the operation to be directed and managed by a democratically elected group of surgeons. Democracy, then, and liberal democracy in particular, is "not suited to deal with crisis care situations."

Already, as Shearman and Smith contend, we are headed for a form of authoritarianism. Liberal capitalism, they argue, is a force that already produces a form of authoritarian rule by corporate elites. Although enmeshed with liberal democracy, its ultimate goals are antagonistic to it, and in the long term act to undermine it. They thus predict that democratic regimes, much like their communist counterparts, will prove to be only a very brief moment in the history of humankind. Their transformation into authoritarian rule is likely to be catalyzed by their failure to deliver solutions to the environmental crisis. The solution is to be found in a centralized system that can direct a technological revolution as vast as the Manhattan Project and the NASA space endeavors. Requiring an unleashing of the potential of technical expertise, a project that must proceed in all developed countries, not just America, it needs to be put forward with the socio-economic foresight and political legitimation of a Marshall Plan.

And they take the argument a step further. Where most writers in this vein merely make the case in relative normative theoretical language, Shearman and Smith name an exemplar. For them, Singapore, an authoritarian country with a strong techno-managerial orientation, offers a model of governance for the future. Indeed, for them it is an example of benevolent technocratic governance. Many will find this disturbing, but it is important to note that various Asian governments, China in particular, have taken a strong interest in this system of governance (Ortmann and Thompson 2016; Beeson 2016).

Shearman and Smith are an exception to the general rule. Rather than advancing a bold call for rule by experts, technocratic arguments are usually stated less directly. Mostly such writers tend to lodge their technocratic assumptions more in the arguments underlying their positions. In many ways they gear their arguments, one way or another, to specific concerns that have more generally emerged in the course of environmental debates. Often, moreover, they are more optimistic about the role of technology and the ability of liberal capitalism to come to grips with the environmental crisis. They typically make the case for an ecological Manhattan project in search of new sources of energy, if not geoengineering of the atmosphere. A representative of this more subtle technocratic orientation is Anthony Giddens. Although Giddens would most likely eschew any reference to technocratic thinking, his work is illustrative. Toward this end, we turn to a brief examination of his widely discussed strategy for dealing with climate change.

Decoupling Environmental Protection from Social Change

In Giddens' (2009) work, in comparison with that of Shearman and Smith, we encounter a more subtle example of the way technocratic thinking undermines democracy without actually criticizing it. Offering in no uncertain terms a technocratic rejection of ecological politics and the environmental movement, Giddens finds the solution in "ecological modernization" rather than changes in capitalism and the lifestyles it has generated. It is the view that wins favor in the centers of power.

One of the most influential sociologists in contemporary times, Giddens has taken the position that climate change is different from other environmental problems and should not be considered in light of other social and political problems—in particular, it is not a problem of political economy or social justice in his view, or what Eckersley (1992) has referred to as "emancipatory environmental theory." He even argues that we should not seek to reduce carbon emissions at the expense of further economic growth. Developing countries should also not be asked to make major contributions to carbon reduction until they have a chance to catch up economically. Economic growth and environmental protection, in this view, need to be disconnected and treated separately. The answer to the problem of climate change is to be found in a crash program in technology development and innovation. This might require forms of planning, but such planning would not be of the traditional socialist variety. Rather it would involve approaches that would assist and direct markets with the transition to ecological modernization, an approach to which we return.

In support of his position, he criticizes the relevance of the environmental movement, contending that it is value-oriented instead of focused on the problem at hand. In a view that contradicts aspects of his earlier writings, Giddens complains that environmentalism is grounded in "life politics," which he affirmatively defined in previous writings as a reflexive understanding of humankind's relationship to nature (Thorpe and Jacobson 2013). But now such life politics is seen to be problematically in tension with the capitalist nation-state. Essentially, he advocates for what can be described as a rather narrowly defined instrumental approach based on market-based strategies. Questions and issues related to lifestyles and social values posed by the environmental movement should be cordoned off from the discussion and analysis of climate change. In the name of urgency, he advocates a politics grounded in a belief in technological progress and reliance on expert systems. Devoting little attention to democracy and citizen participation, he rejects the importance of grassroots movements and the call for sustainable lifestyles. Ethical questions about how we should live together sustainably need to be replaced by pragmatic strategies from the top down. Toward this end, the agents of change will be a mix of governmental, business, and technical elites. Unabashedly, he sees the solution to climate change to be found in the "technological fix."

Techno-Environmental Expertise: Politics and Power

Finally, we need to take up the question of power as it pertains to technocracy. Many writers have argued that the technocracy critique is a "straw man" thesis standing on thin propositions (Brint 1994). More often than not, the thesis is rejected because experts remain subordinate to the governing elites. Given the usual definition of technocracy as the assent to power of technical or knowledge elites, the judgment is not on the surface of the matter false. The assessment, however, neglects other significant dimensions of the technocratic phenomenon. For one, as Laird (1990) has written, the role of technocracy rests in important ways on the fact that it shields—or can shield—governing elites from political pressure from below. It does this in part by employing technical languages that frame problems in ways that make it hard for normal citizens to engage with the issues, especially in public decision processes.

This concern is all the more important insofar as technical experts increasingly define the issues, which, as Schattschneider (1960) long ago pointed out, is one of the basic sources of power. Particularly significant here is the techno-managerial tendency to turn social and economic problems into technical issues that lend themselves to administrative rather than political solutions. This apolitical strategy not only undercuts democratic politics, it also serves to narrowly frame the issues in ways that distort understanding. This can be understood on the level of managerial strategy, but it can also be understood as a deeper social phenomenon associated with modernity. As Foucault made clear, society over the past centuries has been increasingly organized around expert knowledge—that is, around forms of professional knowledge that define and shape the issues that confront modern societies. In Foucault's theory of "governmentality," this is described as a form of power increasingly exercised by the professional disciplines. As this is a strong form of technocratic power, we need to consider it further.

For Foucault (1972) and his followers, the "political" in the contemporary world can no longer be adequately understood in terms of dominant elites and centers of power. Power now has to be seen as multiple and diversely decentralized (Foucault 1980). Manifested in multiple, ubiquitous forms, political power no longer just belongs to the state alone; it is dispersed and in effect everywhere. It is at work among psychiatrists who determine the social and medical status of homosexuality, street-level social workers who interpret the needs of the poor, or risk assessment experts who decide the levels of "acceptable risk" that should form the basis for environmental policymaking. From this perspective, these newer forms of "disciplinary power" are fundamental to the professional disciplines themselves. As the agents of expert discourses, the professions constitute the techniques and practices that disperse power and social control away from the formal centers of governance (Fischer 2000: 24–8). Such bodies of knowledge and their discourses are the foundations of extensive administrative forms of regulation. Combining knowledge with profit and power, this new disciplinary order provides a way of controlling large numbers of people, rendering their behavior stable and predictable, without using uneconomical and ostentatious displays of sovereign power, in particular police or military force which can risk open protest or rebellion on the part of mass publics.

Expert disciplines thus take shape at the intersection of words and things, power and knowledge (Foucault 1973). The knowledge produced is part of the discursive practices by which rules are constructed, objects and subjects are defined, and events for study are identified, framed, and thus constituted. Such disciplines function in ways that can be appropriated by particular institutions, from armies to hospitals and schools. At the same time, they remain irreducible to—or identifiable with—any particular institutional form of power in society. Disciplines, in this way, are seen to shape and "colonize" modern institutions, linking them together, strengthening their instrumentality, and extending their grip on social processes (Ferguson 1985).

Professional disciplines, operating outside of (but generally in conjunction with) the state, are thus seen to predefine the very worlds they have made the objects of study (Sheridan 1980). Because this power is *exercised* rather than *possessed* per se, it is not the privilege of a dominant elite class actively deploying it against a passive, dominated class. Disciplinary power, as such, does not exist in the sense of power in the ordinary usage of the concept. Instead, it is lodged in a complex network of "micropowers" that permeate nearly all aspects of modern social life. For this reason, modern power cannot be overthrown and acquired once and for all by the destruction of institutions or the seizure of the state apparatuses. Such disciplinary power is "multiple and ubiquitous"; the political struggles against it then tend to be local forms of resistance that seek to combat interventions into particular sites of civil society. Because such power is constructed through organized networks rather than easily identifiable points of control and decision-making, each localized resistance can only be effectively leveled against the horizontal links between various points in the network. But this can induce effects on the network as a whole.

Foucault's theory has been applied specifically to environmental governance. Focused on the regulation of social interactions with the natural world, the theory of "eco-governmentality" examines the way government agencies, in combination with expert knowledge, construct our understanding of the environment. The resultant construction is viewed both in terms of the creation of an object of knowledge—i.e., the environment—and a sphere within which certain types of intervention and management are developed and deployed to further the government's larger aim of managing the lives of its constituents. Public environmental management is thus seen to be dependent on the dissemination and internalization of this knowledge/power relationship among individual actors, creating a decentered network of self-regulating elements whose interests become integrated with those of the state.

Implicit here is the claim that the types of "knowledge/power" relationships produced in the process of making nature intelligible to the state have an important influence on the evolution of state rationality itself. Through this process of environmental subject formation (i.e., defining citizens' relationships to the environment), and the creation of environmental subject positions (as socio-environmental relationships among citizens), such systems operate to normalize certain ways of acting while marginalizing others. The way different theorists approach these subject positions varies. Darier (1999), for one, focuses on the construction of the environmental subject (for example, environmental citizenship) as a site for resistance to consumerism and the commodification of the relationship between the individual and the environment. Agrawal (2005), for another, explores the ways the concept of "environmentality" denotes an acceptance of nature by the individual as an object to be managed and their own accompanying involvement in the process. Such definitions are seen as the key to the analysis of state management of the environment.

The traditional state-centered theory of power, according to this theoretical orientation, has hindered our ability to recognize these discursively based expert powers dispersed throughout the social system (Ferguson 1985). The standard emphasis on the political position of the technocrat misses the more fundamental nature of discursive power. The recognition that the most significant power of the environmental professions is lodged in basic conceptual categories of thought and language opens the door to a discursive understanding of the role of the environmental disciplines in modern society. If power is everywhere, it can legitimately be investigated at the margins as well as the center of society. This permits a shifting of attention away from state politicians to civil society actors, in particular environmental movements.

Works by Rutherford (1999) on U.S. Environmental Impact Assessments methods and by Agrawal (2005) on local forest governance in India are examples of this method of analysis. Both illustrate how the production of specific types of expert knowledge (statistical models of pollution, or the economic productivity of forests) coupled with specific technologies of government (such as Environmental Impact Assessments or local Forest Stewardship Councils) can bring individual interests in line with those of the state. It is thus the creation of frameworks that rationalize behavior in particular ways rather than just the analysis of specific outcomes that becomes the focus of inquiry. In this regard, one could perhaps conceptualize the Intergovernmental Panel on Climate Change (IPCC) as a form of disciplinary power in response to the crisis ahead.

From a similar perspective, Luke (1999) has provided a good picture of the philosophy, methods, and discursive practices of these forms of environmental expertise, which he describes as "technocratic eco-managerialism," by looking at the production and reproduction of such knowledge in the modern university. As he explains (Luke 1999: 103), the modern research university becomes a "vital site for generating, accumulating, and then circulating such discursive knowledge about nature, as well as determining which particular human beings will be empowered to interpret nature to society." In this view, "these educational operations now routinely produce eco-managerialists, or professional-technical workers with the specific knowledge—as it has been scientifically validated—and the operational power—as it has been institutionally constructed—to cope with 'the environmental crisis' on what are believed to be sound scientific and technical grounds."

The mission statements of these schools and programs speak of a need to produce leaders necessary to deal with environmental problems. By and large, the focus is reform-oriented with emphasis on practical eco-knowledge, workable solutions, and the management of resources. Such programs present environmental problems as in need of experts trained in various aspects of ecological modernization. Implicit is an understanding that the problems are too complicated for lay citizens to deal with. Indeed, there is little about the environmental movement, civil society, or participatory governance to be found in these programs. Rather the emphasis is placed on scientific ecological training and the placing of such ecological professionals in positions where they can advise and guide modern state-corporate institutions confronted with critical problems.

It is a view embedded in the primary techniques that the students learn, in particular, ecological economics, cost-benefit analysis, risk assessment, resource management, and nature conservation (especially for the use of public lands and parks). As such, professional technical knowledge replaces the traditional naturalist's humanist understanding of the environment with an academic discourse that emphasizes positivist knowledge about ecological causes and consequences that is instrumental for the design and implementation of solutions to particular problems. As Luke put it, the approach evolves from a managerial perspective which empowers experts (such as engineers and scientists, corporate managers, and financial specialists). Eco-managerialism, with the assistance of central governmental guidance, now extends this logic by imposing corporate managerial practices upon the ecology in an effort to provide the corporate economy with the materials it needs. This involves a better coordination of economic production patterns, resources utilization, and consumer consumption. Toward this end, the goal is not only to protect the environment, but also to learn enough about the natural system in an effort to redesign and reassemble it in ways that assist in efficiently producing and redistributing resources at levels capable of supporting the contemporary system of economic relationships (Keulartz 1999).

While many students surely enter these programs with an interest in protecting the environment on its own terms, the curriculum primarily trains them for the instrumental—largely technocratic—tasks of eco-managerialism geared to adapting and managing the requirements of an ecologically modernized economy. To a considerable degree, as Luke explains, these programs serve—intentionally and unintentionally—to deal with a wide spectrum of eco-rationalization projects that seek to redesign land, forests, and water in ways that permit engineers and managers to administratively guide them. Knowledges about ecological systems "are not objective timeless verities, but rather are the operationalized findings of continuously evolving practices for those forms of homogeneous engineering that have been constructed by major research universities" (Luke 1999: 118). These schools are places where knowledge about production, distribution, regulation, and circulation are produced by environmental scientists to help guide the decision-making bodies of liberal governments and capitalist enterprises.

Conclusion

This chapter began by pointing to the evolution of technocratic practices in the development of modern-day environmentalism. This set the stage for a more specific examination of the basic arguments for technocratic policy-making, in particular as a call for an apolitical alternative to democratic policymaking. We then examined the more fundamental tensions between the rationalities or logics of ecology and democratic governance. This included Goodin's oft-cited analysis of substantive environmental goals and the political processes by which they might be decided. In this view, as we saw, a green theory of political agency cannot be derived from the substantive theory of green values. Green goals and values are seen to have priority, independent of political opinion.

The discussion then turned to the writings of contemporary theorists such as Shearman and Smith who emphasize the failure of democratic systems to take necessary actions in the face of the coming climate crisis. Alluding to Plato and his philosopher king, they offer an explicit call for rule by experts. Differently, Giddens was seen to offer a more subtle case for subordinating democratic practices to an emphasis on technology. In view of the seriousness of the coming crisis, and the limited time available, he has argued for decoupling questions of eco-technological development from other social issues often advanced by environmentalists, in particular the need to change modern lifestyles. This is, in short, a call for a crash program in ecological modernization with a heavy emphasis on technological development. Finally, the chapter turned to the question of technocratic power, examining the sophisticated theory of disciplinary power put forward by Foucault and his followers. Along the way, this outline of the broad technocratic thrust has set the stage for a more specific examination of technological strategies, to which we turn in Chapter 3.

Notes

1. Remarks at the U.S. Department of State, Washington, DC. September 28, 2007.
2. For example, it might be discovered that a power generator located in a particular community would spew toxic chemicals into the air that would lead on average to one death for every million local residents per facility per year and would offer power for $0.11 per kilowatt hour of electricity. Another type of generator, it might be determined, could lead to an average of two deaths per million community member per facility, but would offer power for $0.08 per kilowatt hour of electricity. For the risk-benefit analysis, these two types of impacts—death per million and price per kilowatt hour—are said to be "objective categories," as their actual levels are taken as empirical facts (Hiskes and Hiskes 1986).
3. Putnam (1997: 385–7) has explicated from the technocratic literature six fundamental tenets basic to this ideology and the "mentality" it shapes. (1) Technocrats believe "that 'technics' must replace politics and define their own task in 'apolitical' terms." (2) They are "skeptical and even hostile toward politicians and political institutions." (3) They are "fundamentally unsympathetic to the openness and equality of political democracy." (4) They believe that social and political conflict

is, at best, judged to be "misguided, and at worst contrived." (5) They "reject ideological or moralistic criteria, preferring to debate policy in practical, programmatic terms." (6) They are "strongly committed to technological progress and material productivity" and are "less concerned about the distribution questions of social justice." Politics is seen as a process that "can and ought to be reduced to a matter of technique, that is . . . political decisions should be made on the basis of technical knowledge, not the partial interests or untutored value preferences of politicians."

4. Shearman, it can be noted, is a medical doctor.

3

Technocratic Strategy as Central Steering: From Sustainable Development to Transition Management

In this chapter we continue the discussion of technocratic theory and practice by examining its implications for governmental steering. We explore it as an approach for technological developments to deal with climate change and examine the innovative Dutch strategy of transition management designed as a "new mode of governance for sustainable development" (Loorbach 2007). Developed as an attempt to identify socio-technical options and to move them into the policy decision processes, the strategy illustrates the way in which technocratic thinking can unwittingly seep into projects with a wider set of goals.

Sustainable Development and the Technocratic Challenger: Ecological Modernization

The idea of sustainable development, as a theory and practice, has organized and guided much of environmental politics since the late 1980s. Put forward by the United Nations' World Commission on Environment and Development (also known as the Brundtland Commission), the concept of sustainable development has been the official response to the problem of excessive economic growth and its implications for environmental protection, especially as it manifests itself in terms of the economic tensions between the wealthy industrial and poverty-ridden developing nations (World Commission on Environment and Development 1987). According to the advocates of sustainable development, and its later variants, the answer to the problem is to be found not so much in limiting growth, but rather in inventing a new kind of growth, "sustainable growth." Instead of calling for general reductions in the rate of growth per se, as was the issue for environmentalists at the outset, the stated goal is to pursue guided growth capable of meeting the requirements of the present without compromising the ability of future generations to meet their own needs (Carter 2010: 207–37). Adopted by the United Nations and thousands of nongovernmental organizations as the basic framework for global protection, adoption of the concept scarcely stopped there. Sustainable development became the environmental creed more generally. Environmental experts speak of sustainable cities, sustainable corporations, sustainable agriculture, sustainable communities, sustainable population growth, and so on (Fischer and Black 1995).

The concept's strengths, however, are at the same time its weaknesses. Beyond the rhetoric of sustainable development, a closer examination of the idea shows it to blur the underlying ambiguities upon which the concept has been constructed (Torgerson 1995). Because the relationship between environment and economy in sustainable development was not clearly specified, the concept was never without its problems, or at least not for long. At the outset, it had the advantage of bringing environmentalists and industrialists back to the table after the stalemate resulting from the "limits-to-growth" debates of the 1970s. The fact that different interests can read competing meanings into sustainability offered a place for both of them, opening the way for new discussions. But both environmental groups and industrial leaders soon fought over the meaning of the term. For the environmentalists it was about environmental protection; for the developing world and corporate industry it was about growth. Even though sustainable development combined environmental protection and economic growth, it has never answered the hard question—namely, how to strike the balance between these two very different goals. This failure has proven particularly difficult for deliberations between the developed and developing countries. While the ambiguities were precisely what made the concept's unusually wide acceptability possible, they also rested on the same assumptions—even contradictions—that have made it possible for different proponents to support the idea. In short, all can read their own competing meanings into sustainability (Fischer and Black 1995).

But there is more involved than ambiguity. Typically overlooked is the fact that sustainable development subtly weaves in a powerful political bias that favors the developers. The effort to build a bridge between environmentalists and industrialists, whether in the First or the Third Worlds, clearly plays to the advantage of industrial interests. Insofar as it resituates environmental discourses within the frame of economic growth, or at least requiring an accommodation between them, it shifts or softens argumentation away from the movement's earlier calls for "limits to growth." In doing so, it serves to privilege the already powerful interests of industry over the harsher environmental critics. Not only did it set the discourse on a new footing much more favorable to industrialists, the industry advantage is in many ways built into technical-analytical models that emerged to guide sustainable development as both a theory and a strategy, such as cost-benefit analysis.

What we also see upon a closer look at these conceptual tensions is that the development bias is also built into the technocratic interpretation of sustainable development that came to dominate environmental policymaking. As a programmatic strategy for global environmental protection, the refinement of sustainable development in the hands of planners and managers—both academics and practitioners—increasingly emerged as a new techno-managerial discourse for guiding socio-technical environmental change. This was especially the case with regard to a variant or revision of sustainable development known as "ecological modernization," a thinner approach which stresses the role of markets, technological innovation, and government–business partnerships largely at the expense of socio-ecological change. Much of contemporary environmental policy, particularly in the corridors of power, is based on this alternative interpretation put forward under ecological modernization.

As a weaker approach to sustainability, the theory of ecological modernization gained enthusiasm for several interrelated reasons. In part, it has been a response to the fact that sustainable development has run into theoretical or ideological problems (Carter 2010: 207–40; Young 2000; Brand 2010). And in part it is a response to the fact that sustainability has not taken hold in practice. In any case, sustainable development, as conceived by progressive environmentalists, has planted few deep-seated roots. Indeed, there are no countries that have fully adopted sustainable development as a systematic practice. This led, for differing reasons, to a retreat to the weaker version of ecological modernization. Most significantly (or most problematically, some will say), ecological modernization has sought to avoid the kinds of fundamental economic and social structural changes called for by sustainable development (Hajer 1995: 32). Rather than fundamental greening of the system as a whole, ecological modernization limits its reforms to a policy subsystem. As such, environmental policy is only one policy domain competing with the other domains—business, transportation, social welfare, and the like.[1] It attempts to sidestep the very issues that the critics of sustainable development have found problematic with the approach (issues that often have also made it difficult to adopt and implement sustainable development). For others, it has simply been the fact that sustainable development has failed to take hold—or even gain much ground—in the practices of either the state or corporate industry. Ecological modernization thus emerged as the "reform" alternative more compatible with the existing system of liberal capitalism.

More recently it would seem that ecological modernization has been replaced or augmented by calls for a "green economy" or "green growth." But these, it can be argued, are to a considerable extent only newer rhetorical flourishes in the face of the failure of specific formulations to take hold. Further, a closer look shows that these variants have more or less only repackaged the technical components of ecological modernization. In all of these formulations the way forward is seen to depend on new eco-technological innovations with business playing a central role in bringing that about. Thus, if ecological modernization, as a concept in environmental discourse, might appear to be fading, it is still alive and well in the call for eco-technological innovation strategies for green growth.

Whether ecological modernization or eco-technological innovation, this technocratic strategy can be interpreted as a victory for liberal capitalism over traditional understandings of environmentalism. It is essentially a more linited effort to ameliorate environmental problems by building more modest or manageable ecological criteria into the existing production and decision processes. What is more, its theorists—such as Giddens, as we saw earlier—now typically elevate the role of economic growth (including the role of profit-making) to a central place in ecological renewal, arguing that it is necessary for environmental protection. Growth, in short, is said to provide the resources to finance the required environmental transitions. Rejecting the earlier limits-to-growth argument, the emphasis on technological innovation is seen to be the specific precondition for environmental revitalization. In this view, only growth driven by technological change can provide the kinds of resources needed to tackle the challenges ahead. That is, instead of treating environmental amelioration as an economic burden, a technologically driven environment transition is seen as a potential source of further growth. In this conceptualization, environment and technology, as well as environment and economic competition, are closely related. Little, however, is said about how this new source of wealth would be turned into environmental protection. This political question is largely left open.

Without necessarily rejecting the earlier "end-of-pipe" technical solutions, the emphasis turns to technological design in an effort to push research into the development of ecologically sound technologies. "Technological fixes" are designed to reduce air pollution and waste, as well as to reorient the processes of manufacturing. Of particular significance is the emphasis on green technologies such as pollution abatement equipment and cleaner forms of energy, the minimization of packing materials, recycling, and the reduction of harmful chemicals in the production processes (Carter 2010: 227–37).

Consistent with free market principles, this eco-modern emphasis on technology is accompanied by the use of cost-benefit analysis to guide decision-making. The theory sees the market as the main mechanism for organizing these efforts. The market is to assume the primary role of conveying environmental ideas and specific practices, with manufacturers, financial institutions, and consumers all making contributions to the effort. Fundamental is the need to internalize the costs of production—that is, building in the cost of environmental destruction (Carter 2010). Thus, the basic market assumptions upon which democratic capitalist societies are organized remain untouched, while solutions such as carbon sequestration, biofuels, nuclear energy, and geoengineering move to the fore. In this conception, technical expertise—both as engineering and economics—take center stage. The solution, as Bjorn Lomborg explains it, is efficient technology.[2]

Toward this end, the state relinquishes its centralized function as regulator and assumes a more decentralized flexible role as facilitator of cooperative government–business partnerships. The state shifts toward various policy instruments to help facilitate the efficient changing of production and consumption in more ecologically benign ways (Carter 2010). To the degree that it acts independently, it uses market-friendly techniques such as tradable pollution permits and eco-taxes.[3]

Relying on capitalist production and the liberal state, ecological modernization strategies exclude basic concepts related to environmental struggles, north/south tensions in particular. Gone, most significantly, is the emphasis on social justice and democratic participation. Indeed, the argument tends to focus explicitly on the need to avoid such controversial issues. In this view, the critical question of how to re-balance the ecological system should not be burdened with complicated questions about social justice and democratic rights that want for solutions under the best of circumstances. In the process, the approach holds the existing distribution of power and relations of social inequality in place. To the degree that they are considered important, they should be dealt with in other contexts.

In short, both sustainable development and eco-technological alternatives have placed emphasis on science and technology in various ways. Since the emergence and evolution of the pursuit of sustainability, in all forms, it has been increasingly defined and administered by professional experts. The result, despite the cultural diversity and complexity of the global sphere, has tended to be a narrow framing of environmental degradation as a technically based problem amenable to ecological planning. Basic to the approach is the idea that the knowledge necessary to implement a strategy for sustainability can be obtained—knowledge about what constitutes basic needs, how to define aspirations for a better life, ways to manage sustainable growth, methods for judging the ability of the biosphere to absorb human pressures, which technologies to improve, and how to marshal environmental resources (Fischer and Black 1995). Nature, in this perspective, can be analytically reconstructed as a biophysical cybernetic-like system in which experts can conceptually dismantle, redesign, and reassemble it to coordinate adequate resources, without straining the global carrying capacity. In the process, nature's energies, materials, and sites are empirically "operationalized" in ways that lend themselves to predictable year-in year-out monitoring of levels of rains, soil creation, timber growth, fish populations, or agricultural output. Through the efficient application of such "eco-knowledge experts can calculate the carrying capacities of regional and local eco-systems" (Luke 1999).

Conceptualized in this way, sustainable development and eco-technological modernization are both more than normative goals; they also constitute the basis of a planning methodology for the management of an industrially sustainable future (Taylor and Buttel 1992). Indeed, this is especially the case for the more industry-friendly interpretations of this conception. To take up the assignment, both in national and international organizations, cadres of university-trained sustainable development experts and environmental planners have rapidly emerged (Luke 1999). Insofar as the knowledge needed to carry out this project is seen to exist—or is at least obtainable—all that would seem to remain is the mobilization of the moral-political will to introduce the strategy.

In the real world of environmental politics the assumptions upon which a sustainable society is built are estimable. Most problematic is the fact that the available environmental knowledge is nowhere sophisticated enough to reveal the limits of nature, thus permitting us to exploit resources safely up to that limit. We are, for example, only barely beginning to develop the capability to measure the phenomenon of global warming. The idea that we might literally monitor and manage such environmental change with the kinds of precision suggested by sustainable development experts is more a scientific ideology than a certainty within our reach, especially within the critical time frame posed by global warming.

But these limitations haven't stopped the hope for the technical fix (Huesemann and Huesemann 2011). The search for such technologies to deal with global warming, even when they border on science fiction, are receiving increasing attention in both the technical world and the political realm (*Economist Magazine* 2010; 2015). These include strategies to invent organisms that consume CO_2, the capture and sequestration of carbon by-products from burning coal, the building of nanoparticles that provide new fuels, and the geoengineering of the atmosphere, among others (see Appendix at the end of this chapter for a more detailed picture of these technologies). All hold out the possibility of finding ways to circumvent the need to fundamentally restructure our way of life. None of them offer support for environmental democracy.

Nuclear power serves as a good illustration. Although nuclear power has suffered one setback after another, including the Fukushima tragedy in Japan, its supporters do not give it up, the Conservative government of the U.K. being perhaps the best example. Despite exorbitant costs and uncertain outcomes, and even perhaps foolishness, the Tories have pushed ahead. Such political and economic leaders do their best to keep nuclear energy alive, believing that later when supplies of energy are in short supply, many citizens will prefer it to limited electricity per day. Indeed, they are now experimenting with mini-reactors which can be distributed around a country in substantial numbers. Interesting, nuclear power now receives support from a number of influential environmentalists, such as Hanson (Vaidyanathan 2015), Brand (Woody 2009), Nordhaus and Shellenburger (2013) and Monbiot (2013). Relatively recently, it is important to note, this was evident in the hallways of the Paris COP 21 meetings, where an active effort to promote nuclear energy and other technological solutions was very much a part of the discussions, driven in significant part by Bill Gates (Green 2015).[4]

Eco-Technological Innovation as Transition Management

An especially significant strategy that has captured the attention of many seeking to deal with the challenge posed by sustainability has been a focus on "transition management." While approaches such as ecological modernization have identified goals and tools for achieving a more technocratic conception of sustainability than envisioned by sustainable development more generally, such perspectives do not offer a theory of agency for carrying out the strategy, other than assumptions that business, government, and industry will—or should—pursue them, generally voluntarily. However, other eco-technologically oriented groups concerned with how to shift toward a more sustainable societal system have in recent years taken up the question of agency and action strategies, especially as they relate to the persistent issues of sustainability and the pressures of climate change (Shove and Walker 2007). Most important has been a group based in the Netherlands that developed and experimented with, and in part implemented, what they call "transition management." It is offered as a new "mode of governance for sustainable development" designed to shape governance networks in ways that avoid the limitations of both top-down and bottom-up approaches (Loorbach 2007).

The transition orientation seeks to understand and guide a "socio-technical transition" of modern society to a sustainable system for the future.[5] Employing a mix of theoretical orientations, it is most fundamentally grounded in systems theory. Drawing as well on empirical studies of the relationship of science and technology to innovation, the study of complexity, and a focus on multi-level systems of governance, transition management has opened up and advanced an important new line of social-scientific exploration. Pioneered by Jan Rotmans, a systems-oriented mathematician at the University of

Rotterdam, this work on transitions has had the unusual distinction of being not only a research project, but also a component of the program of top-level government agencies. In conjunction with policymakers in the fields of energy and environment, Rotmans and his research team managed impressively to move the transition research project beyond the academic realm. Receiving support from various ministries and political leaders, it became an approach adopted as part of the national environmental policy plan and the country's strategy for energy transition. As a five-year-long project focused on the theory and practice of transition management, it has been presented as a new paradigm for multi-disciplinary research, technological innovation, and governance (Grin, Rotmans, and Schot 2010).

One can fairly raise the question as to whether large-scale, complex systems can in fact be managed, but the transitions researchers take this as a given. The transition approach is designed to facilitate a socio-technical transition to sustainability, emphasizing resource management and ecological modernization. Indeed, Smith and Kern (2009: 78) have described it as an attempt to "reinvigorate ecological modernization." As such, it seeks to understand and advance a more efficient use of resources to meet existing economic and social needs by rationalizing the systems and subsystems that deliver them. It is driven by an interest in learning how to promote future innovations in sustainable technology that can, in turn, become institutionalized as part of the modern-day system of governance designed for ecological steering.

Taking a long-term perspective, transition management seeks to identify multi-level pathways for achieving the goals of sustainability. It has done this first by commissioning a large body of research that has sought to uncover the keys to understanding earlier innovations that have profoundly influenced and changed the world, such as the invention and development of railroads, earlier forms of communication such as the telegraph and telephone, the shift from coal to oil, and the like (Geels and Schot 2007). But even more importantly for the present discussion, it calls for the establishment of action-oriented "transition arenas" that bring together leading players to think about and deliberate on the directions and dynamics a transition to sustainability should take—scientists, engineers, policy consultants, historians of technology, industrial leaders, and government officials, among others. As Loorbach (2007: 131–56) explains, the "transition arena is an institutional space where innovators with various backgrounds, perspectives and ambitions are brought together to develop shared long-term perspectives and a transition agenda that increasingly will influence regulatory policy and our societal systems." Located outside of established policy networks, these "niche" arenas are set up to avoid capture by powerful and hierarchical vested interests, especially from the powerful energy sector (but also the sectors of transportation, agriculture, water management, education, and health care, among still others). Removed from the established subsystems, or functional "regimes" in transition management language, these groups are encouraged to "think outside of the box." They can put aside the standard criteria and modes of thought that traditionally govern thinking in the respective subsystems regimes, including the policy methodologies typically employed to measure success and failure, and the kinds of political and social pressures exercised by the dominant organizations. Toward this end, these arenas are organized and guided by "transition managers" who help them develop transition "storylines" grounded in a specific vision of a sustainable future. They do this by facilitating a process of learning through "reflexive" deliberations designed to inform the "societal landscape" and in the process destabilize existing regimes, such an energy regimes in need of transition.

At the risk of oversimplification, the participants in the deliberative arenas begin by exploring the considerations related to the goal of guiding societal arrangements and how they might apply to the pursuit of sustainable socio-technical systems (Loorbach 2007). The visions and goals are generally based on broad political and policy understandings related to major long-term challenges such as carbon emissions reductions. The participants selected to be in the arenas deliberate and debate the transition options for achieving the long-run goals that visions of sustainability would require. Scenario-building techniques are used in developing the sustainable goals into socio-technical visions.

Special emphasis is placed on the identification of the co-evolution of technologies, social processes, and practices. The "social" in socio-technical focus, as such, attends to the configuration of institutions, markets, culture, knowledge bases, material interests, and the relations that co-evolve with technological development. By relating these to the corresponding technical factors, the participants seek to understand the ways that societal institutions must change to effectively introduce and implement new sustainable technologies. In the process, emphasis is placed on mutual learning, consensus building, and developing a shared problem perception in relation to the goals.

With the visions providing a sense of direction, participants deliberate over potential pathways toward the realization of these visions. In this phase, processes of learning and adaptation provide the essential links between long-term goals, socio-technical pathways, and short-term actions. Working collaboratively with stakeholders, transition pathways can be explored through experimentation with alternative socio-technical practices in sustainability niches (e.g., organic food, eco-housing, solar heating). The associated learning informs the design of sustainability niches, is used to re-evaluate guiding visions and transition pathways, and helps to improve strategies for achieving deeper intuitional change. The institutionalization of transition pathways is generally the final discussion, although this dimension is less developed than the others.

Especially important for the learning process is the emphasis on experimentation. Transition management theorists call for conducting "transition experiments" to test and gain knowledge of alternative sustainable technologies and the practices they would give rise to, especially experiments in the energy sector. Beyond practical learning from these experiments, the goal is to put pressure on contemporary governing regimes to bring about sustainable change. According to the model, government needs to foster the diversity essential for the creative niches that carry out innovative experimentation, but refrain from "picking winners." If the selection at the regime level is shaped to move toward sustainability, it is argued, winners emerge in an "evolutionary way."

Strong emphasis is placed on the understanding of non-technical factors such as institutions and cultural factors, considered to be preconditions for sustainability. Although technology is seen as pivotal, there is a need for a goal-oriented, strategic, co-evolutionary governance strategy, which stresses the dynamic interrelations between cultural, structural, and technological innovation (Weaver et al. 2000: 286). The transitions management model thus stresses learning processes rather than technology-push policies (Smith and Kern 2009).

There can be no question that the approach grapples with a critical issue of our time, namely how we aim to make a very fundamental shift toward a more sustainable society, in the interest of future well-being, if not survival. In this regard, this transition group has sought to take the "bull by the horns." As a member of the group's external scientific advisory board, I can say with plenty of first-hand knowledge that a heroic effort was made to learn about technological innovation processes and how they offer insights to facilitate the kind of change that needs to be taken up today. Having said that, though, it is not difficult to uncover implicit—and not so implicit—technocratic features of this effort.

One astute Dutch observer described transition management as a "benign form of technocracy."[6] Similarly, Scarse and Smith (2012: 61), both sympathetic to the transition management project, have spoken of its "technocratic overtones." As they write, "an *implicit* political model is revealed...in the assumed relationship between transition 'goals' and their achievement." Once the goals are established, they tend to be "beyond the reach of actors at the level of any particular regime but...by implication, instantly changeable by policy-makers, who are informed by a vanguard of transition managers acting on society's behalf, with popular support for the goals simply being assumed." The approach, as such, involves a "rationalistic model" of policy-making; the lessons learned in the niches will be consensually adopted and carried through. In their view, "the model assumes that true social values can be identified..., that policy emanates from values, that policy determines outcomes, and that social values can be decided upon, institutionalized and translated into policy action by policy makers." The approach basically never "considers how to build a wider political legitimacy for authoritative leadership that can overcome political and economic resistance to its visions and niches."

These points were at times raised by critics at the various conferences organized by the transitions management group to present both progress along the way and later the final research findings. These conferences at times gave rise to discussions that were awkward, perhaps even somewhat unwelcome. An underlying issue was the assumption that sustainability is an "objective" category about which there is wide agreement. The initial direction of the project, as well as the development of the guiding sustainability vision, were largely taken for granted. As outspoken critics such as Shove and Walker (2007) explain, it was largely assumed that there is an orienting vision in the field of environmental policy "that is defined and shared by a constituency of institutional actors who are by implication and example located within national or regional organisations." In fact, though, one of the problems with sustainability discourse has been the ambiguity and often outright disagreement about what a sustainable society would look like. Indeed, the perspective of ecological modernization represents one of them, having emerged to challenge the progressive environmentalists' understanding of the need to more fundamentally change the Western way of life. In short, a major issue that has defined the politics of sustainable development was in significant part removed from the table. The point was sometimes conceded by transition management writers, but the research continued as if it didn't make a critical difference. The general argument put forward in the face of the critique was that the challenge of a crisis is impending, that there is little time to waste, and there has to be a starting point.

This, of course, is true—time is short and modern societies need to start moving on this problem if there is any chance at all of avoiding an unprecedented crisis. But this position is taken at the expense of overlooking the fact that transition management has a politics of its own based on the choice of basic definitions, the path dependencies of particular choices, its acceptance of the legitimacy of particular strategies, and the ability to motivate the relevant actors to implement transition management. Indeed, one can argue that the neglect of the other social and political perspectives was not without consequences. Over time a certain sense emerged that transition management was itself driven by a political agenda, particularly as it related to the further funding of the project. Moreover, there were particular suspicions that led more than a few in the larger social-scientific community to view the project skeptically, a skepticism that the arguments in defense of the research framework were unable to quell. In the end, the project was not renewed as much because of these differing opinions as it was related to empirical findings and normative prescriptions, although the two were clearly interrelated. When it was over, despite the heroic promises, the most Rotmans and his associates could claim without much objection was that they had successfully introduced a transitions management discourse. This was, in fact, true. Toward this end, it also gave rise to a journal and a network to further promote research and deliberation about environmental transitions. But how to go about such a transition remains an open question among a broader group of social scientists and policymakers. Others merely rejected the strategy as an undemocratic, misguided effort on the part of a group of scientists to guide society according to particular visions (some would say their own vision).[7]

In this latter regard, it is important to note that the word democracy never—or almost never—appears through a very large body of papers and reports as a component of the project. Nowhere, for example, is it discussed as an element in construction of the deliberative arenas. The need for participation, to be sure, is fully represented in the discussions, but it is the participation of selected stakeholders, not a broader range of groups drawn from the public. One can easily see that the researchers updated the systems perspective with language that is associated with deliberative governance—including "reflexive governance"—social learning, and perhaps even work on deliberative democracy. But these concepts are carefully situated in the broader systems framework that dominates the transition discourse. The participants deliberate but they do it within the established parameters of the project, determined by the criteria of the system (whether the existing system to be changed or the new one to be brought about). The broader, more open—democratic—question of what to deliberate about was substituted by specific transition tasks the participants were asked to reflect on and discuss. In this regard, the approach could easily be understood as an accommodation to an enlightened form of participatory managerialism, or techno-managerialism—but not participatory democracy. Underlying this conceptualization was an implicit view—and even explicit for some—that the topic was too serious for democracy per se. Sometimes this was supported by the perception—not incorrect—that the public has failed to act. Others viewed the public as unable to grasp complex issues and thus not particularly suited for talking about the environmental challenge. One major figure, expressing exacerbation, said he didn't care—it was time to act before it was too late.

Questions of when and how the goals are to be managed are not entirely ignored, but they receive little careful scrutiny. Although there is a focus on assessing winning and losing approaches, the broader political implications for the socio-political system as a whole are largely neglected—that is, the classic questions of who gets what, when, and how are clearly subordinated to the goal of sustainability. There was some attention devoted to how transitions managers might generate public support of the long-term implementation of such change, but it was not a major focus.

Shove and Walker (2007: 3) argue that the transition managers assume a privileged position that obscured transition politics, smoothing over conflict and inequality. As they put it, "there is a politics to the very processes of abstraction involved in defining something to manage... and to the implication that there are managers... who sit outside the boundaries and who can apply managed tools including levers, niche-building machinery and engineering devices from a privileged, knowledgeable and above all, external position." There is little attention paid to the fact such abstractions involve political constructions rather than issues of technical analysis. As they point out, there is little or no politics of problem or policy formulation involved in this governance process. Moreover, no serious analysis of such a system can leave out the social actors "who are cast as managers from the systems they seek to intervene and of which they are a part." Nor is there much sense that the stakeholders, identified by the transition managers, are neutrally exempt from strategic behavior and the play of power.

They also find little sense that the visions of a sustainable future are shaped by the social contexts to which they pertain. In short, as they write, "it is necessary to recognize that provisional templates for transition are political statements that can only be partially inclusive (when there are ever more actors on the social stage), contingent (when conditions are dynamic) and potentially unstable as material forms and practices evolve over time." There is, in short, a politics to transitions management, a playing out of power concerned with when and how to decide and when and how to intervene, which cannot be hidden beneath the temporary illusion of "post-political common interest claims of sustainability" (Shove and Walker 2007: 5).

Thus, despite the official rhetoric of the project, the politics of construction and choice of visions and images of the future were either missing or underplayed. While the goal was commendable, one could easily judge the project to be an apolitical form of politics, albeit subtly so—that is, a modern-day form of technocratic thinking. Never does the approach announce the superiority of the scientific cadre, but rather it does this in various ways by way of excluding others.

Technocratic Strategy and the Environmental Response

It is not difficult to see that the turn to experts and technology will have substantial appeal to many, both members of the public and the governing classes. Moreover, it is hard to imagine overcoming or surviving these environmental consequences without major technological innovations, in particular in the field of alternative energy. As we have seen, many argue that it is our only way out of the crisis. And current technological efforts do make progress, even if too slowly (Davenport 2016a). Open, however, is the question of implications of such technical expertise for the processes of governance. While this is difficult to judge in advance, it seems fair to say that it will most likely be at the expense of democratic politics. This technocratic orientation thus poses a serious challenge to environmentalists who emphasize public participation and environmental democracy. In many ways, it is a challenge that has all too often failed to receive the attention it requires.

This is not to say that environmentalists have somehow failed to recognize the technocratic challenge. Indeed, this turn to and reliance on technical expertise has not been without environmental critics. Many environmentalists, unsurprisingly, see this as an attempt to take technical control of the problem in ways that avoid major changes in the economy and the way of life that have created the problem. For these critics, the reliance on technology will at best achieve a "technological fix" that fails to tackle the problem at its roots (Huesemann and Huesemann 2011). Although it would not be without social consequences, such fixes are seen to occur along the lines of existing economic and social structures, in particular those of corporate capitalism. It can be equated with fitting the car with a new tailpipe while ignoring more fundamental problems with the motor.

Fixing the motor, of course, is also a task for the technical domain. But the process is not as straightforward as some might think. Under liberal capitalism the process of technological change is mediated by the profit incentive. In large part, as the environmental community points out, the technical community is funded by the corporate sector or its representatives in the political system, including the military. Geared to the productive imperatives of the industrial system for which these technologists work or are funded, namely to turn a profit, such basic technological change has often not been the major focus of the techno-scientific community. As such, technological development has been driven mainly by economic (often military-related) interests rather than those of the larger public. By and large, the efforts of scientists and engineers have been harnessed by the industrial capitalist system motivated by profit-making, typically oriented to near—rather than long-term—financial gains. Engineers, in short, develop new technologies that are integrated into and marketed by the capitalist economic system, a point seldom problematized by the technical community. Under this arrangement, in the environmentalist's view, advanced industrial capitalism can continue to do what it does best, create material goods and services, while scientists seek to "geo-engineer" climate change consequences. In the name of ecological modernization and green growth, the economic engine can continue to turn out the goods for high consumption-oriented societies through market-oriented, piecemeal processes of fixing the global climate. In short, unsustainability is perpetuated. Or as Blühdorn (2013: 16; 2016) puts it, the "politics of unsustainability" sustains the general dominant environmental paradigm.

Many environmentalists see this to be prophylactic and thus irrational in the face of the consequences. First, there is for them no reason to believe that science and its technologies can achieve these changes, some of which would require minor miracles in view of time limits imposed by the coming emergency (Revkin 2010: 2). Even if such technological changes prove to be possible, they will not necessarily be in time, roughly defined as the next thirty to forty years. In any case, they will not be available given present levels of research funding.

But environmentalism is in many ways stronger on critique than solutions. Eco-democrats, to be sure, argue that genuine sustainability involves more than the introduction of new technologies. It also means rethinking our way of life. This not only requires political deliberation about basic goals and values, but also increased forms of social involvement at lower levels of the society. Innovation in business firms, for example, requires new forms of participatory management, if not economic democracy. In addition, various echelons of government will need to deliberate, develop, and coordinate sustainable policy strategies. Democratic decision-making, for environmentalists, is required to forge legitimate decisions and to motivate a wide segment of the population to assist in their implementation, especially at the local level where the problems have to be dealt with.

It is nonetheless hard not to recognize that such democratic participation is difficult to find and we aren't always sure that it is up to the task, particularly when it comes to matters related to science and technology. Compared to the technocratic project and its "scientific" decision techniques, environmental democracy is much more of a theory than a well-developed practice. In comparison it is weak, if not underdeveloped. Which is not to say that it is wrong. Rather, the contention is only to acknowledge that a lot more work would have to be done—theoretically and practically—before democratic environmentalism might be able to defeat a powerful technocratic politics in the corridors of power, presented most likely as a form of ecological modernization.

Conclusion

In this chapter we have explored the technocratic emphasis on ecological modernization and its emphasis on technological solutions, including the concept of the technological fix. While technological solutions will be necessary, this approach ignores or downplays their social and political consequences. We then turned to the Dutch model of transition management put forth by Rotsman and his colleagues. This can be seen as a genuine effort to develop a socio-technical systems approach that attempts to explore technological development in the social context and to find a more discursive format for selecting options that might be adopted by governmental decision-makers and the larger public. But for a variety of reasons this approach seemed in the end to lean more on experts than a broad swath of the public, despite the good intentions and best efforts on the part of its theoretical and programmatic leaders.

Bringing these discussion together, we again encounter the important question as to whether we can bring technical expertise and democracy together. The standard theoretical response to this question in political and social theory is deliberation. As writers such as Baber and Bartlett (2005) argue, the way to bring technical knowledge and social norms and values together is through democratic deliberation. In their words, "the theory of deliberative democracy offers the foundation for a possible and practical reconciliation of rationality, strong democracy, and demanding environmentalism." It is a point made by others such as Dryzek (2000), Smith (2003), Baber and Bartlett (2005), Hajer (1995), Gundersen (1995), and Eckersley (2004), as well as myself (Fischer 2009). Given the central importance of such deliberation to the questions at hand, we thus turn in Chapter 4 to an examination of ecological citizenship and environmental deliberative democracy.

Appendix

Climate Change: Searching for the Technological Fix

Insofar as climate change poses the greatest threat to future societies, some would even argue survival, the search for new technologies to dramatically reduce carbon emissions is well underway. While there is no "Manhattan Project" as such, the need is as great as the challenge. But there is a wide range of projects under discussion, ranging from rather conventional strategies to move away from fossil fuels to new technocratic strategies that have an air of science fiction about them.

The most obvious of the approaches focus on the promotion of solar, wind, and water energy technologies. These constitute the most green of the various strategies; all are being advanced but at a pace too slow to meet the challenge, thanks mainly to opposition from the traditional energy companies. Beyond these technologies, there have been significant efforts to turn to biofuels to replace gasoline. Here significant progress has been made, but use of the technology has run up against the problem of world hunger. The switch from using crops such as corn for food to fuel has driven up the price and availability of food supplies for many of the poor people around the world, putting the technology in a holding pattern. Others focus on natural gas to fuel cars and heat homes, which is cleaner than coal and oil and the supply is abundant. Technologists are busy trying improve it in ways that might dramatically facilitate further use of this energy source.

Economic and political leaders, however, tend to put their emphasis on big technological projects, such as nuclear power, carbon sequestration, and geoengineering. Nuclear power was regaining a new life in the face of climate change, up until the nuclear power plant catastrophe in Fukushima, Japan in 2011. But many engineers still hold out hope for a safer form of nuclear power, seen by many as one of the most impressive technological feats of the modern age. The hope is that new safety features can ease the worries of an anxious public. Despite the fact that Germany has decided to discontinue nuclear power, other countries like the U.K. have continued the effort. Most stunningly, engineers in Japan even after Fukushima have advanced plans to build more nuclear power plants. This technocratic community, known as the "nuclear tribe," argues that Fukushima should be thought of as a "learning experience."

Beyond nuclear power, another approach deals with the decarbonization of coal, or "carbon sequestration and storage." This is an especially important consideration because coal remains very abundant and accounts for 50 percent of energy production in the U.S. and other advanced countries, even more so in China. Some argue today that, given the increasingly high-level needs for energy, which will only grow in the future, finding a way to deal with coal is the only viable solution (Farrell 2010). Despite the high cost and technical difficulties that remain, inventing this solution thus becomes a top technical priority.

In addition to reducing carbon levels, biological engineering will most likely play a major role in dealing with the consequences of too much carbon. One of the most devastating effects of global warming will be a loss of agricultural output. Already the world food supply is beginning to experience serious shortages in many places. To deal with this problem, innovative agricultural technologies will be needed, especially those that can make new crop varieties available on a worldwide scale. This leads many to call for an intensification in research on genetically modified organisms (GMOs), reducing the need for pesticides. (A question raised here is whether the problem is one of too little production or maldistribution.) Others scientists point to potential benefits through the genetic modification of trees, which would help to both reduce CO_2 in the atmosphere and promote forest biodiversity.

A new set of ideas comes from the emerging field of synthetic biology, which seeks to build organisms by constructing new species with information from genetic codes. More than just a pipe dream, such organisms have already been developed. This movement seeks to replace the petrochemical industry by manipulating chromosomes to create an energy-producing insect—"a bacterium that will ingest CO_2, sunlight and water, and spew out liquid fuel that can be pumped into American SUVs" (Zakaria 2008). In this version of a "brave new world," large-scale processes of bacteria-processing fermentation are seen to produce complex molecules with high-level energy contents that can also be easily adapted to the established energy infrastructures. Its leading advocate, Graig Venter, speculates that biological mechanisms could be developed that could suck up the excess amounts of CO_2 and thus represent an important solution to global warming. In his words, through what he calls "environmental genomics," his laboratories "are trying to create a new value system for life" (Pilkington 2007). Further out, and highly questionable, others have argued that we will in time be able to re-engineer the human body to accommodate the new climatic conditions (Mead 2012). According to a few, we might even be able to accommodate the body to climatic conditions on another planet. Indeed, some argue that the ultimate solution will be to move to other planets.

Another technology in development is nanotechnology, seen to hold out the possibility of weaning us away from fossil fuels, oil in particular. Although nanotech is only in its infant stages, and not always easy to understand, it is seen to have potential for decreasing our consumption of oil while making our technologies more efficient. It offers the possibility of new pathways for a transition to renewable sources of clean energy, solar power in particular (Mahajan 2010). It could also include the development of new porous nanoparticles that can capture carbon before it is released into the air, as well as permit the development of new and more powerful automobile batteries, facilitating the development and use of electric cars. Indeed, the number of proposed possibilities that further development of these minute particles might bring are too many to mention here. It wasn't all that long ago that nanotechnology sounded futuristic, but today large amounts of money are being poured into research and development and various applications are already emerging in an entire industry of eco-manufacturing based on nanotechnologies.

Beyond this, we move into a realm that raises new technological possibilities that could conceivably be developed, some of them sounding science-fiction-like. Most important in this respect is geoengineering. Geoengineering refers to an approach designed to fix the climate change problem after the carbon has been released into the atmosphere (Kreuter 2015). Instead of attempting to stop or reduce the emissions in the first place, the most prominent approach involves efforts to reflect the sun's rays and the heat they generate back into the outer atmosphere. Technically this is possible but proves at this point in time to be very complicated, both technically and politically. Some have shown the way that this technological possibility has been advanced by coupling it with the language of crisis (Lederer and Kreuter 2015).

The technology of geoengineering is highly contested, as it could cause or trigger any number of unknown, unanticipated, problematic consequences, both ecological and political (Hulme 2014). There is no shortage of unknown dangers and risks associated with this approach, and maybe many more that are yet unknown. For one thing, nobody knows if it would actually reduce the carbon problem. Some of the unexpected consequences might even create irreversible damage. However, a growing number of scientists and engineers are today working on this alternative (*Economist Magazine* 2010: 99–100), another sign geoengineering has begun to be considered by organizations such as the United Nations. As the *Economist* put it, "What is sometimes called Plan B seems to be taking shape on the laboratory bench—and seeking to escape outside" (*Economist Magazine* 2010: 99).

The political implications are also highly uncertain. Altering the atmosphere would require a degree of international cooperation that we have seldom seen. Among the questions it poses are: What would happen if one country tried geoengineering without the consent of the others? What if one country could unintendedly—or even intentionally—alter the atmosphere of another country? Thus, some see this "brave new world" as a last-ditch strategy as a matter of survival, a last resort to deal with the problem (Victor et al. 2009). Others argue that we must begin now.

For the critics of geoengineering, such an approach to altering the atmosphere is immoral. Some see it as a distraction from the task at hand: the need to reduce carbon outputs. This has led to the argument that geoengineering is a "moral hazard" insofar as it diverts our efforts away from the more fundamental concern of bringing the Earth's ecology into a sustainable balance. In this view, it is the ultimate technological fix, permitting us to continue with an unsustainable way of life. Instead of developing carbon-neutral technologies, we can continue to push more carbon into the air and then limit our focus to an attempt to clean the air.

The one thing that is certain is that these strategies would bring the scientific community to the forefront. Although research is already underway along these lines, it will surely become more important and prominent as the indications and consequences of climate change become more apparent, especially as they start to become seriously problematic. Not only would it require putting more scientists in charge of important public projects, it would surely give rise to open debates among members of the scientific community. All would surely not agree on the advisability of all strategies. The range of the debate, though, would be more technical than political. It would most likely rule out the know-nothing positions of politicians representing the narrow interests of particular industries.

Notes

1. From a green perspective, environmental values have to move to the center of the political system with all other policy domains coordinating their activities with these green priorities.
2. Interview with Bjorn Lomborg by F. Zarakia, CNN (July 17, 2009).
3. Germany, Norway, Sweden, and the Netherlands are primary examples of countries that have embraced ecological modernization and implemented many of the measures it advances (Carter 2010).
4. http://www.theecologist.org/News/news_round_up/2986746/cop21_leaves_nuclear_dream_drift.html.
5. The concept of "socio-technical" is used here in a general sense rather than a reference to a particular social-technical theory.
6. Thanks goes to Robert Hoppe of the University of Twente for sharing this interpretation with me.
7. Scarse and Smith (2012: 61) put forth an interesting speculation on the project's lack of interest in politics. As they put it, "The relatively late interest in politics among TM advocates may relate, in part, to its origins as a policy-oriented realm of academic thought developed in consultation with policy-oriented elites. Any political challenges inherent in TM prescriptions had to be downplayed in order to gain assent from policy elites."

Part II

Democratic Prospects in the Face of Climate Crisis

4

Environmental Democracy and Ecological Citizenship: From Theoretical Ideals to Practical Alternatives?

In this chapter we take up two basic themes in environmental political thought: ecological citizenship and environmental democracy. Not only are these interrelated theoretical orientations advanced by environmental political theorists to counter the kinds of technocratic eco-authoritarianism discussed in the two previous chapters, they are presented as essential foundations of a sustainable way of life. The future of democratic governance in view of the climate crisis is thus seen in important ways to depend on the viability of the environmental democratic challenge. We agree here with the premises of environmental democracy, but point to a pressing need to give more attention to the relationship of this theory to the realities of both political power and the limited time frame now available for achieving such a challenging societal eco-transformation.

In an effort to assess this challenge, the chapter begins by setting the theory of environmental democracy in the more general political context of which it is a part, namely the long-standing relationship of environmental politics to democratic participation. The discussion then turns to assessments of the literatures on ecological citizenship and environmental democracy. Both theories are recognized to offer important ideal visions, but they are also seen to remain far removed from the eco-political realities of the day. They have, in short, little or no chance of emerging as dominant paradigms during a climate crisis and the accompanying social disruptions that are foreseen later in this century. Given the power relations that govern modern capitalist political systems, calls for environmental democracy will no doubt be further marginalized by political elites, top-level corporate managers, and techno-scientific elites. At best they will support various forms of liberal environmentalism and the ecological modernization project. As we argued in Chapter 3, this eco-modern focus on the technological fix is in fact designed to avoid such a participatory democratic transformation.

The discussion then more specifically takes up the deliberative approach to democratic environmentalism, the work of deliberative democratic theorists in particular. If the more exaggerated claims often associated with environmental democracy are paired off and the focus is shifted to more readily available and in many cases useful practices advanced by deliberative theorists, such as methods for promoting citizen engagement, agonistic democratic practices, pedagogies for policy learning, and deliberative policy analysis, this orientation has indeed an important role to play in thinking about and establishing or extending democratic environmental practices.

How, then, can we bring these elements to bear on the immanent eco-political realities ahead? How might these insights help us to both preserve and promote democratic values during states of environmental emergency when democratic practices might well be shelved to await better times? Although the argument will be developed more fully in later chapters, we suggest that environmental deliberative democracy should usefully reorient itself away from the current emphasis on larger political-institutional structures—e.g., parliaments, administrative agencies, regulatory processes, and deliberative polling, both at national and global levels—and include, even concentrate on, its potential contributions for specific local environmental projects. Whereas deliberative environmental democracy will have little chance of marching through the dominant institutional structures of the capitalist state, we shall later argue that its practices can fit quite well with particular participatory civil society movements and projects outside of this system, the ecological relocation and ecovillage movements in particular. Not only are these projects already underway, they stress activities that generally correspond to the goals and commitments of ecological citizenship and environmental democracy. Indeed, as we shall later see, they have already planted roots that provide fertile soil for such socio-ecological transformation. This connection, in fact, could open up a productive relationship between environmental political theorists and the practitioners on the ground. In particular, we shall suggest that theory and practice of participatory environmental governance, drawing on both the theories of deliberative democracy and emerging practices of deliberative governance, can offer a productive bridge between them.

Environmental Politics: Participation and Democracy

We begin by pointing to the fact that during some forty years or more of environmental political theory, the connection between environmental participation and democracy has come to more or less be taken as an article of faith. This commitment to democratic environmental participation has its origins in both the theory and practices of the environmental movement in the late 1960s, which were also in important ways part of a more general post-material cultural shift in Western societies (Ingelhart 1971). Basic to the emancipatory politics of this post-material orientation was an emphasis on the values of self-expression and self-determination through participatory democracy. From the outset, this involved a focus on decentralized forms of local participatory policymaking. For some it even involved self-governance in civil society. By the 1990s, such thinking came together in both the theory and practices of environmental movements and green political parties. Although participation need not involve deliberation, the emphasis on participation coalesced around the ideas of deliberative democracy and deliberative governance during the early years of this century (De-Shalit 2000: 135–45).[1] Democratic participation and public deliberation are now seen as essential for resolving environmental problems and even more importantly the creation of an ecologically sustainable society (Hayward 2013). This literature is so extensive that the commitment to environmental participation and ecological democracy scarcely needs to be demonstrated. Not only is it enshrined in official environmental documents, the literature constitutes a very long list of leading scholars in the field.[2]

In more recent years this emphasis on democratic environmental participation has been extended to climate change as well, sometimes as calls for global environmental democracy. Most of the writers calling for global democracy to deal with climate change offer a normative argument. They essentially assert that there is nothing necessarily incompatible between ecology and democracy. In general, they concede that democracy today has not been working to solve the climate crisis, but argue that there is no reason why we should give up on it. The problem is seen to be a lack of genuine democracy. By this they refer to a form of democracy that is free of political manipulation by elites and parties and the absence of distorted modes of communication, in particular the modern media. It permits citizens to participate in making decisions that affect their own lives.

Leggwie and Welzer assert that citizen participation must be an essential component of a future climate policy—that the reconstruction of industrial society can only work when members of society can understand and identify with it, which requires that the affected be involved in the "operationalization" of climate policy thinking (2008: 41). In Hayward's (2013: 3) words, for example, "real democracy is for grown-ups who think and act responsibly together in the interest of all [and believe that] the serious threats that climate change brings need a grown-up response." Siller (2010) insists that a solution to the climate crisis can only be made to work through democratic processes.

Gould (2013: 2) writes that "the urgency of the present environmental crisis cannot displace the normative centrality of democracy, which needs to be transformed rather than dispensed with." Citizen "participation in democratic decision making is more required than ever, both as an expression of people's equal agency or their equal right to jointly determine the conditions of their life together (with an adequate environment being one of the paramount conditions)." Stehr (2016) writes that "researchers who flirt with the idea that more authoritarian governance would help us address global warming are badly mistaken. What's really needed is more democracy."

The same view, in more muted tones, has also been taken up in the environmental governance literature, where the talk is more about participation than environmental democracy, per se. Environmental participation need not be democratic, but the implication in much of the literatures is that participation will promote and facilitate forms of democratic governance.

The argument here, to be clear, is not that these writers are wrong per se. It is rather to cast doubt on the realization of such genuine democracy in the time frame imposed by the climate change challenge. We support here the importance of democratic governance, but see the need to more carefully consider the political context that might make it possible. That is, it is not likely to occur at the nation-state level, and certainly not at the global level. We thus have to search elsewhere for other options.

Participatory Environmental Practices

One of the early participatory initiatives was built into the landmark environmental legislative decision in the U.S. in the early 1970s to require an environmental impact assessment (EIA) for all projects with potentially harmful effects on the environment, a practice that also became a requirement for development projects throughout Europe in the later 1980s and 1990s (Staeck, Malek, and Heinelt 2001: 33–42). Such assessments require consultation and input on the part of citizens which—in theory if not always in practice—can occur throughout the research and decision-making processes. EIA is designed to increase public awareness, to assist in balancing competing interests, to minimize public controversy, and to hinder politically biased decisions on the part of public officials.

Here one must also mention the "Right-to-Know" movement. Basic to the participatory thrust have been efforts to supply the knowledge and information needed to make intelligent decisions. At a practical level, the struggle to obtain "Right-to Know" legislation is generally viewed as an essential part of environmental democracy. Such legislation, for example, supplies citizens with ways to obtain information about chemicals dumped in their neighborhoods; what sorts of toxic elements are in the air they are breathing, and so on. As Hazen (1997) puts it: "Right-to-Know programmes provide both an opportunity to participate in environmental decision-making and a responsibility to understand and assess the meaning of the data fully."[3] While the concept is limited in many ways, it has nonetheless been used to empower communities to take charge of their own investigations. This is especially important for methods such as participatory epidemiology, a methodology which permits citizens to conduct their own investigations into environmental health issues (Novotny 1994). As Hazen continues, it has opened the door for the citizens "to influence decisions affecting their own well-being." But given the nature of the crisis, this is not enough. The doors need to be swung wide open.

Since the introduction of such EIA practices and Right-to-Know laws, participation has spread across the whole range of environmental decision-making processes (Beierle and Cayford 2002). But no call for public participation has been more widely circulated than that put forward in the United Nations document, *Our Common Future* (1987), otherwise known as the Brundtland Commission report. Prepared for the Rio Earth Summit in 1992, the report asserts that access in reaching sustainability "will depend on widespread support and involvement of an informed public." It calls for an enlarged role of citizen participation in environmental planning, development decision-making, and program implementation (Stirling 2009). Recognizing that workable policies and programs have to be constructed around patterns of everyday life, the Rio Summit advanced a program for the promotion and support of sustainable development at the local level of the citizen—namely, the Local Agenda 21 Action Program. As a result, participation has come to be considered a "best practice" in environmental policy and implementation. Accompanying these practices have been a range of studies focused on examining how and when environmental participation works (Newig and Fritsch 2009).

Further, the World Bank, which often works closely with the UN, has been influenced by such participatory efforts. It has, in this regard, initiated participatory projects to accompany the assessment of the impact of their development projects around the world, many of which have caused serious environmental degradation, especially the construction of large dams in the developing world. Although widely seen to be an environmental culprit, the Bank has itself commissioned major research efforts to study the role and uses of local participation and has developed methods for engaging civil society organizations in both economic and environment-related projects, described as an upstream effort to better understand both the constraints and opportunities to civic engagement at the level of the country. For many in the environmental movement this has been viewed skeptically as an effort to cast the Bank's projects in a better light. But the fact that the Bank has taken up the cause of participation has not altogether been a negative contribution (Mansuri and Vijayendra 2012). If nothing else, it is a very different orientation to the environment than the one exhibited by the Bank in earlier decades.

All of these efforts, both theoretical and practical, focus on the role of the citizen and the opportunities to participate in environmental problem-solving and, as such, have significantly influenced public environmental discourse. Not least important has been their impact on academic social science, both empirical and theoretical. In the realm of theory, environmental political theorists have devoted considerable effort to theorizing environmental citizenship, the role of citizen deliberation in environmental policymaking, and the democratization of the green state. We turn to a discussion of these concepts in the following sections. Before doing that, however, it is important to concede that the following discussions do not do justice to the extensive literatures on these topic. The point here is to illustrate the ideal character of these discussions—some might say utopian—and in doing so, to underscore the distance between them and current political realities. The discussion seeks in large part to stress that none of the political-ecological transformations these theories call for will be at hand before the onset of the serious and unavoidable impacts of the coming environmental crisis. If the point is not altogether apparent, it is nonetheless ignored or underplayed in this literature. It is also the goal to set the stage for an effort to rethink the ways that elements of these concepts can be employed in a more practical context.

Ecological Citizenship: Cornerstone of Environmental Democracy

In the thriving field of green political theory a great deal of emphasis has been placed on environmental or ecological citizenship during the past couple of decades. Dobson and Valencia Saiz (2005: 157) refer to this as the "turn to citizenship" in the literature of environmental political theory. For Dobson (2003: 206), perhaps the leading writer on the subject, ecological citizenship is "the exercise of ecologically related responsibilities, nationally, internationally, and intergenerationally, rooted in justice in both the public and private spheres."

There is also a need at the outset to clarify the sometimes confusing distinction between environmental and ecological citizenship. Some theorists use the term "environmental citizenship," some refer to "ecological citizenship," while others seem to use the concepts interchangeably. To those who specify a difference, environmental citizenship generally refers to citizenship in the tradition of liberal political theory emphasizing environmental rights, whereas ecological citizenship is more oriented to the civic republican tradition and a focus on obligation and responsibility (Dobson 2003: 5–6). In the discussion that follows, largely in the name of consistency, we will use ecological citizenship, but in a general rather than a specific way.

While citizenship is widely seen as the foundation of a democratic environmental approach, there is no one theory of ecological citizenship. In broad terms, though, there are two basic orientations: a rights-based approach and an approach to responsibilities based on personal duties and obligations. Both can be considered as basic foundations of an environmental democracy.

With regard to rights approaches, much of the theoretical work stems from or is related to liberal political theory, focused on individuals in fixed territories. Given that free-standing individuals in the liberal theory are assumed to pursue their own understandings of the common good, often based on some aggregation of individual self-interests, this approach suffers from an inability to settle on one particular conception of the environmental good, or what a sustainable society might look like. Indeed, from a liberal perspective it is quite alright to altogether question the importance of environmental sustainability—and some surely do. In short, as atomized political actors, individuals under liberalism have no way (and need no way in liberal theory) to come to an agreement on a common good.

But there is another version of the rights-based approach that has more traction among many environmental political theorists. For these writers, often appealing to the classic interpretation of the evolution of rights put forward by Marshall (1950), there can be a social right to a secure environment. In view of ecological interdependencies, human as well as natural, this perspective can better speak to the problem posed by the atomistic characteristic of individuals in liberal theory more generally. Indeed, a substantial literature has emerged concerning the possibilities and implications of establishing the environment as a human right—namely, that a safe and sustainable environment is a condition for the enjoyment of human rights more generally and, as such, a criterion for the pursuit of a good life. Or to put it even more forcefully, that the right to "a livable and sustainable environment" might itself be included as a basic human right (Shelton 1991: 105). It is worth noting, in this regard, that more than seventy nations have already included such an environmental provision of one sort or another in their constitutions (Hayward 2000).

Given the global nature of the environmental problem, this literature has often extended environmental citizenship across national borders, which presents another problem for liberal political theory. Essentially, rights theorists tend to speak mainly of rights as they are understood and practiced in liberal democratic countries, rather than to a more transboundary cosmopolitan understanding of the global challenge. Ecological citizens, in this view, also need to think of themselves as part of the planetary environment as a whole. The concept, as such, can be understood as a horizontal relationship between citizens rather than a vertical link between citizens and the nation-state (Dobson 2003). This way of thinking about the citizen clearly involves a more global understanding of the problem. It implies a form of "universal political subject" (Stewart 1991: 74). Or, as van Steenbergen (1994), puts it, we move from the national citizen to the "earth citizen."

The potential power of the concept of the "earth citizen" can be substantial, just as was "spaceship earth" in the earlier days of the environmental movement. The discursive importance of a powerful metaphor and the images it can project should not be underestimated. Although the idea of a post-national environmental citizenship largely remains metaphorical (Christoff 1996), such images are basic to the processes of long-term social change. While the idea has sometimes emerged in the political rhetoric of global environmental struggles, it largely remains in the realm of academic theorizing, outside of practical environmental politics.

One way to give substantive meaning to the concept of ecological citizenship, as Dobson (2003: 119–20) argues, is to connect it to the concept and measurement of the "ecological footprint," referring to the amount of ecological resources—natural capital—each individual or group consumes. Though difficult to measure in any exact way, a focus on the impact of the ecological footprint is theoretically credible as a substantive foundation for a more robust concept of ecological citizenship. Dobson, in this regard, notes that environmental impact is nonterritorial. Rather than a narrow focus on rights, as he argues, earth citizenship should focus on how each person impacts on the environment and the ways this interconnects us across the planet. In his words, the concept of the footprint "is both an expression of the space of ecological citizenship and a way of framing decisions as to the direction of citizenship responsibilities" (Dobson 2003: 115).

This brings us to the issue of environmental responsibilities more generally. According to van Steenbergen (1994: 146), the thing that differentiates the environmental movement from many other movements is its emphasis on responsibility. This also underscores the tensions between the environmental movement and the liberal state. As we have seen, a liberal-oriented rights-based ecological focus largely fails to consider collective responsibilities for the common good of all, an issue essential to environmentalism. Specifically, obligations and responsibilities thus remain missing dimensions in the liberal understandings of citizenship generally. Given the importance of ecological interdependencies, it is difficult to talk about environmental protection without recognizing our obligations to the planet. Indeed, the rights perspective is all too closely aligned, even if unintentionally, to the very push for the satisfaction of self-interests that has contributed so heavily to the overuse of resources and the forms of degradation it has brought with it. For this reason, some environmental theorists seek to correct the underlying limits of rights-based theories generally by pushing for a more active theory of citizens' responsibilities. Toward this end, they typically appeal to some form of civic republicanism approaches to ecological citizenship.

Citizens, in this view, need to feel an active responsibility to the environment. Such responsibility requires engaged deliberation in pressing questions of sustainability, as they pertain to the private as well as the public sphere (Dobson 2003). In contrast to liberal theory, which limits responsibility to activities in the public realm, the theorists of citizen responsibility recognize private acts to have public implications. Responsibilities in the private realm—and their corresponding duties and obligations—are understood to be non-reciprocal, rather than contractual. As such, the environmental citizen "does the right thing not because of incentives, but because it is the right thing to do" (Dobson 2003: 129). As such, environmental democracy in this conceptualization is ultimately anchored to the virtue of the citizens.

Such a turn to the ethics of virtue can be related to Aristotle's understanding of citizens, habits, and ethics. As Powell has put it, we "might consider citizenship as the collection of habits we develop to participate in our democracy and fulfil our civic duties."[4] Environmental citizenship, as such, might then be defined "as the collection of virtuous habits that emerge to attend to our obligations to the planet and to one another by ensuring a clean, healthy, and sustaining environment." The virtuous citizen would regularly engage in decision making processes to ensure a sustainable environment.

To conclude this discussion of ecological citizenship, it is hard to find fault with the idea of ecological citizenship; it is surely something to strive for in the struggle for a sustainable society. But a candid assessment should proceed on two fronts. On the theoretical level one can argue, as both Torgerson (2001) and Garside (2013) do, that ecological citizenship needs a stronger political interpretation. In this view, shared by Kenis (2016), an emphasis on individual responsibilities is not strong enough, as it lacks a specific sense of political commitment. The ecological citizen should be a political actor struggling to reduce the overall footprint rather than just a virtuous citizen who tries to reduce his own footprint. Torgerson also takes issue with Dobson's emphasis on civic republicanism, pointing to a potential conservatism lodged in the concept. Civic republicanism, as he argues, mainly involves a responsibility to the existing polity and thus can mean a responsibility to the status quo. For him, the issue should be more a commitment "to a principle that arises within the existing order but is not realized by it." For Torgerson, a better alternative would be "participatory environmental governance," an approach we take up in Chapter 6.

It is also necessary on a political level to acknowledge the very large distance between the ecological citizen and the dominant world of the consumer citizen, who is an essential contributor to the environmental crisis. Citizens in the West, but increasingly so in the East, are socialized into the culture of the "consumer society" and its ideology (Urry 2011: 54–60). They are, as such, fully removed from a meaningful concept of ecological responsibilities.

How to move in this direction is a difficult question. A standard response is environmental education and proposals to integrate ecological responsibilities into the school curriculum, a process that has been underway in many places. An unresolved problem here is, what should be in the curriculum? Much of the discussion assumes that a common, uncontested environmental good can be identified and agreed upon. But experience shows this to be a difficult agreement to reach, a point to which we return.

There is, moreover, little clear evidence that such educational programs can produce ecological citizens per se. Most students are found to be for protecting the environment until it actually impinges on their own lives, for example the need to stop driving their cars. Some research shows that such learning depends on lived experiences with the natural ecology, but this has to be much more than a walk in the woods.

Environmental learning thus remains as uncertain as it does important. What brings about a personal transformation is the pressing question. We know from theories of transformative learning that this usually requires a break with an experienced reality, often in the form of a crisis of one form or another (Fischer and Mandell 2012). A real worry in the case of climate crisis is that an intellectual transformation will be too late to avert tragedy. In short, we again confront the problem of time. There is little reason to believe that such a large-scale transformation can occur in the increasingly limited amount of time available to make a difference in the effort to ward off catastrophe.

At this point, we shift to the discussion of environmental deliberation which is, as we have noted, an important part of ecological citizenship, especially as it pertains to ecological responsibility and environmental learning in civil society. The basic issue is straightforward: democracy requires citizen deliberation. Democratic citizens have a responsibility to discuss the environmental predicament of which they are a part. Important here, however, are such questions as whether ordinary citizens are up to this task and whether the institutional conditions for such undistorted deliberation are available.

Deliberative Environmental Democracy

Before taking up deliberation in an environmental democracy, we need to first ask, what is environmental democracy? The question is complicated by the fact that there are various meanings attached to the concept of environmental democracy, which extend across quite a broad spectrum of arrangements. What is more, sometimes the term seems to often be used to refer more to social justice than democracy per se (Faber 1998). While there is an important relationship between them, they are not the same thing. Even though many tend to speak as if democracy and social justice involve something of a choice between one and the other, social justice can only be sustained and extended through democracy. As the famous theologian Niebuhr put it, "man's capacity for justice makes democracy possible; but man's inclination to injustice makes democracy necessary."

Few theorists have sought to offer an explicit definition of environmental or ecological democracy. Worker and Ratte (2014) of the World Resources Institute point out that "environmental democracy can be defined in somewhat different ways, but all of them involve the belief that citizens affected by environmental concerns should have equal rights in participating in the environmental policy decision-processes." Along these lines, Mitchell (2006) refers to a democratic alternative that first seeks to include interested or concerned publics in environmental policymaking processes and, second, is not characterized by structures and processes that systematically distribute ecological amenities to some groups while burdening others with environmental degradation. Hazen (1997), for another, writes that environmental democracy reflects the "recognition that environmental issues must be addressed by all those affected by their outcome, not just by governments and industrial sectors." As such, "it captures the principle of equal rights for all those in the environment debate—including the public, community groups, advocates, industrial leaders, workers, governments, academics and health care professionals." For people, she continues, "whose daily lives reflect the quality of their environment, participation in environmental decision-making is as important as in education, health care, finance and government."

One could understand these definitions as statements about conditions for democracy generally applied specifically to issues related to environmental decision-making. From this view, environmental decisions would require no particular exceptions to the usual requirements for democracy. That is, there are no particular issues related to environmental decisions that are different from those confronted by other types of issues. The approach, in short, recognizes no need to treat environmental decisions differently, owing to high levels of complexity, uncertainty, or the need to decide and act swiftly in matters related to environmental crisis. A widely accepted definition, it could be understood in terms of interest group competition and bargain.

But does it go far enough? Others think that more needs to be involved. This is especially the case when it comes to theorists of deliberative politics. As Baber and Bartlett (2005) have written, environmental politics involves "some form of deliberation, some form of collective agreement about how to manage our social relations." But what form of deliberation remains a subject of much theoretical debate. In their view, as well as many others, it should take the form of deliberative democracy. This leaves the question, what form of deliberative democracy? What would it look like; how would it operate? Perhaps the most theoretically rigorous effort in this direction has been the attempt to work out environmental deliberative democracy in terms of Habermas's theory of communicative interaction (Dryzek 1995; 2000; Eckersley 2004). In general, though, there are competing perspectives, all raising difficult questions (Smith 2003).

The emphasis on environmental deliberation resonates with the dominant focus on citizen participation and deliberative democracy more generally in political theory since the early 1990s. Deliberative democracy, as Smith (2003: 53) puts it, "has established itself as a new orthodoxy within contemporary democratic theory" and it is thus "no surprise that it has been the subject of much debate in green political theory." One of the first, if not the first, to advocate a deliberative turn in green political theory was John Dryzek. His work, in fact, remains an important source of insight in the development of this body of theory.

By and large, deliberation and deliberative democracy have emerged to challenge the belief that citizens are geared only to their own self-interests, a view long influential in many circles in political science. In sharp contrast, deliberative democratic theorists have sought to revitalize a more classical understanding of democracy focused on citizen participation, moral reason, and discussion of the public interest (Gutmann and Thompson 2004). These concerns are especially poignant for environmental struggles, as environmental protection for future generations is one of the public interest issues par excellence.

Deliberation, in this view, refers to "debate and discussion aimed at producing reasonable, well-informed opinion in which participants are willing to revise preferences in light of discussion, new information, and claims made by fellow participants" (Chambers 2003: 309). It is a view that has been embraced by environmental theorists such as Smith (2003), Dryzek (2000), Baber and Bartlett (2005), Dobson (2003), Gundersen (1995), and Barry (1999). These writers have advanced a concept of deliberative environmental democracy that seeks to bring environmental politics in close contact with the theory of deliberative democracy, including rigorous efforts to formally integrate them.

Citizens engaged in environmental deliberation, in this perspective, still have their own interests but they are expected to advance reasons as to why they take their views to be in the interest of the other participants as well (Fischer 2009). As Warren (2007: 272) writes, the primary contention is that "deliberative approaches to collective decisions under conditions of conflict produce better decisions than those resulting from alternative means of conducting politics: coercion, traditional deference, or markets." The resulting decisions "from deliberation are likely to be more legitimate, more reasonable, more informed, more effective, and more politically viable."

Acknowledging that citizens make many—or perhaps even most—decisions based on their personal or group interests, deliberative theorists insist that they are also capable of submitting their own interests to more reflection. And sometimes, importantly, they act in the public interest. Indeed, the emergence of the environmental movement itself can be taken as primary evidence of this reality. At times, moreover, this involves efforts to create new interests and values, such as sustainable development.

Deliberative democratic theorists also find that liberalism implicitly rests on a restricted conception of reason largely influenced by the dominance of scientific rationality. In this way, as is often seen in environmental deliberation, that which counts as legitimate argumentation is problematically narrowed from the outset. The point is particularly important in environmental politics given the prominent role of technocratic forms of expertise (Fischer 2009).

Often neglected are the distinctive viewpoints of groups at the margins of the dominant culture, in particular those who employ other modes of reason and expression. This becomes especially important in the context of global environmental politics, where other cultures come into play, including indigenous groups in the developing world. It is an argument that feminist theorists have also leveled against neo-positivist modes of science and reason (as well as against some deliberative theorists who emphasize particular modes of "rational" reason). The answer to these charges put forward by deliberative theorists tends to be a call for a more open, democratically inclusive approach to discourse and deliberation. It is an argument that relates directly to environmental struggles, especially to those that raise questions about the nature of reason and argumentation. Much of environmental theorizing has placed a good part of the blame for the crisis on a distorted mode of technical reason associated with the industrial revolution and the developmental path to which it gave rise.

The theory of deliberative democracy, however, has not been without its critics. These range from issues about scale (how far can deliberative democracy be extended) to questions related to social quality among the deliberative participants in a pluralist society, the role of interest groups in deliberative processes, questions of complexity and the role of expertise, the compatibility between citizen participation in deliberative projects, deliberation in representative democracy, and more (Elstub and McLaverty 2013). Of particular importance has been deliberative democracy's emphasis—or sometimes overemphasis—on processes designed to forge consensus. A strong argument can be made, as Machin (2012) maintains, that environmental politics is more about disagreement than consensus. In fact, one can ask to what degree it is possible to find a common environmental good that all can agree on. As such, an agonistic theory of democracy would appear to be better suited to environmental deliberation.

A second issue has to do with the nature of political representation: how can deliberation be designed in ways that allow participants to be considered representative of the larger public? Another question concerns the degree to which the average citizen is capable of participating meaningfully in complex decisions. And yet another one is the failure to fully consider the role of political power, social influence, and protest movements (Mendonca and Ercan 2015) that circumscribe deliberative settings, thus introducing hidden influences. In this latter regard, many find that deliberative democracy is too far removed from the real-world political processes and thus consider it to be a normative project rather than a ready alternative. Pointing to the distance between the orientations of the citizen and the deliberative theorist, Blaug (1999: 134–5) writes that "if deliberative theory is to be of real use, if it is to be a pragmatic and earthbound practice, it will need to address democracy not just as it appears in the elevated view of the political theorists, but also as it is actually encountered in the everyday world of ordinary people."

Another even more challenging argument put forward by various sociologists concerns the process of "individualization" associated with late modernity. This work points to the fact that as citizens focus more on the creation and fulfillment of their own identities and subjective interests, they become less inclined to engage in processes of deliberative participation with fellow citizens in the search for a common good. In this view, citizens' senses of their own identities are based more on their personal life experiences than connections to their fellow citizens. More or less, then, this undercuts an essential precondition for deliberative democracy (Middlemiss 2014; Beck and Beck-Gernsheim 2002), a point to which we return in Chapter 5.

Taken together, these concerns point to a heavy theoretical agenda. The point here is not to answer these questions; that is beyond the scope of the current discussion. In the main, the objective is to indicate the nature of deliberative democracy and the issues it raises. As it currently stands, it largely remains a theoretical project with too few connections to real-world political decision-making processes, short of serving as a counterfactual against which specific existing arrangements can be judged. Judged against practical criteria, some—if not many—of its critics might describe it as a quasi-utopian project.

While deliberative democracy can have something of a utopian ring to it, especially given the time frame imposed by climate crisis, this approach is not without a number of concrete projects. Indeed, some of them can have special significance when it comes to focusing on the local level and small communities. In this regard, it is important to consider the work of "mini-publics," an effort to address deliberative theory in empirical fashion through a range of experiments. Numerous activists and scholars in the U.S. and Europe have developed deliberative projects and procedures that bring citizens' preferences to bear on policy issues, including environmental issues (Gastil and Levine 2005; Fishkin 1996; Bohman 1996). In various ways, these projects can be taken as practical experiments that can test different contentions of deliberative democracy theory (Chambers 2003).

Deliberative Environmental Forums as Mini-Publics: Citizen Juries and Consensus Conferences

Deliberative mini-public research programs focus on finding ways to bring forth citizen views on complex social, economic, and environmental issues that bear directly on policy decision-making (Groenlund et al. 2014: 12). Such projects have evolved from efforts to redesign standard approaches to gathering public opinion—citizen meetings that employ methods such as focus groups, deliberative polling, national issue conventions, televoting, and study circles—on to more sophisticated scenario workshops such as citizen juries, planning cells, citizens' assemblies, consensus conferences, and approaches to public budgeting (Rowe and Frewer 2005; Gastil and Levine 2005; Fishkin 1996; Andersen and Jaeger 1999; Hendriks 2005; Joss and Durant 1995). Concerned with what Fishkin (1996) describes as "considered judgment," such work attempts to comprehend the ways in which citizens develop informed public preferences.

Two especially important deliberative approaches are the citizen jury and the consensus conference, both of which have been used to explore environmental issues. The citizen juries (or panels) allows for a significantly higher degree of unstructured citizen involvement. They offer citizens the opportunity to deliberate in detail among themselves (usually for several days) before coming to decisions on the question they are asked to judge; they do this after listening to expert presentations, putting forth their own questions and concerns to the experts, and deliberating the issue among themselves (Crosby and Nethercut 2005).

Whereas citizens juries have generally focused on relatively narrowly defined local problems (e.g., how to reorganize public transportation routes to reduce cars on the road), the consensus conference was institutionalized at the national level by the Danish Board of Technology and has addressed broader economic and social questions (Fischer 2009). Implemented in 1987, consensus conferences have focused on issues such as air pollution, energy policy, risky chemicals in the environment, sustainable agriculture, food irradiation, the future of private transport, gene therapy, and the cloning of animals.

In many ways, the consensus conference is similar to the citizen jury but it differs in several important respects. One is that the participants are given a topic but no specific charge to answer; they generally decide for themselves which questions they want to address and answer. Another is that the citizens more actively cross-examine the experts. The deliberations also tend to be longer than in the citizen jury. And further, where most forms of citizen panels operate behind closed doors, the Danish consensus conference has been more open to the public as a whole, typically being held in the Danish Parliament. Indeed, one of its primary purposes has been to inform and stimulate broad public debate on the given topic.

The Danish Board developed a model for a "citizens' tribunal" designed to stimulate broad social debate on issues relevant to parliamentary-level policy-making. In an effort to bring lay voices into technological and environmental inquiries, the Board sought to move beyond the use of narrow expert advisory reports to parliament by taking issues directly to the public. The Board, in short, developed a framework that bridges the gap between scientific experts, politicians, and the citizenry (Kluver 1995). Not only has it been widely credited with invigorating contemporary democratic practices, it has built understanding and trust among citizens and experts as well. The consensus conference reflects, as such, an enlightened techno-environmental approach to bridging the gap between scientists and citizens, namely a turn to deliberation. It recognizes the need for discursive interaction between science and the relevant stakeholders, if not the public generally.

Based on this successful innovation, the consensus conference process has been widely imitated in many other places in the world. In the U.K. examples of consensus conferences include topics that range from plant biology to radioactive waste management. These efforts have received positive reviews by the Royal Commission on Environmental Pollution (1998). All of this has made the fact that the conservative politicians in Denmark closed down the Board of Technology all the more astonishing. They preferred to see the discussion of environmental problems restricted to the narrower frames that better suit traditional economic interests.

Such deliberation is essentially an effort to expand discussion beyond the restricted conception of reason inherent to liberalism, especially as reflected in environmental struggles. By broadening the spectrum of views that bear on an environmental problem, deliberation can extend the boundaries of rationality. Toward this end, deliberative theorists have also developed theory and methods for the practice of deliberative policy analysis (Hajer and Wagenaar 2003), as well as a new deliberative role for public administrators, to which we turn in the remainder of this chapter.

One of the main criticisms of mini-publics is that they remain small, one-off projects. They make clear that citizens can participate in innovative forums and that they can be important for motivating citizen engagement and learning. But they seldom have a direct influence on policy processes. Thus to situate these more limited benefits in a larger understanding of deliberation, other theorists have more recently located these projects within a concept of a larger "deliberative system" as a whole, which we take up below.

Environmental Democracy and Citizen Deliberation in the Deliberative System

The deliberative system refers to the networks of communicative deliberation throughout the political system—whether parliamentary debates, mini-publics, deliberative policy-analytic inquiry, or discussion around the kitchen table (Mansbridge 1999; Parkinson and Mansbridge 2012). It is an attempt—albeit ambitious—to understand how they all play a role to one degree or another in shaping public opinion and political discussion. This has largely emerged in the context of efforts to locate or institutionalize deliberative democracy, such as in the case of the citizen jury. It is an interesting theoretical move as it allows for the inclusion of a wide range of different types of communications at different levels of society. As Elstub and McLaverty (2013) have put it, "rather than simply focusing on the extent to which particular types of institutions do or do not meet standards of deliberative democracy, the focus is now on their interdependence, interaction and therefore how to combine these institutions with other processes to ensure the norms of deliberative democracy are prevalent across the deliberative system as a whole."

This deliberative perspective has also been applied to climate change by Stevenson and Dryzek (2014). In *Democratizing Global Climate Governance*, they set out to examine the role of environmental discourses related to climate change policy. Analyzing an extensive range of venues—public spaces, environmental networks, the United Nations' Intergovernmental Panel on Climate Change (IPCC), and more—they investigate the progress of the primary discourses in the global deliberative system of climate change, with an eye toward discovering openings to promote global deliberative democracy. What they find is at best a mixed picture. Although there is no shortage of deliberation, it tends to be fragmented into networks and niches that generally do not talk to one another, or when they do, seem to have little impact on one another. At best, as the authors acknowledge, deliberative environmental democracy related to climate change is seen to remain a goal to strive for rather than an ongoing reality. And, to assess the portrait candidly, it is hard not to conclude that we have a very long way to go. One could easily judge it to be unlikely, certainly in any meaningful time frame related to climate crisis.

By including a broader focus on deliberative politics, which need not be democratic, this orientation implicitly acknowledges problems for moving forward that result from the narrower deliberative democracy perspective. In the deliberative systems perspective the deliberative democracy projects can be located within the larger communicative system, permitting as well as mapping their relationships to other communicative processes. In doing so, this highlights—perhaps unintentionally—the fact that deliberative democracy in the real world of politics represents little more than a small number of experiments here and there. In this regard, at least at the higher systems level, it can remain a normative ideal, serving as a counterfactual in the assessment of ongoing deliberations in the same way that Habermas's theory of speech acts was designed to serve. Without overlooking a few interesting projects in deliberative politics at the national level, especially in Brazil (Pogrebinschi 2013), many still consider the theory of deliberative democracy to mainly be relevant to the local level. But this need not be a hindrance for present concerns, given that there will likely be a retreat to the local level on the part of many during the climate crisis, a topic to which we will return. It is at this level that the struggle for democracy has the best chances, even if still difficult. Social movements and small communities such as transition communities and ecovillages might well continue and further develop democratic practices. They are, in any case, well positioned to do so. And the deliberative systems perspective arguably provides a way of situating them in the larger political systems. Later we argue that the theory and practice of participatory governance at the local level offers a home for this effort.

The deliberative systems perspective also offers the advantage of focusing more specifically on the relation of technical discourses to other forms of communications in societies. This becomes especially important in the struggle to engage scientists and engineers in climate crisis politics, as we shall also see later.

Finally, another concern has to do with what some theorists see as a shifting tendency in deliberative systems theory toward an implicit elitist perspective. Writers such as Owen and Smith (2015) and Hendriks (2016) argue that citizens are ironically being left behind in deliberative systems theory (also see Rothstein 2013 on the problem of institutional bias). Here it would appear that the move upward to the level of the societal system, and certainly the global system, threatens to shift attention away from the deliberative role of the citizen. According to Owen and Smith (2015), the dominant theories of deliberative systems tend to ignore or underplay the actual deliberation taking place between citizens. It is an argument that underscores the importance of conceptualizing the local level on its own terms, a concern to which we return later in the book. Citizens, it can be said, need to find their own means of democratic expression through their own local activities.

Extending Environmental Rationality: Deliberative Policy Analysis and the Facilitation of Deliberation

None of this is to suggest that deliberation is less important. Independent of deliberative democracy and mini-publics, deliberation remains an essential part of reason and social learning more generally. Deliberation, as Dryzek (1987) has perhaps illuminated more than anyone else, can help us to expand the rationality of environmental decision-making processes. Democratic environmental deliberation, if not deliberative democracy specifically, moves to the heart of the matter by addressing the classical concept of "bounded rationality." Originally introduced as an aid rather than hindrance to policymaking, bounded rationality is a way to explain how and why policymakers and analysts typically bite off more manageable segments of a problem, instead of pursuing a comprehensive approach to a problem as a whole, as they are deemed to be beyond the analysts' intellectual capabilities (Fischer and Gottweis 2012). Unlike conventional policy analysis—which employs a rigorous yet overly simplistic empirical-deductive logic—deliberation can expand our ability to reason, if not be rational per se.

Because both our imaginations and empirical-analytic capabilities are limited and thus easily prone to errors, deliberative argumentation makes it possible for politicians, policy decision-makers, and citizens to expand their thinking beyond their own familiar, limited understandings and outlooks. Deliberative engagement offers the advantage of gaining knowledge, abilities, and experiences of other participants. In short, it can lead to policy solutions that individual participants would not have brought to the fore.

Deliberation speaks, in fact, to the characteristics that set off environmental problems from other policy issues, namely higher degrees of complexity, problematic levels of uncertainty, nonreducibility, and often the need to act sooner than later, as in the case of climate crisis. These are, to be sure, the characteristics that lead the eco-authoritarians to call for a central role for the experts. On the other hand, Dryzek, along with Fischer and Gottweis (2012), argues that it is just these features that require the kind of broader perspective that deliberative policy analysis can bring to the fore. Insofar as environmental policy experts draw on established knowledge, they are often unable to see new and unexpected dimensions associated with ecological complexity and uncertainty. Moreover, because of their general adherence to technical and instrumental forms of reason, conventional administrative agencies, regulatory law, and market mechanisms are unable to manage the needs for high degrees of coordination under conditions of complexity. It is thus necessary to turn to an alternative form of reason, practical reason, which seeks to coordinate both means and ends. Not only is such reason oriented to political and ecological learning through communicative processes, it combines the "collective cultivation of virtuous behaviors rather than administration of people" (Dryzek 1987: 200). It leads, as such, to better arguments for particular courses of action (Fischer 2003: 133–6).

With regard to environmental complexity and the problem of coordination, when confronted with problems such as the Fukushima catastrophe or the BP oil disaster in the Gulf, deliberation permits the inclusion of negative feedback—defined by Dryzek (1987) as "the ability to generate corrective adjustments when a natural system's equilibrium is disturbed"—and to "facilitate coordination across different problems" (so that solving a policy problem in one policy field doesn't just create other issues and problems in others).

The task of environmental governance, from this perspective, involves designing democratic institutions that enhance the opportunities for full and equal participation on the part of citizens as they seek to clarify and revise social understandings and values in collaborative problem-solving and decision-making (Rosenberg 2003). For environmentalists this has from the outset meant not only moving deliberation beyond the dominance of techno-industrial interests to include the less powerful but affected interests, but also the creation and dissemination of a new environmental ethics and the values that would support it. An important method for fostering this kind of policy discussion has been the development of deliberative policy analysis, a method designed to facilitate argumentation and discussion among the participants (Hajer and Wagenaar 2003; Fischer and Boossabong 2016). Essentially, it is a method that seeks to reorient the theory and practice of policy analysis to facilitate public deliberation, the practice beyond its traditional emphasis on public managerial decision-makers.

What is needed from the perspective of deliberative policy analysis and the argumentative turn in policy inquiry more generally (Fischer and Gottweis 2012) is a reconceptualization of the role of the public servant as facilitator of public engagement. The goal is to create "communities of participation," in the words of Feldman and Khadermian (2007). The challenge confronting those working in the public sector is to interactively combine knowledge and perspectives from separate domains of knowing—the technical, political, and local/experiential domains. Bringing about more inclusive practices of governance in general—and environmental governance in particular—involves inventing participatory contexts in which the representatives of these forms of knowing can discursively share their perspectives in the common pursuit of problem-solving (Fischer and Gottweis 2012).

Beyond merely identifying and disseminating information from these various ways of understanding and analyzing environmental policy problems, such work involves translating ideas and arguments in ways that facilitate mutual ecological understanding and deliberation among the participants and discursively promote a synthesis of perspectives that helps to simulate different ways of knowing relevant to the problem at hand.

Conclusion

In this chapter we examined the emphasis in environmental discourse on democratic participation and deliberation, in particular as it has emerged in environmental political theory. Toward this end, we considered leading theories of ecological citizenship and environmental deliberative democracy. Missing from these otherwise forward-looking theories, it was argued, is a close relationship to the political possibilities that will—and will not—be available in the face of an immanent eco-political crisis. Not only do they tend to neglect questions of power, they are seen to fail to take into account both the ecological and political timetables. Given the pace at which we are speeding toward climate crisis, we are quite plausibly running out of time. As things stand, we find here little reason to believe that there is going to be a sweeping societal transformation that ushers in environmental deliberative democracy, in any case not before the arrival of the crisis. By seeking a more emancipated political world through a theory that is quite removed from current realities, these efforts tend to create a theoretical vacuum between a contemporary politics of necessity and a future point in time when social arrangements might be organized in more sustainable ways.

We have nonetheless advanced the possibility of dealing with this situation by connecting deliberative elements of these theories to a conception of participatory environmental governance, a theory which operates between democratic theory and concrete policy practices. We pointed, in this regard, to the ability of deliberative practices to enhance environmental learning, both for ecological citizens and environmental policy analysts. We noted that deliberative projects such as the consensus conference and the citizen jury can both help to inform citizens on their own terms as well as lead to greater political engagement. Further, the practices of deliberative policy analysis can facilitate the understanding and analysis of environmental problems defined by high degrees of complexity and uncertainty. In addition to detecting aspects of environmental problems that go unnoticed, deliberative policy analysis can assist in forging action-oriented consensuses needed to deal with both environmental goal-setting and policy coordination.

Facilitating these connections, however, requires reorienting these theories away from an emphasis on larger political-institutional structures and refocusing them more on participatory civil society projects developed by progressive environmental movements. Where environmental democracy has little chance of altering the institutional structures of the capitalist state, we argue later that many of its specific practices sit quite well with local movements and practices outside of the dominant political-economic structures. As we shall see in the final chapters, these theoretical orientations can be particularly relevant to smaller-scale projects such as decentralized governance, ecological relocalization, and bioregional movements, all significant movements which have been largely neglected, ignored, or downplayed by most of these theorists. Not only are all of these movements already underway, as practices of ecological citizenship and democracy they can be expected to expand as the climate crisis imposes itself more harshly on ordinary life. As such, we shall argue that these socio-political projects offer fertile terrains for both the theoretical advance of these concepts of deliberative participation and the new avenues for practical ecological applications. We argue that ideas about ecological responsibilities and deliberative decision-making practices can be interwoven with both a conception of participatory environmental governance and the small but growing movement toward small-scale sustainable communities that it can serve.

Before taking up these issues, however, it is important to more carefully examine this institutional gap between environmental political theory and the eco-political crisis that these practices might help to narrow. Toward this end, we examine in the next chapter various theoretical efforts that attempt to "green" state institutions with the assistance of eco-democratic theory.

Notes

1. It is important to distinguish between participation and deliberation, an issue that has given rise to considerable debate (Pateman 2012; Mutz 2006; Curato 2014; Berger 2013). All deliberation is participatory, but participation is a much broader concept. In addition to deliberation, especially as understood by deliberative theorists, participation includes a range of other action from voting to working on a political campaign, attending meetings, helping to implement a project, and more.
2. See, for example, Dryzek (1990; 2000), Gundersen (1995), Shiva (2005), Baber and Bartlett (2005), De-Shalit (2000), Paehlke (1995), Eckersley (2004), Smith (2003), Mason (1999), Bäckstrand et al. (2010), Beierle and Cayford (2002), Williams and Methany (1995), Faber (1998), Demirovic (1994), Purdy (2010), Mitchell (2006), Morrison (1995), Hester (2010), Mathews (1996), Barry (2001), Christoff (1996), Lundmark (1998), Barns (1996), and Demirovic (1994), among numerous others.

3. Hazen further writes that the release of information "is not enough, by itself, to characterize the impact a facility may have on a community. It must be combined with information on exposures and hazards to assess whether there is a potential risk; and in addition, more sophisticated data must be collected and assessed to identify the risk's magnitude. The obligation to use information responsibly is crucial to the programmes' continued success."
4. Personal communication with Greg Powell, September 27, 2013.

5

The Green State as Environmental Democracy? Political Power, Globalization, and Post-Democracy

In this chapter we extend the preceding discussion about environmental democracy to the question of the "green state." There is a growing literature on the greening of the state, but much of it today largely takes a functional perspective oriented to environmental regulatory capacities and policy transitions (Duit, Feindt, and Meadowcroft 2016; Paterson 2016).[1] In the early 2000s, however, there were important debates about the possibility of a green democratic state. Although these discussions have tended to fade in recent years, the issues they raised are still relevant for an assessment of democratic environmental prospects. For this reason, we first examine the theories of three leading political theorists: Eckersley, Dryzek, and Barry. Although their works largely fall far short of identifying practical political openings for restructuring existing state institutions and practices, the issues and problems they raise remain instructive. In the second half of this chapter we assess these concerns against broader contemporary political trends, in particular concerns about "democratic deficits" and the theories of "post-democracy" that have accompanied them. Even more specifically we examine the "ecological paradox" that a post-democratic politics poses for a sustainable transformation. These issues, as we shall see, suggest that in pursuit of environmental democracy we might best look for alternative locations outside of the state.

The Green State and Environmental Democracy

Ecological citizenship and deliberative environmental democracy, examined in Chapter 4, are generally seen by their theorists as basic components of the "green state." Some such as Wilson (2006) have argued that the focus on the green state can in fact be an antidote to the abstract theorizing of ecological citizenship and deliberative environmental democracy. Although this suggests a useful turn, even a casual reading of the literature on the green state indicates that it has not been the case. Theories of an eco-democratic green state are largely as removed from the ongoing world of politics as its elements, sometimes more so.

We do not need to examine this work in detail here. A basic outline can illustrate the gap between theory and practice, permitting us to more clearly understand the challenge at hand. Toward this end, we begin by introducing the arguments of Dryzek, Eckersley, and Barry. Dryzek's position, one can say, comes the closest of the three to the more traditional orientation of the state in both environmental political thought and the politics of environmental movements. For the most part, greens have been skeptical about the role of the state, given the state's continuing material emphasis on economic growth, coupled with a paternalistic style of governance. Rather than relying on the state, green theorists—radical greens in any case—have typically called for decentralized forms of governance. Often the writers emphasizing environmental citizenship and environmental deliberation in civil society fall into this category. Eckersley and Barry, as we shall see, deviate from this orientation and call for an explicit emphasis on transforming the state. For them, the state is the only actor with the power to restrain environmental degradation. They see the greening of the state as a way to *enable* strategic action on the part of movements and activists.

We turn first to Dryzek (2000; Dryzek et al. 2003), as his contribution expresses various concerns about a focus on the state and thus can set the stage for a discussion of the work of Eckersley and Barry. Dryzek, like those both theorizing about and working with environmental movements, advises caution when it comes to involvement with the state. He worries about co-optation and marginalization, given the power of the state and its goals and priorities. In his view, the active locus of environmental politics has been and must remain in civil society and the public sphere must be located outside of the state. This is even more the case, he argues, when it comes to democratic environmental action.

After examining the possibilities of movements being included and excluded from the state at various levels of policymaking, Dryzek takes the view that the best position for environmental movements under contemporary circumstances is active exclusion. That is, to maintain as much distance as possible from the state. Worried about the possibility of co-optation of the public sphere by the state, for which there is surely enough reason to justify the concern, he fears the loss or weakening of the environmental movement's oppositional leverage. Although the movement is weak in comparison to the state, it has nonetheless managed through its civil society activities over past decades to bring environmental issues to public attention and onto the governmental agenda. As states have often opposed these initiatives, either directly or covertly, the need to protect the future of civil society is seen to be essential. In just this regard, in a seminal work on the factors contributing to successful environmental policy, Jänicke and his research group found that "it is not primarily the institutional set-up of representative democracy which is advantageous for positive policy outcomes, but rather the... participatory, legal and informational opportunity structures available to proponents of environmental interests... which appear to be most decisive."

State opposition to environmentalism stems, according to Dryzek, from the need to fulfill basic imperatives that often clash with ecological concerns. Indeed, Dryzek's analysis of core "state imperatives" is a useful framework for helping to understand these conflicts, especially as they pertain to climate change policies. A functioning state must, in his view, fulfill several basic or "core" imperatives—secure economic growth, protect the nation internally and externally, and collect revenue and resources needed to carry out the others. These imperatives, in particular the imperative of economic growth, often stand directly in the path of environmental protection. If basic environmental regulations cannot be formulated in ways consistent with state imperatives, they are likely to either be rejected or watered down.

This is a political reality which environmental movements need to acknowledge, despite the often promising enticements of the state. To preserve both their credibility and their ability to act strategically, such movements, it is argued, should keep a healthy distance from the state. This is especially the case, Dryzek argues, if the goal is environmental democracy. Only a free and open civil society can ensure space for democratic environmental deliberation on courses of ecological action. In his view, political "power can be exercised from and in civil society in several ways," which include changing the terms of public discourse and debate, legitimating collective action, the constitution of deliberative forums, and enabling protest movements (Dryzek 2000: 101–2). The worry, he maintains, is that all of these can be significantly truncated, if not quashed, when civil society groups are too closely aligned with the state, a contention borne out by a considerable amount of evidence. It is an argument associated with well-established traditions in green political thought, particularly the call for localized, decentralized forms of governance which return us to both a closer relationship with nature and each other. Eco-anarchism would be the most radical of such positions, but scarcely the only one, as Dryzek's work makes clear.

At the same time, though, there is no convincing reason to believe that the expansion of the public sphere has to be a positive force for environmentalism, as indicated by the rise of anti-environmental groups in Western countries, particularly the U.S. and U.K. It is a mistake to assume civil society to be a united or unified realm, when in fact it more typically reflects the divisions in society more generally. Indeed, the public sphere has also given rise to active, often aggressive opposition to environmentalism, every bit as worrisome as the weak environmental record of the state. Civil society does, as Dryzek argues, provide greens with an open space for democratic struggle, as well as a sphere for environmental innovation and experimentation. But greens are not alone in this realm and other countering forces can and do work to stymie such efforts. In this regard, civil society offers no magic formula, often only reflecting the broader tensions and conflicts in the society as a whole.

Eckersley and Barry, in contrast, have rejected this circumscribed focus on the public sphere, arguing that the solution rests with changing the state. It is not that they reject the public sphere per se, but rather they stress its limitations. In particular, they emphasize its distance from political power, the very thing that many environmentalists have more traditionally counseled. For the power to introduce and implement environmental policy change, they argue, environmentalists need the state. Thus the better strategy, as they see it, is to environmentally *condition* the state—that is, to green the state.

Conditioning the State?

For Eckersley (2004) the state has to be the focal point of green politics, including democratic environmental politics. Her argument, which is complex and not easily summarized, adds new dimensions to the discussion, namely the international realm now basic to many environmental problems, climate change in particular. In a globalized world, only the state has the power to ecologically reorient the productive apparatus, especially in agreements with other industrial nations.

Notably, Eckersley pays much less attention to civil society. For her, civil society lacks the power to make the necessary transformations. Agreeing that the liberal-democratic state is unable to adequately confront the ecological crisis, she largely views civil society environmental movements in terms of their ability to instrumentally push the state in ecological directions. While she has nothing against ecological experimentation in civic society, such movements, in her view, are most useful when they confrontationally engage the state in the face of the urgency of the environmental crisis, climate crisis in particular. That is, the public sphere can influence the state by extending or withdrawing legitimacy, but it is not an independent site for ecological sustainability. The goal of grassroots environmental movements in her work is to advocate for the "ecocentric state," with which Dryzek would not disagree. But rather than doing this from a distance, for Eckersley this means actively engaging the state, even infiltrating its institutions. This reminds one of what Rudi Dutschke referred to as "the long march through the institutions" (Dutschke and Gollwitzer 1980). Acknowledging that this view runs counter to the anti-statism of most green political theory, Eckersley argues that only the state can counter the power imbalances that lead to environmental destruction, especially the economic imbalances created by large transnational corporations. Only the state institutions can establish authoritative rules and regulations capable of halting the pace toward environmental climate crisis. For these reasons, environmentalists must in her view enter the realm of state activities and engage in an active struggle to reshape the state institutions.

In other words, instead of worrying about the ways the environmental movement can be restrained by the state, in particular the imperatives of economic growth, she argues for removing the constraints by elevating sustainability to the level of a requirement in the "ecocentric state." For her, Dryzek's core imperatives are too strong and too instrumental. She takes state functions to be more flexible, thus holding out the possibility of changing them in environmentally friendly ways. If such efforts do not eliminate the state's requirement of sustaining capitalist market arrangements, they could modify them with environmental conditions. In some ways, she even seems to be suggesting something like an irrevocable commitment to ecological sustainability, as if to add another imperative to Dryzek's list. But most importantly for the present discussion, there is little on how this is to happen politically; the specifics of a strategy remain altogether unclear. In short, her discussion remains at a high level of theoretical abstraction that often sounds as improbable as it does unrealistic.

In addition to her skepticism about civil society, Eckersley also has little interest in the sorts of decentralist approaches that have often appealed to radical environmentalists. Democratization, she argues, is not about "decentering authority." In this regard, Eckersley's focus moves in the other direction, namely to the international sphere, and toward this end she speculates about the extension of democracy into the global realm. While this direction follows a certain theoretical logic advanced by a number of democratic theorists (Gould 2013), it does move the argument further from practical political realities. While she offers a vision of the green state, this work is more of a long-range blueprint than a strategy. The same can be said of Dryzek's theory, but his is closer to a strategic orientation, at least from the perspective of environmental movements that need to decide and establish their relationship to the state.

Transforming Ecological Modernization?

Barry (1999; 2008) too argues for a focus on the state. He is sympathetic to eco-anarchist and bio-regional theories, but he sees no inherent reason to reject the state. Where radicals have often portrayed the state as inherently problematic, he takes a different stance. Like Eckersley, he emphasizes the need for state power to bring about a fundamental transformation, the sort which is necessary to deal with the climate change crisis. The environmental movement, while avoiding co-optation, needs to work with the environmental authorities of the state.

Especially interesting for the present discussion, Barry acknowledges the gap between theory and real-world practices and calls for an "immanent critique." Theory needs to start with an analysis of the existing political realities. For him, this means, in line with Eckersley, that the state has to be taken seriously, as that is where the power to make changes is located. He advances this critique by arguing for a theoretical understanding that can integrate agency into what is otherwise an institutional structure approach. Here he is on good footing, as this could offer a theoretical link between environmental movement politics in civil society and state policies and practices.

Toward this end, Barry advances a specific strategy for the environmental movement. The integration of agency and state structures should take place through the theory and practice of "ecological modernization." He acknowledges that this will come as a surprise to many environmentalists, but argues that ecological modernization can be transformed. It is, as he maintains, the one available approach to environmental renewal that currently has traction, even if flawed.

Barry clearly understands the limitations of ecological modernization, as outlined in Chapter 3, but contends that it can be improved, if not subverted, in ways that more fully address the issues of sustainability. Despite its technocratic limitations, examined in Chapter 4, it is a model of sustainability that has the blessings of power, having been created by policy elites in and around the state. Moreover, it does contain elements necessary for a more sustainable approach, particularly its focus on essential technologies. Thus, he takes ecological modernization to be a concrete project with which environmentalists can and should work. The goal would be to incrementally add the missing social and cultural dimensions that are basic to a more progressive version of sustainable development. That is, to move from questions of economy and environment to a more socially sensitive discussion of those linkages that includes questions of human welfare more generally. In this regard, environmental protection could be advanced as a component—or better, extension of—the social welfare state (Meadowcroft 2012). This would also resonate with the argument that a healthy environment is a basic requirement for the good life.

Tempering the Argument

Later, in an edited volume, Barry and Eckersley join forces to update and further advance an appeal to "'reinstate the state' as a facilitator of progressive environmental change rather than environmental destruction" (2005: x). In their concluding chapter, much in line with Barry's earlier argument, they seek a strategy that could co-opt ecological modernization—still seen as the only game in town—thereby turning it into a strategy for a more progressive program capable of guiding environmental change. But this time the tone seems more cautious. Rather than hoping to co-opt the existing conception of ecological modernization, the call is for the introduction of a "stronger" form of it. In my reading, however, as with most of the essays in the book, the argument remains based more on the hope of a possibility rather than any clear sense of a concrete political strategy to realize such an approach. After an extensive exploration of the possibilities, it seems fair to say that they come up with little that holds out genuine hope for the future of such transformation, near or far term.

While Eckersley and Barry can be commended for their theoretical efforts—the state surely remains important—there are some unavoidable problems here that need more attention. At the most general level, many environmentalists will argue here that Eckersley and Barry accept but then tend to sidestep the fact that the state has played a very large role in environmental destruction. The modern state itself is responsible for a great deal of environmental destruction. Not only does it support and facilitate the capitalist imperative and the ongoing thrust to expand economic growth, it also engages in ecologically disastrous military adventures (Thomas 1995). In this regard, the state has taken a contradictory position between capital accumulation on the part of business and the public legitimation of public decisions. This means, in effect, that state support of business has to be portrayed as a contribution to the public interest. But this is often not the case, especially in environmental affairs. To bridge this tension the state has often had to engage in a form of symbolic politics, which we take up below.

Turning more specifically to the argument about ecological modernization, during the years subsequent to Barry's argument advanced in 1999, ecological modernization has proceeded apace with no such modifications, despite the environmental movement's objections. And there is little apparent reason to assume that this will not continue to be the case. A basic reason for this, I would aver, has to do an underlying dimension of ecological modernization that Barry failed to fully appreciate—namely that as a technical fix it exists more or less to avoid these other social and cultural questions. That is, it emerged in significant part to counter the very kind of environmental politics that Barry is calling for. As power elites support ecological modernization for this reason, there will be little interest in altering it in the direction Barry suggests. There is then no compelling or necessary reason to believe that the strategy is workable. And second, as if we could miss it, these same political elites—largely of neo-liberal bent—have sought with considerable success to reduce rather than expand the social welfare state. There is, in short, little afoot to suggest a receptivity to such programmatic alterations, even if incremental.

To this one might also add the obvious. The environmental movement does not have the kinds of power and resources needed to effectively challenge modern-day political leaders, who for the most part enjoy heavy support from the powerful corporate world. This might suggest, though, a different approach to the same goal. The one group that might have some impact on such changes would be the technical community. An environmentally conscious group of scientists and engineers, reaching a critical mass, might have enough social or political standing to move in this direction. It is an old argument without much of a track record. But given their importance to ecological modernization, with its bets on technology as the way out of the crisis, it is still the case that some engineers might have enough influence to introduce more socially compatible alternatives in the advance of the model. There is nothing certain about this, to be sure. But if it would prove to be a possibility, this would require a different set of moves on the part of the environmental movement (Fischer 1990).

Post-Democracy: Implications for Eco-Democratic Politics

We also need to examine the theory of the green democratic state against another influential literature that has emerged over the past decade. Although there have long been complaints about the thinness of democracy in Western democratic systems, it has until more recently been discussed in the mainstream literature on democracy as a malady that should and can be remedied. That is, the core political practices could still be presented as relatively democratic, while acknowledging the need to bring various political activities back in line with core principles. Campaign finance reform would serve as an example. In this case, money plays too big a role in political elections, but this can be fixed by passing stricter laws controlling campaign funding.

Relatively recently, however, other writers have begun to advance the idea that more is involved than dysfunctional practices. In this view, to stay with the issue of campaign finance, the role of money is now seen to have corrupted the system to such a degree that it can no longer be portrayed as roughly democratic. The problem, as such, has now distorted the very core of the political system and cannot simply be reformed away. Many now say that the system is better described as a plutocracy. In this alternative view, such political systems are seen to have crossed over into a non-democratic realm, or what Crouch has theorized as "post-democracy." If we can no longer speak of democracy, the idea of environmental democracy would seem to be even further removed from the realm of political possibilities.

There are now influential theorists who ask if democracy remains a realistic goal. This question, of course, moves against the grain. But increasing numbers of writers think there is good reason to ask if democracy has become anything more than modern-day myth (Wolin 2008). Despite all of the fine words offered in the name of democracy, some argue that democracy, in the true sense of the idea, is already a thing of the past and may no longer be within reach.

One might ask: Is this question a serious concern? Before addressing it directly, it is first important to be clear that we are not saying that there is no political participation, a foundation of democratic politics. That is, the argument here is not that citizen participation is altogether disappearing. Rather it is to take note of the fact that it tends too often to be relatively marginalized, when not simply ignored. It is also not to overlook the energetic activities of advocacy-oriented interest groups. Instead, it recognizes that the activities of these groups are often part of the problem. As is well documented, the interest group process is not particularly democratic. When citizens join these groups, they discover hierarchical structures with little interest in their active participation.

Although people in the West take democracy for granted, often without really thinking about what it means, it is important to remember that democracy is not a given (Meaney and Mounk 2014). For one thing, modern democracy is a relatively recent phenomenon in the larger context of human history. Moreover, most of the people in the world still do not live in democratic countries, and even where democracy has flowered, progress has often been slow. The history of democracy has been a story of ebbs and flows.

Put in this context, is it still meaningful to talk about participatory forms of deliberative democracy, let alone environmental democracy? The answer, to be sure, depends on what we mean by democracy. Clearly, there are democratic forces in play. But, at the same time, there are few contemporary political systems that could be called strongly democratic in terms of their basic structures and practices. For this reason, it is reasonable to contend that the sort of strong democracy that green political theorists call for is quite far removed from anything that passes today for democratic government. This should not be taken to mean that normative prescriptions—blueprints—are unimportant. But they should not be confused with practical realities. Forward-looking thought—even utopian thought—has an important place in the consideration of political action. Such prescriptions, however, only point us toward the horizon in political struggles that can take long periods of time. As already noted, no other issue is more important in this matter than the question of time.

In examining the criticisms of democracy, it is important to begin with a few words about the condition of contemporary democracy. To be sure, democracy, depending on definitions, maintains a high level of support around the world. Such support, though, tends to vary in different places. Despite the enthusiasm, it is important to keep in mind that only about 45 percent of the world's population (living in 46 percent of the world's governments) live under political systems that can be described to one degree or another as democratic. Many on the list can only at best qualify as thin democracies.

Hard to overlook is the fact that levels of frustration with democracy are high. Surveys show that citizens support the principles of democracy in theory but tend to be frustrated with the practices of contemporary democratic governments. There is a widespread concern that these systems are not solving the problems confronting their citizens and societies generally. The trend is particularly strong in the United States, where political participation is low and citizen disappointment high. In the case of the European Union, surveys find that not many more than half of the EU's citizens are satisfied with their own governments, and even fewer with the EU itself (Hobolt 2012). In 2016 the U.K., in fact, voted to terminate its membership. And there is even less enthusiasm to be found in many of the newly democratizing nations of Eastern Europe. One unmistakable consequence has been the rise of populist groups, such as the Tea Party and various militias in the United States. In Italy, for another example, there is the Northern Alliance, in France one finds the Le Pen movement, there is the Party for Freedom in the Netherlands, and in Germany one finds the Alternative Future for Germany party. Beyond populist movements, there are any number of more authoritarian right-wing groups such as neo-Nazis in both Europe and the United States, fueled in significant part by the growing waves of migration around the world.

Authoritarian trends, in short, can be identified in many countries of the world (Cohen 2016). It is easy for people in the West to forget that over half of the world's population live under authoritarian regimes of one sort or another. Further, many see these countries becoming more prominent on the world stage at the expense of the U.S. and Europe. Indeed, in many quarters there has been a growing enthusiasm for such systems, if one can describe it as such, particularly in the less developed and developing worlds. It is now often reported that such countries find that they have as much or more in common with China than they do with countries like the United States or United Kingdom. One of the main contentions that comes up in such discussions is that authoritarian regimes can deal with problems in a more rational and decisive fashion (an argument that often overlooks or neglects the fact that such regimes can be authoritarian forms of state capitalism). In this regard, Singapore is often cited as an attractive example for these countries.

Interestingly, those who cast a favorable light on Singapore single out its technocratic form of government, described as a benevolent form of authoritarianism. Even China has taken an interest in the Singapore model (Ortmann and Thompson 2016). And, as we have seen, Shearman and Smith invoke it as the best bet for dealing with the environmental crisis.

For these reasons, it should be obvious that the sort of strong democracy that green political theorists call for is quite far removed from anything that passes today for democratic government. In addition, there appear to be few openings in this new world of governance that offer democratic political traction. From this perspective, one might argue that green political theory unintentionally misdirects our attention. Such an argument becomes clearer, and perhaps more convincing, when we consider the contributions of Colin Crouch (2004) and Ingolfur Blühdorn (2013). Their work moves beyond frustration with the state of contemporary democracy to the development of theories of "post-democracy" and "simulative democracy."

The concept of post-democracy, advanced by Crouch in 2004, has received wide attention. Writing against the backdrop of increasing levels of voter apathy and citizen distrust, a worry that leading writers refer to as a "democratic deficit" (Müller 2016) and a "political crisis of representative democracy," Crouch takes this concern further and argues that we have entered a period beyond democracy, or "post-democracy." This, he argues, raises worrisome questions about the relevance of conventional thinking about democratic politics. In his view, we now live in a post-democratic period which renders the pursuit of democracy even more difficult, even perhaps out of reach. Toward this end, he points to the fact that an active enthusiasm on the part of the citizenry has largely given way to "boredom, frustration, and illusion." Powerful minority interests have everywhere pushed ordinary citizens aside in the political policy processes. Typical "top-down publicity campaigns" are now designed to persuade disinterested or disaffected citizens to vote. As Crouch describes it, politics now has become "stage-managed by spin doctors and PR people." To the degree that we find calls for more democracy, they typically take the form of "an ongoing struggle between the demand for transparency and professional well-funded efforts to avoid that transparency."[2]

In Crouch's provocative view, we have begun "a move beyond the idea of 'rule by the people'" to a challenge of the "idea of rule at all." Politicians regularly turn to political marketing and public relations techniques designed to manipulate public opinion. Such techniques provide political leaders with a way of discerning public views without offering citizens the opportunity to organize and control their own communicative engagements. Rather than genuine public deliberation of the issues of the day, we regularly encounter distorted communicative processes in media-dominated political processes.

Crouch speaks of the "growing incapacity of modern citizens to work out what their interests are" (2004: 28). Under neo-liberal post-democracy the manipulative powers of the corporate world have transformed citizens into unreflective consumers (2004: 49).

Through an examination of the underlying economic and social dynamics that have created this modern-day political malaise, Crouch goes beyond contemporary complaints about the failings of democratic governments, in particular those of countries such as the United States and United Kingdom. His main argument is that contemporary "democratic" governments only hold up a facade of democratic principles as they slip "back into the control of privileged elites in the manner characteristic of pre-democratic times" (2004: 6). Particularly problematic here are neo-authoritarian tendencies ushered in by the politics that emerged with the deregulatory politics of Reagan and Thatcher in the 1980s before sweeping across an increasingly globalized post-Cold War world. The result, as widely chronicled, has been states that have focused on market freedoms at the expense of the socio-economic concerns of their citizens.[3]

Liberal democracies in the West, Crouch argues, have moved into a phase in which institutions are still described as democratic but the civic culture of participatory engagement that supported and sustained democratic politics is largely exhausted. Whether or not it is altogether exhausted is up for debate. But there are, in any case, no civil society actors who have anything near the kind of political leverage that the corporate world possesses, or the governmental institutions which cater to their interests.[4]

There are, in this view, very few forces that can counter this co-optation of the political process. Owing to the decline of industrial manufacturing and the working classes, coupled with the globalization of economic activity, this disproportionate corporate takeover of the political sphere has become a defining trend over the past decades. The process is seen to have cut back the democratic spaces and processes needed for public participation, leaving political leaders disconnected from the people or publics they still purport to represent. This is reflected in the fact that the public reports a widespread lack of confidence and trust in their politicians and political institutions. Regardless of regular elections, citizens regularly express little faith that things change with the replacement of one set of elites with another, with no shortage of evidence to support such claims.

Without suggesting Crouch's argument to be any less important, it largely corresponds to the well-known left-oriented critique of modern political economy and the realities of political life to which it has given rise. Although writers such as Mair (2013) and Streeck (2014) have advanced more detailed and analytical versions of the same collapse of democracy argument, Crouch has succeeded in supplying this reality with a new name—"post-democracy"—that has had enough resonance to move this critical discussion of the current political malaise beyond a relatively marginalized radical critique into mainstream academic circles, even at times onto the public media.[5]

Whereas contemporary political theory is focused on democracy, deliberative democracy in particular, Crouch asserts that democracy is more a legacy of the past than part of our future. There are, moreover, no simple bromides to offer. Progressives, he argues, will not be able to reverse the course of post-democracy. To be sure, they can "learn to cope with it—softening, amending, sometimes challenging it—rather than simply accepting it" (2004: 12). But democratic transformation is not within reach in this perspective, although Crouch hedges on this point in later discussions. We thus have two lines of political investigation that largely proceed independently of one another. One is more or less an analysis of existing political trends and the other an exercise in normative theory. All things considered, the post-democracy argument appears to prevail when it comes to considering the future of democracy, at least as we generally understand it. One thus seems entitled to raise questions about the contemporary relevance of much environmental political theory. Deliberative democracy in a green state could easily be portrayed by their critics as theories groping in the dark.

Post-Democracy and the Politics of Simulation: Explaining the Ecological Paradox

This post-democracy argument becomes even more poignant for environmental politics and the greening of the state when we turn to the writings of Blühdorn (2013), who examines the post-democratic decline more specifically in terms of what he describes as the "ecological paradox." Extending the argument that democracy is now little more than a facade for a very different kind of political process, Blühdorn seeks to extend the post-democracy discussion with an even more critical conception of what he refers to as the post-politics of "simulative democracy." Where Crouch and others see post-democracy to be the consequence of deeply anti-democratic practices imposed on society by those with wealth and power, Blühdorn argues that while the inequalities of wealth and power are central to the problem, the story is even more complicated. In his view, the withering of democracy is also the product of an "emancipatory-progressive struggle" inherent to the very processes of individualization that accompanies modernization. As such, he pursues the analysis at a more fundamental theoretical level, identifying deeper democratic assumptions that have become problematic, perhaps even outmoded. Toward this end, he focuses on the modernist norm of the "autonomous identical individual" capable of democratic deliberation devoted to articulating "a general will and common good" (Blühdorn 2013: 25). In this view, "the existence of an externally demarcated and internally homogenous *demos*" required for such democratic discursive participation is being chipped away by the modernist processes of progressive differentiation, fragmentation, and fluidity that characterize contemporary societies, consumer societies in particular. As he puts it, "the bourgeois-modernist idea of a unitary, stable identity which supposedly evolves and matures through a person's lifetime has been superseded—or at least supplemented—by the intrinsically contradictory ideal of a multiple, fragmented and flexible identity" (2013: 26). To state it overly simplistically, citizens are increasingly unable to identify with one another and thus to discover common interests, thus undercutting the possibility of participatory solutions.[6]

There are, he argues, two different competing and generally contradictory forces contributing to this fragmentation. The most important is the constant change created by the dynamics of corporate capitalism and the lifestyle consumerism it promotes. Rather than bringing citizens together, these processes work to distance them from one another, often in the name of achievement and the conspicuous consumption typically associated with advancing up the social ladder. Such self-interest-based status systems are enabled and promoted in significant part by capitalist control of the media and the practices of marketing and advertising.

The other contribution to this social fragmentation comes from a quite different direction, one generally critical of the contemporary ideologies of capitalism and consumer society. Here Blühdorn moves beyond Crouch to point to the emancipatory politics of social movements that emphasize subjectivity and "identity politics." That is, many of the movements that Crouch would point to as primary forces struggling against the decline of democratic politics are seen themselves to be part of the problem. Such emancipatory movements, as well as postmodern theories that reflect such subjectivity in the academy, advance the cause of individual self-realization and self-determination—e.g., women's, minority, and LGBT liberation movements in the larger emancipatory struggles inherent to contemporary societies. As a fundamental challenge to elitist politics, the cause of these movements emphasizes self-expressive values pursued through both civil society and established political institutional channels, often in the name of participatory democracy. These groups demand that democratic governments acknowledge their interests and make available the conditions and resources for the advance of the issues and concerns of their members.

While these efforts have made gains in opening up important political debates about social and political inequities, even leading to policy changes in numerous cases, Blühdorn underscores the ways they also contribute to the further fragmentation of the larger political community. Although these struggles help to liberate particular repressed social groups, they provide no mechanisms for bringing groups together. The result is increased demands—often inflated demands—on existing political systems by newly activated groups that now assert their autonomy and independence. As he puts it, "in line with modernisation's trajectory of individualisation and subjectivisation, contemporary citizens are ever more assertive in their demands for participation, representation and government responsiveness" (Blühdorn 2013: 27).

This means that as the democratic subject (individual and collective) of the democratic political system melts away, the system also needs to support and even intensify its purported commitment to representative democracy and the demands of its political subjects. This "post-democratic paradox," as Blühdorn (2013: 27) writes, materializes as the dilemma of having citizens conceiving "of themselves . . . as autonomous subjects, an end in themselves" and, as such, the "political principal and the democratic sovereign . . . while at the same time they also want to (and must) embrace the freedom, promises and imperatives of the *opportunity society*." They thus seek to "liberate themselves from restrictive democratic norms so as to optimize their strategic position in . . . a relentlessly competitive modernity."

Examples of this paradox abound. To take just one illustration of such social fragmentation, consider the fact that some college campuses now cater to such demands by providing separate dormitories and kitchens for students of different ethnic groups. These groups are encouraged—implicitly when not explicitly—to move forward in society independently of other groups. And, more often than not, they do so competitively with little or less concern for the common realm per se. Indeed, concerns for the common good can at times be seen by such groups as getting in the way of individual self-realization.

Another example concerns citizens who want limited government and fewer taxes but at the same time desire and expect more public services. In fact, such demands can converge with the lure of the "opportunity society." The advance of such cultural value shifts under neo-liberalism often easily merge unintendedly with the self-determination of the lifestyle politics of consumer society. When this happens, according to Blühdorn (2013: 30), democratic governments become platforms "for the individualised struggles to secure an optimal strategic position for succeeding in liquid, competitive and inherently unstable modernity."

In this view, contemporary lifestyle consumerism, coupled with demands for self-realization, undercuts the ability of similar citizens to identify with one another and to discover common interests. With little common ground—or *perceived* common ground—to connect them, the possibility of authentic democracy becomes meaningless, even slips away altogether. But this happens at a time when the calls for more democracy grow only louder. That is, the same people who demand greater self-realization and autonomy call at the same time for the extension of democracy based on common interests.

How do modern societies and their governments manage the conflict between these two centrifugal forces? The answer for Blühdorn is "simulative democracy"—that is, holding out the appearance of democracy without actually supplying the substance of democracy. As he explains, under conditions in which the processes of modernization have "exhausted the project of democracy" but have also "radicalized democratic demands, the practices of simulation perform or make imaginable the validity of democratic values." Such simulations make the "commitments and trajectories" which contemporary individuals take to be "indispensable and non-negotiable" appear meaningful at the same time that they are "outdated and counterproductive" (Blühdorn 2013: 28).

Simulative democracy, Blühdorn (2013: 28) writes, is the "performance of democracy" in the post-democratic society. As a performative process, it involves substituting the signs of democracy for democracy itself. It relies on the ability to create the appearance or pretense that the needs and values expressed by otherwise freely engaging citizens are attended to by their governments. Simulative democracy, as he continues, "is about stabilizing the idea that there is a more or less homogenous demos which can generate and articulate something like a general will [by] reassuring citizens that despite the overwhelming power of global corporations, credit rating agencies and systemic imperatives, democratically elected governments are still in control of the social order." Expressed more generally, "simulative democracy is about producing narratives and arenas which respond to the democratic needs of contemporary citizens—but without compromising... equally powerful non-democratic needs."

To the degree that Blühdorn is correct, this is not good news for the prospects of environmental democracy. Moreover, he employs the model to explain a particularly problematic paradox of environmental politics—namely, why awareness of the environmental crisis is widespread, along with regular declarations of the urgent need to pursue a course of ecological sustainability at the same time that governments and citizens continue to engage in the very practices that generate the crisis. This "ecological paradox," he argues, is grounded in and advanced by these same modernity-driven cultural/value shifts that plague democracy more generally. The demand for environmental protection, in this respect, is in all ways one of the many citizen demands on government. The paradox, in this case, involves the ability of governments to find ways to facilitate the continuation of material production and accumulation in the face of the ecological limits to rampant consumerism. Once a force for social and political emancipation, modernization, under the consumerism of techno-capitalism, turns out be a force for unsustainability with democracy playing a central role in its reproduction, or what Blühdorn calls the "politics of unsustainability."

As a clean and safe environment is one of the requirements for a high-quality lifestyle, citizens place demands on the democratic process to maintain high standards. That is, as Blühdorn explains, consumer societies and their citizens mobilize democratic norms in support of value choices and lifestyles that are clearly socially exclusive and ecologically unsustainable. Governments serve as "a tool for the stabilizations and legitimization of lifestyles which, more visibly than ever, can be sustained only at the cost of increasing social injustice and accelerated environmental degradation" (Blühdorn 2013: 30). To illustrate the point, Blühdorn offers the example of the U.S. government marshaling democratic values to legitimate strategic military interventions that have next to nothing to do with the advancement of democracy, and very much to do with securing the materials required to sustain the styles of its consumer society, in particular resources such as oil and other scarce minerals in geographically remote regions of the world (Blühdorn 2013; Keane 2009: 807).

The politics of unsustainability thus relies on simulation. Through the media and deliberative spaces, including policy networks and decentered forms of governance, individuals and collective actors are able to continue their consumer-oriented lifestyles with few compromises and still see themselves as environmentally minded. Examples of such narratives of reassurance are found in the policies of ecological modernization and the Green New Deal put forward by the UN during the global financial crisis. In these stories the "science of sustainability" is seen to provide the foundation on which the structures and processes of a sustainable future can be built without sacrificing material growth, or at least not too much. In concrete policy terms, they are most specifically articulated in the narratives of ecological modernization. It serves, as we have seen, to render modern-day consumer capitalism as socially and ecologically benign.

In reality, liberal governments have long served to facilitate the consumer society. However, Blühdorn's theory is new in the sense that it offers a compelling and provocative conceptualization for rethinking the political-ecological character of consumerism confronting the twenty-first century. It is also new in the sense that it helps us grasp a number of basic internal contradictions that make the possibility of ecological reform and the greening of the state far more complicated than most realize. That groups have used their leverage to gain advantage for themselves has always been inherent to democratic capitalist systems, despite an earlier political science that saw government as a neutral actor merely mediating the process. What is different today is the pace of this consumerism and the fact that it has run up against natural ecological limits. The concept of simulative democracy, to be sure, is similar to the process of "symbolic politics," identified earlier by Edelman (1985). But this does not detract from the issue at hand. Blühdorn's innovative use of symbolic simulation raises serious issues about the possibility of a green democratic state. If we have entered a period of post-democratic, simulative politics, we have surely moved to post-environmental democracy.

But this should not be the final word. While it seems clearly the case that we are in something of a post-democracy phase, with simulation helping to explain the ecological paradox, there is room to ask if these theories capture the story as a whole. Although they contribute to our understanding of the workings of the modern state, both parliamentary and presidential models of government and their electoral processes, there are at the same time significant counter-forces at work that constitute strong democratic challenges to the existing system, emancipatory movements in particular. Although Blühdorn considers these forces, he largely examines them against their implications for the existing governmental structures. Against this background, they involve apparent contradictions. To argue, however, that they make for contradictory complications is in need of some qualification. In many ways, to be sure, they pose contradictions for the existing political order. But these movements are out to change the existing order and exploiting contradictions is one of the levers of change.

In addition to their challenges to liberal-capitalist political structures and the limits to electoral politics, social and political movements and the theorists who labor to advance their causes and concerns have also given rise to calls for new forms of governance, which constitutes a broader understanding of political activity. While such challenges look weak against the entrenched power of the existing governmental systems, this does not mean that change can altogether be written off. In this regard, it is significant to note that large numbers of people increasingly share the view that the "system" does not work. This is particularly clear in the U.S., where a solid majority shares this opinion. Such a lack of credibility and trust is potentially dangerous for contemporary political systems, especially if they cannot be held within tolerable limits. Unfortunately, however, the forces behind these perceptions are politically divided over the explanation of these system failures. For example, the Occupy Wall Street movement pointed to the capitalist system, while the Tea Party has blamed government. Missing here, of course, is the fact that these two explanations are inherently related. But as long as they remain politically divided, the thrust of these movements is to a considerable degree held in check.

This may only mean that it will take a long stretch of time for such liberal-capitalist political systems be restructured, either through regular processes of policy change or as the result of successful political struggle. History, as we know, is complicated and unpredictable. One need only consider the former

Soviet Union. It was widely thought by many that the Soviet Union would last forever, at least a lot longer than it did. And when it collapsed, it happened quickly. On the other hand, it is the element of time that motivates this particular discussion of environmental politics. All things considered, such changes are unlikely to occur before the climate crisis becomes critical.

Finally, we also take a different tack and posit an alternative perspective that we will return to later in the book. This involves another way to understand the democratic crisis by rethinking the concept of democracy. In this respect, one can argue that Crouch, if not Blühdorn, holds up the failure of democratic politics against a somewhat idealistic conception of what is seen to have existed in the past. For one thing, there is nothing new about political elites manipulating the decision processes to benefit their own elite constituents, in particular their wealthy constituents. It is rather that the techniques available to them are more sophisticated than in earlier times. Even here, however, a media-oriented politics based on polls and political advertising is not without its surprises and contradictions, as the use of social media has shown. What is more, Crouch and Blühdorn do not argue that opposition has disappeared; rather they only point to the difficulties its activists confront. In this regard, there is another useful interpretation consistent with the analysis of post-democracy that is perhaps more realistic and, in some ways, less pessimistic. This involves a turn to an insurgent understanding of democracy and Wolin's (1994) theory of "fugitive" democracy and the sorts of "outbreak" politics that emerge in the course of political struggles, such as the environmental struggle. We take this up more directly in Chapter 11.

Outbreaks of democratic resistance typically take place in civil society, as theorists ranging from Dryzek to Absensour emphasize. Indeed, we turn in Chapter 6 to an examination of just one such civic society political project called participatory governance. Given that it generally stands outside of the state institutions analyzed by Crouch and Blühdorn, it can not only stand alongside their work but also potentially offers a ray of democratic hope against the otherwise bleak picture they offer.

Conclusion

In this chapter we continued the discussion of environmental democracy, this time through a consideration of the theories of the green state. Here again it was difficult not to notice the distance of these theories from real-world politics. Eckersley's theory, one has to concede, is offered as an exercise in political theory. Her approach is to look for openings through which a progressive environmental politics can move. But the discussion leaves little room for optimism. Barry, despite his welcome call for immanent critique, does not do much better but for a different reason. The strategy he suggests—a struggle to reshape ecological modernization—appears to have already shown its political limitations. And to make matters worse, the theories of post-democracy and simulative democratic politics make credible cases that modern-day politics is moving more away from the democratic state than toward it.

We are thus again left with the sense of a vacuum. As the probability of the climate crisis comes to more fully penetrate the minds of both politicians and citizens, there will be calls for swift action—even in many cases calls for dire action—and the system will not wait for the ecological citizen to step up to the challenge. Indeed, elite forces, including the militaries, will turn to science, engineering, and political control—often authoritarian—in the face of fundamental change. Democracy, as we have argued, may well be put on the shelf in many places, as has often been the case in times of economic and social emergency. This leaves then the question of how long such a state of emergency will last and whether there will later be a return to democratic values and practices. The longer the crisis lasts, the less likely it will be that future generations will opt for a return to a world they did not know. Indeed, living with the consequences of democratic failure, they may perhaps decide that it was not such a golden era, as it is still presented today. Moreover, one should even acknowledge the possibility that the technocratic state under the circumstances might be greener than the liberal-capitalist state.

The best hope for democratic governance in the face of the emergency may be that some groups, even if only on the edge of society, can hold together and maintain these values in one form or another. One candidate for this possibility is the relocalization movement and the practices of participatory governance. Participatory governance, as a response to democratic failure, might hold out some possibilities of maintaining democratic practices at lower levels of government, being closer to the earlier environmental movement, the civil society politics, and decentralized governance. Toward this end, we turn in Chapter 6 to an examination of the theory and practices of participatory governance as they have emerged in recent decades. As a democratically inspired form of governance, it offers some of the more important projects involving strong citizen participation at the local level, including forms related to environmental citizenship. Indeed, following Torgerson's (1999: 126) view, participatory environmental governance might serve as a more politically committed form of environmental citizenship. In the process, this discussion opens the way for the presentation of field research focused on two informative case studies of innovative approaches to environmental participatory governance, namely the community forest movement in Nepal and the global ecovillage movement.

Notes

1. Writers such as Meadowcroft (2012) have sought to explore the emergence and evolution of the "environmental state," particularly through comparisons with the evolution of the welfare state. The environmental state, or "ecostate," however, tends to refer to that subsection of the larger state that deals with environmental regulation and policy. Moreover, this environmental state is not necessarily specified as a democratic environmental state (Druit, Feindt, and Meadowcroft 2016). Indeed, much of the effort in this direction would easily be described as the advance of ecological modernization. Others such as de Geus (1996) have examined strategies for the ecological restructuring of the state, but these discussions also do not prioritize democratic strategies.
2. The discussion of transparency should also include the secret surveillance of nearly all forms of public and private communication by the U.S. National Security Agency and its counterpart in the U.K., not only those of American and British citizens but also citizens, armies, and corporations around the world. It is also a serious concern in Germany, after it was learned that the German secret service was cooperating with the U.S. agency, which was discovered to be spying on the German Chancellor.
3. In the process, the state has become colonized by powerful corporations that impose their own interests, as the financial crisis of 2008 and the politics of Wall Street that followed made clear. This, of course, is not new, as Marxists have long argued, but it becomes even more problematic in a globalized world of transnational corporations. As the primary institutions in the post-industrial age, the dominance of these global enterprises results in a dramatic and unapologetic "return to corporate political privilege under the slogan of markets and free competition" (Crouch 2004: 51).
4. Although there are regular calls for political accountability and transparency, as well as demands for the extension of political and social rights, the energy of such political systems, Crouch (2004) maintains, has exited the formal political sphere. So, while on the one hand democratic processes and practices would at times appear to be stronger—thanks to activities in the public sphere—democratic politics has lost much of its content. Politics, in reality, has become a matter for negotiations among economic and political elites, which can operate with high degrees of independence from civic society. Here one can find no better example than the elite politics that prevailed after the disastrous global financial collapse in 2008. Despite movements such as Occupy Wall Street, as well as efforts by progressive politicians, little in the way of fundamental reform has taken place. Indeed, those responsible for the collapse were allowed in nearly all cases to continue unscathed.
5. While in previous decades one heard about the demise of democracy from critics such as Noam Chomsky, today it is even expressed by political leaders such as former President Jimmy Carter. As Carter recently stated, "America has no functioning democracy" (McHugh 2013). The role of money in American politics is surely a major contributor to this failure of democratic politics. Although one or two decades ago many would have been hesitant to openly speak of plutocracy, today there is

nothing unusual about the use of the term. And there are seldom any major objections. It is simply acknowledged to be what it is.

6. Research shows that while this contention is not wrong, it requires some qualifications. Not all people respond in the same way to this process of individualization (Middlemiss 2014).

Part III

Environmental Democracy as Participatory Governance

6

Participatory Environmental Governance: Civil Society, Citizen Engagement, and Participatory Policy Expertise

Democratic participation, as we have seen, has been a central theme in discussions of the environmental movement and the greening of state from the outset. But, as we have also seen in the previous two chapters, these discussions have often been as much idealistic as practical or realistic. Despite the contemporary rhetoric about democracy and democratic governance, genuine public engagement is seen to be in decline. Much of the academic focus on participation is on small mini-projects that are basically experiments that take place here and there. It is not that there is something wrong with these efforts; rather they will not be up to the powerful challenge that will be advanced by technocrats and eco-authoritarians. At best, such efforts will remain on the margins of society, if not ignored and pushed aside altogether. Without in any way downplaying the need for normative democratic theory, we should not run the risk of losing a clear picture of just what we are up against. In view of the democratic deficit examined in Chapter 5, it is hard to overlook the disjunction between contemporary democratic political theory and real-world political realities. As there is not much of a bridge from one to the other, certainly not at the present time, democratic political theory needs to be strategically oriented as well. In the search for alternative avenues, this effort turns to the theory of participatory governance.[1]

The discussion that follows focuses on the theory and practice of a citizen-based form of locally oriented participatory governance that can incorporate key elements of deliberative democracy but at the same time speaks more specifically to ongoing political practices. Toward this end, the first half of the chapter surveys the rise of governance and its emergence in environmental politics. It then examines the claims for governance, in particular the more democratic form of governance, "empowered participatory governance." This is followed by a discussion of participatory governance as "deep democracy" put forward by Fung and Wright (2003), in particular their effort to sort out prescriptive principles for empowered participatory governance. Rather than drawn from theory per se, this approach emerges as much or more from an analysis of real-world struggles. Several concrete examples of such democratic struggles are briefly sketched—participatory budgeting in Brazil, people's planning in India, and community forestry in Nepal—and the new models of participatory or collaborative expertise that have emerged with them. The chapter, in the process, sets the stage for a more detailed examination of the participatory governance in the community forest movement in Nepal and its discursive political role in the effort to democratize Nepalese government and society. Grounded in actual political struggles against hierarchy and injustice, participatory governance is presented as a response to the sorts of conflicts that climate change will increasingly usher in.

Participation, Governance, and Sustainability

From the outset in the 1960s, as we saw, participation was a core component of the environmental movement (Beierle and Cayford 2002; Lafferty and Meadowcroft 1996). This, however, was followed by an increasing technocratization of environmental policymaking (Fischer 2000). Amid concerns for the citizens' role as a result of this technical turn, the Earth Summit in Rio in 1992 initiated a significant shift back to more transparency, local-level decision-making, and nongovernmental public–private partnerships that brought public engagement back to the fore (World Commission on Environment and Development 1987). In addition, the Arhus Convention (on Access to Information, Public Participation in Decision-Making and Access to Justice in Environmental Matters) echoed most of the same themes.

This call for public involvement converged at the same time with the kinds of approaches being both theorized by the new "governance" scholars and in some cases put into practice. Calls for governance have found their way into the policy recommendations and practices of a significant spectrum of prominent international and nongovernmental organizations. For example, various forms of governance have been embraced by major organizations such as the World Bank, the U.S. Agency for International Development, U.N. Habitat, and the European Union. All have put money and effort into the development of participatory processes.[2] Civil society organizations, such as Oxfam, Action Aid, and the International Budget project, have actively disseminated information and promoted participatory practices. Despite the fact that some see governance as having become over time something of a buzzword, it does represent more than a new political rhetoric. It speaks to commitments widely reflected in participatory environmental projects of national environmental ministries and their local agencies around the world.[3]

Environmental scholars followed suit and called for participation in environmental governance. Indeed, it became basic to all environmental initiatives. The enthusiasm for governance, in short, rapidly influenced the field of environmental policy and management (Bäckstrand et al. 2010; Chhotray and Stoker 2010: Chapter 10). This has happened in particular via the search for sustainable development. As Adger and Jordan (2009) show, governance entered environmental debates after the Brundtland report and became one of the primary topics in research on sustainable development broadly conceived. Indeed, it has evolved as something of a small industry unto itself (Kates et al. 2005). Governance for sustainable development, or "sustainable governance" as some refer to it, came to be seen as an essential component of any realistic prospect for a successful transition to a sustainable society (Adger and Jordan 2009: 4).

At one level, sustainable governance is easy to describe. In the broadest of terms sustainability can be understood as a state of affairs in which human activity is in balance with the natural world. But the issue quickly becomes complex when we turn more specifically to the question of what that relationship should look like—for example, what should be sustained? Surely we need to sustain essential elements of the physical world and key processes of economic development, as well as social and cultural life. How, though, do we balance these goals? This proves to be even more difficult as we recognize that the idea of baselines becomes more dynamic and unstable with accelerating climate change (Dryzek 2016: 12). It is here that the importance of governance enters the equation. How do we organize the processes that decide these priorities?

Basic now is the recognition that sustainable governance involves more than an objective focus on physical and ecological systems. It also involves both "human interaction with nature... and human interaction with others with respect to their interaction with nature" (Bromley 2005: 201). As such, it raises challenging questions for a wide array of criteria, such as social and economic vulnerability, equity and inclusion, and general well-being. Given the complexity of the challenge, sustainability will not just happen as part of a pre-arranged process. Moreover, there is a need to understand that it will be interpreted differently in different socio-ecological contexts. This means that governance strategies for sustainability require careful reflection and deliberation (Arias-Maldonado 2013). As Adger and Jordan (2009: 7) put it, "the processes of deliberation, argumentation and discussion are constitutive of governance."

Governance as Theory and Practice

What, then, more specifically, is "governance" about? Essentially the concept of governance emerged to challenge the more traditional focus on government. Although there are different views on "governance," it broadly refers "to the changing boundaries between public, private, and voluntary sectors" in relation to "the changing role of the state" (Rhodes 2012: 33). "Network governance" theory, as the most prominent variant, thus proposes a polycentric network of inter-relationships among public and private actors in which the boundaries between the two sectors are fuzzy and blurred. In many or most discussions the emphasis is on horizontal networks of actors seen to deal better with the cross-cutting complexities of more and more policy problems that have to be addressed. Network governance, as Jessop (2003: 101–2) points out, is based on the belief that networking can "overcome the limitations of anarchic market exchange and top-down planning in an increasingly complex and global world."

This has given rise to extensive and ongoing debates, especially with regard to its implications for democracy. For some, it raises the question as to what extent and in which ways we need to understand democracy and its practices differently. The supporters of network governance argue that its potential to mobilize a wide range of actors and to gather and assimilate information is a precondition for the practice of modern-day democracy. Its critics, on the other hand, contend that a lack of transparency and accountability too often associated with network governance undercuts democratic government.

With regard to transparency and accountability, governance is often presented as a way to develop a trust-based equitable consensus through pluralist forms of deliberation capable of solving policy problems. Such networks are seen to open up and extend public participatory spaces between states and markets that can inclusively empower citizens and communities to engage in public problem-solving (Deakin and Taylor 2002). Torfing, Peters, Pierre, and Sorensen (2012: 188–91) argue that they offer new avenues for politically mobilizing citizens, bringing them into decision processes and offering them a wider array of options to choose from a larger spectrum of public service providers, as well as greater chances for the citizen consumers of such services to have an influence on their delivery.

For Klijn (2008: 506) governments "have become more dependent on societal actors to achieve their goals because of the increasing complexity of the challenges they face" and it is thus only through the collaborative processes of governance that "society's policy problems can be resolved." Rejecting traditional bureaucratic government, Skelcher (2010: 164) states that governance reflects the involvement of citizens, civil society groups, and businesses with governmental institution and agencies in the development and implementation of public policy. Involving deliberative mechanisms, partnerships at various levels, and forms of co-production, it is a response to the fact that stakeholders and citizens are demanding more significant roles in the exercise of governmental or public authority (Hajer and Wagenaar 2003: 2). Governance, in short, means that citizen inputs become a much more central part of the policymaking processes, even potentially a central component.

There has been no shortage of research in recent years that examines the ways that public policies and service delivery are being carried out by new patterns of participation and governance, including public, voluntary, and private market arrangements, either in conjunction with the state or at its expense. This research has focused on democratic citizen participation, including projects such as citizen juries and public budgeting, to examine the degree to which citizens can in fact meaningfully participate.[4] It terms of citizen competence, empowerment, and capacity-building, it has often focused on human development and social capital.[5] Capacity-building, as the development of a community's ability to deal collectively with problems such as environmental degradation, can contribute to a sense of social togetherness. Research shows, for example, that places with participatory community governance have higher levels of capacity for dealing with disasters (Aldrich 2012). With respect to service delivery, a range of environmental experiences demonstrate that community participation can improve the efficiency of programs (in terms of uses of resources) and the effectiveness of projects (that achieve their intended outcomes) in the provision and delivery of services, in both the developed and developing worlds (Newig and Fritsch 2009; Koontz and Thomas 2009).[6] Regarding social equity, participatory practices can also have the potential to combine efficiency with equity.[7] Research indicates that decisions made through the participation of community members rather than by traditional elites or unaccountable administrators can offer less powerful groups in the community better chances of influencing the distribution of resources or delivery of services (Issac and Heller 2003; Fischer 2000). In terms of political representation and the distribution of power,[8] participatory governance has the potential to correct the failures of representative government (Chhotray and Stoker 2010; Heinelt 2010; Fischer 2012).[9] But while the potential is there, much of the empirical investigation has produced mixed assessments.

Although there are no hard and firm guarantees, participation in governance can lead to important payoffs (Dahl 1994). While there is nothing simple or straightforward about it, these new practices do hold out important possibilities that can be extended. While it cannot be said without qualifications that decentralized participation leads to greater efficiency, equity, or representation, experiences suggest that the conditions of success depend on conscientious effort and design. As various environmental struggles show, especially those over the siting of environmental hazards, positive outcomes depend on the degree of political representation and the distribution of power it reflects. All too often, when not properly designed and implemented, such projects can be co-opted by other groups in the interest of less obvious or hidden objectives (Malena 2009). Even when the efforts are successful, research shows that such projects are frequently rewarded by being governmentally institutionalized, at which point they are often adjusted or manipulated to serve purposes other than those intended. From a critical perspective, this instrumentalization of participation can be seen as a "political technology" employed to control processes and projects in ways that hinder or narrow the possibilities of popular engagement. For some, this is seen in the fact that it has become part of a general call for "good governance" by prominent organizations, including the United Nations and the World Bank.[10] In these cases, it has led critics such as Cooke and Kothari (2001) to describe participation as "the new tyranny."

The widely spoken rhetoric of "best governance practices" aside, there is then a substantial need to connect these prescriptions with more empirical verification. In too many cases, as Offe (2009) has argued, the focus on governance in the mainstream of political science has become something of an ideology seemingly unconnected to political realities, suggesting in his view that it is an "empty signifier." Davies (2011), for similar reasons, argues that the transformative potential of governance has been overstated. Governance theory and its emphasis on horizontal networks is, as he sees it, fully compatible with neo-liberal reforms that push responsibilities away from government to individuals and groups, typically with fewer resources and less power. As such, Davies underscores the fact that a progressive approach cannot be tied to the dominant power relations.

Even some of governance's strongest supporters have tended to modify their views on governance, particularly as they pertain to democracy. Stoker (2011) and Torfing et al. (2012) have acknowledged that many of the earlier writings were overly enthusiastic. Although governance generally refers to a new space for decision-making, it does not, in and of itself, indicate the kinds of politics that take place within them. Both theory and empirical experience with governance show that there are numerous patterns of participation and non-participation, from elitist top-down forms of interaction to radically democratic models from the bottom up. Indeed, in some cases, such participation has been manipulated or misused to obtain elitist nondemocratic goals. This is especially the case for governance projects that have evolved hierarchically, unintendedly or otherwise (Griggs, Norval, and Wagenaar 2010). They have tended to reproduce existing social and political equalities, even if in somewhat different ways.

This has led Torfing and colleagues (2012) to retreat by arguing that governance is not intrinsically undemocratic and that under some circumstances it can augment democratic practices. In an attempt to reorient and recapture the enthusiasm with which governance was initially greeted, these writers have suggested a reformulation they call "metagovernance," or "governing governance." In this new "second-generation" conception of governance, multi-level interactive governance networks are seen to require new tools but now operating in tandem with the state. Given that the state in question is not especially democratic, the new formulation does little to offset many of the democratic critiques of governance. Rather than replacing a withering state, it appears now that participation in governance helps the existing state to function better, a point relevant to the effort to green the state. In this regard, Davies is correct to argue that governance and metagovernance are fully compatible with the neo-liberal state.

Not all activities related to environmental governance are geared to democratic participation but it has in general been one of the important components (Newig 2007). Although there are various approaches to environmental participation, with extended discussions on how to categorize them (Fung 2006), practices involving public consultations such as deliberative polling, citizens' advisory bodies, and citizens' juries, among others, are taken here to be thin forms of participative deliberation compared to stronger, more robust forms of participatory governance (Beierle and Cayford 2002).

If Torfing and colleagues are correct in arguing that governance can be made more democratic, particularly under specific circumstances, then we argue in this work that they are the circumstances associated with the variant of governance referred to as participatory governance. Whereas the concept of participatory governance is employed in different ways without a clear definition, we use it here to refer to an empowering form of governance that draws on theories of participatory democracy, various governance practices, deliberative politics, oppositional social movement theory, and counter-hegemonic politics. From the theory of participatory democracy, it attempts to build on Pateman's (2012: 10) argument that participatory democracy requires moving the theory and practice of participation out of the realm of political institutions and processes to a broader conception of a participatory society as a whole. From governance theory it draws on efforts to extend deliberative decision-making out from the center to include civil society actors. And from social movement writers it recognizes that this theoretical formulation needs to be situated in a critical understanding of the dominant power relations that shape the hegemonic politics of the capitalist state and the nature of local struggles—concerning both self-determination and inequality—against the centers of power and local environmental struggles in the case at hand (Laclau and Mouffe 2001; Harvey 2001; Fung and Wright

2003). Merging these theoretical foci offers a way of situating deliberative participatory democracy in the context of a specific set of governance practices that extends not only beyond official decision-making processes, but also to the agonistic arenas of political struggle. Moreover, beyond theory, as we shall see later in this chapter, it descriptively captures the progressive participatory methods and deliberative decision practices of a number of important socio-political movements.

Participatory governance, in this formulation, includes elements of various deliberative approaches, but is distinguished by its move beyond giving advice to shaping or even determining actual policy outcomes, binding or non-binding (Chhotray and Stoker 2010). In significant part, as we shall see later in this chapter, this generally results from a relative independence or distance from the dominant power and influence of traditional bureaucratic governmental structures (Fischer 2006; Dryzek 2000). Focusing attention in this work on participatory governance in ecological affairs, we explore the way the theory and its concepts can be applied to environmental policymaking. In this and the next chapters, we examine important experiences with it in the field. What kinds of experiments have been carried out (Sabel and Zeitlin 2011)? And how should we assess them? In particular, how do they address the democratic deficit?

Participatory Environmental Governance

Whereas citizen participation in the government process has traditionally focused on measures designed to support and facilitate increased public access to information about governmental activities, efforts to extend the rights of the citizens to be consulted on public issues which affect them, and to see that the broad citizenry will be heard through fair and equitable representative political systems, participatory governance seeks to deepen this participation by examining the assumptions and practices of the traditional view that generally hinders the realization of a genuine participatory democracy in environmental politics (Gaventa 2006). It reflects a growing recognition that citizen participation needs to be based on more elaborate and diverse principles, institutions, and methods. These begin with an emphasis on a more equal distribution of political power, a fairer distribution of resources, the decentralization of decision-making processes, the development of a wide and transparent exchange of knowledge and information, the collaborative establishment of progressive partnerships, an emphasis on inter-institutional dialogue, and greater accountability. All of these measures seek to create relationships based as much or more on trust and reciprocity than advocacy, strategic behavior, and deceit. It involves as well the provision of means to engage individuals and environmental organizations outside government through political networks and institutional arrangements that facilitate and support genuinely collaborative-based discursive relationships among public and private sectors. And not least important, its theorists have sought to think of ways to politically institutionalize these civil society processes as regular practices of government with a direct influence on policy decision-making.

In contrast to governance generally, local participatory governance seeks to introduce a more active and empowering form of participation in these spaces, or an effort that Santos (2005) has called the "democratization of democracy." It does this through an emphasis on participatory democratic practices based not only on public deliberation, but also the politics of struggle. Instead of attempting to influence policy through enhanced governance networks working with or in conjunction with governments, the kind of participatory government envisioned here takes a more oppositional stance to governmental power, anchoring its structures and practice more firmly to civil society and the tasks of citizen empowerment. Applied to environmental politics and policymaking, participatory environmental governance resonates with the earlier more radical mode of environmentalism. Emphasizing civil society rather than government policy processes, it assumes a skeptical stance toward state power and the possibility of greening the state, reflected in Dryzek's analysis presented in Chapter 5.

Participatory governance thus focuses on the deliberative empowerment of citizens and aligns itself in varying degrees to work on deliberative politics, in particular agnostic approaches, in political theory and deliberative experimentation in policy-related fields of contemporary political and social research. Although it acknowledges the citizen's role as voter and/or watchdog, it moves beyond this limited function to set the practices of direct deliberative engagement within the political contexts of the pressing issues of the time. As such, it has as much or more resonance with agonistic democracy and its emphasis on conflict as it does the search for consensus through deliberative democracy per se.

In the case of participatory environmental governance, it is discovered to have given rise to different types of civil society actors and new spaces for them to inhabit. In both the developed and developing countries, these have involved important shifts in problem-solving and service delivery, including more equitable forms of support for economic and social development. Along the way it has often meant a transition from professionally dominated processes to more citizen-based environmental activities, frequently taking place within the new civic society organizations, a point that will be illustrated in detail in Chapter 7.

Empowered Participatory Governance

In the search for a prescriptive theory for "empowered participatory governance," Fung and Wright (2003) examine a range of cases—including environmental cases—designed to promote the active political involvement of the citizenry in an effort to sort out what works. Most of the cases involve citizen-based social movements struggling to develop new patterns of interaction with government institutions, engaged in what they call collective countervailing power. Without power to counter opposition from entrenched interests, they argue, participatory governance projects are unlikely to prevail. Thus, while they attempt to constructively move existing institutions in more deliberative directions, they remain oppositional movements anchored to outside forces in civil society.

The first step toward a prescriptive theory, in this view, is to acknowledge the problem of complexity. As they explain, complexity makes it difficult for citizens to deliberatively participate in policy decision-making. They speculate, however, that "the problem may have more to do with the specific design of our institutions than with the task they face." Toward this end, they have examined a variety of empirical experiences in the participatory redesign of democratic institutions, including Porto Alegre and Kerala, that involve innovations supported by movements that elicit the social and political energy of citizens—especially those from the lowest strata of society—in pursuit of solutions to problems that beset them.

Even though these reforms vary in their organizational designs, the policy issues to be deliberated, and their scope of activities, they all seek to deepen the abilities of ordinary citizens to effectively participate in the shaping of programs and policies relevant to their own lives, environmental protection being one of the most important. From these common features they isolate a set of characteristics that Fung and Wright define as "empowered participatory governance." The principles they draw from these cases are designed to enable the progressive "colonization of the state" and its agencies. Relying on the deliberative participatory capabilities of empowered citizens to engage in reason-based action-oriented decision-making, the strategy and its principles are offered as a radical political step toward a more democratic society.

As a product of this work, they isolate three political principles, their design characteristics, and one primary background condition. The background enabling condition states that there should be rough equality of power among the participants. The political principles refer to (1) the need of such experiments to address a particular practical problem; (2) a requirement that deliberation relies upon the empowered involvement of ordinary citizens; and (3) that each experiment employs reasoned deliberation in the effort to solve the problems under consideration. The institutional design characteristics specify (1) the devolution of decision-making and the powers of implementation to local action-oriented units; (2) that these local units be connected to one another and to the appropriate levels of state responsible for supervision, resource allocation, innovation, and problem-solving; and (3) that the experimental projects can "colonize and transform" state institutions in ways that lead to the restructuring of the administrative agencies responsible for dealing with these problems.[11]

While this work is an important step forward, a theory of the design of deliberative empowerment still requires greater attention to the cultural politics of deliberative space (Fischer 2006). Beyond formal principles concerned with structural arrangements, we need research as well on the ways the social valorization of a participatory space influences basic discursive processes such as who speaks, how knowledge is constituted, what can be said, and who decides. From this perspective, decentralized design principles in environmental governance are necessary but insufficient requirements for deliberative participation. We need to examine more carefully how political-cultural and pedagogical strategies can facilitate deliberative empowerment in participatory environmental governance (Fischer and Mandell 2012). The point can be illustrated by the cases below, in particular the case of the community forest movement examined in detail in Chapter 7.

Participatory Environmental Governance: From Porto Alegre to Kerala and Kathmandu

Many citizens' deliberative projects are largely advisory in nature; they supply additional information that can be useful to politicians and the public, but are not built into the governmental structure itself. But there are important exceptions, the most progressive of which have emerged in the developing world, especially in Brazil, India, and Nepal. These innovations, taken here to be exemplary cases of participatory environmental governance, include deliberative processes analogous to citizen juries but which have been more formally integrated into the policy processes of progressive governmental institutions. Of particular importance are the practices of public budgeting in Porto Alegre, Brazil, people's development planning in Kerala, India, and the community forestry movement in Nepal. These innovations have been influenced by social movements, NGOs, and left-oriented political parties, both theoretically and practically. Here we shall briefly outline the experiences in Porto Alegre, Kerala, and Nepal. Turning first to participatory budgeting in Porto Alegre, by all standards one of the most innovative practices in participatory governance, this has become a model widely emulated around the world. Indeed, Pateman (2012) has cited participatory budgeting in Porto Alegre as a quintessential illustration of contemporary participatory democracy.

Participatory Budgeting in Brazil

Under public budgeting in Porto Alegre parts of local budgets, including finances for environmental protection, are determined by citizens through deliberative forums (Baiocchi 2003; Wampler 2009). In Porto Alegre, for example, flooding, waste removal, and road reconstruction have been part of participatory budgeting. Perhaps even more significant, in Recife an entire neighborhood was completely reconstructed in accordance with rigorous environmental standards.

In a state of over 11 million inhabitants, long governed by a clientelistic pattern of political patronage, a left coalition led by the progressive Worker's Party took office in the late 1980s and introduced in 1989 a publicly accountable, bottom-up system of budgetary deliberations geared to the needs of local residences. This involved the development of a multi-level deliberative system shaped by the strategies of political activists, influenced at various points along the way by the socio-cultural pedagogical thought of Paulo Freire. As a result, the city of Porto Alegre has been divided into regions with a Regional Plenary Assembly which meets twice a year to decide budgetary issues. City administrators, representatives of community groups, and any other interested citizens attend these assemblies, jointly co-coordinated by the municipal government and community delegates. With information about the previous year's budget made available by representatives of the municipal government, delegates are elected to work out the region's spending priorities. These are then discussed and ratified at a second plenary assembly. The delegates then put these forward at a city-wide participatory budgeting assembly which meets to formulate the city-wide budget from these regional agendas. After consultations, the executive consolidates a final proposal and sends it to the City Council, which can accept or amend the budget proposal (Gomes and Secchi 2015).

The success of the model is clearly established. The Porto Alegre model of participatory budgeting or variants of it have been put into practice in a large number of countries around the world. In Germany alone, there have been well over 100 exercises in participatory budgeting (or "Buergerhaushalt"), also applied to environmental local environmental decisions (Franzke and Kleger 2012). But most of these projects have been grafted onto existing political processes and have thus been relatively insignificant in comparison to the projects in Porto Alegre.

People's Planning in Kerala

The second case, that of Kerala in India, has involved a full-fledged process of people's environmental and resource planning (Isaac and Heller 2003; Fischer 2000). Located in the southwestern corner of the country, Kerala gained attention in the development community for its impressive economic and social distributional activities in the 1980s. In the mid-1990s, a coalition of left parties led by the Communist Party of India/Marxist decided to extend these activities to include a state-wide, bottom-up system of participatory planning, the goal of which was to develop the Kerala 5-Year Plan to be delivered to the central government in New Delhi.

Pursuing a devolutionary program of village-level participatory planning as a strategy to both strengthen its electoral base and improve governmental effectiveness, the governing party decided that approximately 40 percent of the state's budget would be redirected from the administrative line departments and sent to newly established district planning councils, about 900 in number. The villages, supported by the Science for the People social movement and the Center for Earth Sciences, formulated specific development plans in their own political-cultural context that spelled out local environmental problems and resource needs, development assessment reports, specific projects to be advanced, financing requirements, procedures for deciding plan beneficiaries, and a system of monitoring the outcomes. These plans were then accepted or rejected by vote in village assemblies. The final plans were sent to the State Planning Board and incorporated into the state's 5-Year Plan, forwarded to New Delhi for inclusion in the overall development plan of the national government.

It was a process adapted and reproduced in several hundred places in India, before being eliminated in Kerala by conservative politicians. There have been, however, discussions designed to find a way to bring it back. It is a point that illustrates the need to understand these developments as part of a political struggle.

Community Forestry in Nepal

In the case of the community forest movement in Nepal, participatory environmental governance emerged in civil society owing to initiatives by a national federation of local forest user groups. Evolving in the period after two political revolutions and the struggle to bring democracy to Nepal following centuries of monarchical rule, the government initiated a number of reforms designed to facilitate the process. One of them was the Forest Act of 1993 which permitted the devolution of particular forest activities to the local forest users, including an operational plan to be submitted to the regional

District Forest Office for approval and implementation. Because forest protection has long been in the hands of scientific foresters mainly working for the national forestry ministry, it proved difficult to carry out the provisions of the Forest Act, in particular those related to community engagement in the planning processes. In significant part, it became a struggle between those with scientific knowledge of forests and those in civil society with local knowledge about the everyday realities of particular forests (Santos 2007).

By forming a federation outside of government, the local groups were able to significantly challenge the central ministry. Toward this end, they introduced a system of participatory decision-making in forest governance that succeeded in changing many of the ways that the foresters now relate to them. Not only did it succeed in introducing new laws and policies, it has had a very important impact on the practices of many foresters who have come to support the movement.

Beyond its impact on forestry policy in the country, the federation came to be a major force in the struggle to democratize the political system more generally. In view of the significance of the experience, the community forestry movement offers informative insights for both participatory environmental governance and forestry practices elsewhere. Given the importance of forests to global warming, the resulting model of community forestry became an exemplar for other forest-covered countries around the world to emulate. For this reason, we turn to a more detailed discussion of Nepal in Chapter 7.

As a consequence of such activities in participatory budgeting, people's planning, and community forestry, participatory environmental governance has gained a prominent place on the political spectrum of participatory democracy (Fung 2006). Indeed, it emerged in the 1990s as perhaps the most advanced form of participatory innovations. Promoting decentralized practices, it adds an additional layer of local participatory institutions to an increasingly complex institutional landscape that has in some cases, such as those discussed here, given rise to transfers of both resources and decision-making powers. In environmental governance participatory practices are now ubiquitous and, though the progress is still relatively slow, there is an increasing turn to stronger, more democratic approaches to participatory environmental governance.

In addition to participatory governance's contribution to democratic practices, the political impacts are different compared to other forms of participation. In the case of the citizen jury and the consensus conference, the outcomes are merely advisory. They offer politicians and decision-makers a different kind of knowledge to consider in their deliberations, a form of understanding often more closely akin to the types of thinking they themselves engage in (as opposed to complex technical reports). But in Kerala,

Porto Alegre, and Nepal, by contrast, deliberation has been integrated into the policy decision process. In Kerala, local resource management discussions were channeled up to the State Planning Board for inclusion in the official planning document. In Porto Alegre participatory budgetary decisions were linked into the official governmental budget-making process; the outcomes of the deliberations determined portions of the budget. In Nepal, a strong social movement managed to reshape forestry policy and practices, giving local forest users an important say in the development of forest policy.

Participatory Environmental Expertise: A New Type of Expert

Of particular significance in these environmental projects is a type of NGO working to represent and serve the needs of marginalized or excluded groups. In many of the newly created participatory spaces activists have assisted excluded peoples—such as the poor, women, AIDS victims, and the disabled—in developing a collective presence that has permitted them to speak for themselves. Through such efforts environmental activists and their citizen groups have in many cases succeeded in influencing the policies of mainstream institutions. In some cases, these activities have given rise to a new breed of public servant—frequently schooled in NGOs—devoted to offering assistance to these groups. As government officials or independent consultants to parallel institutions, they have often played an essential role in the development and spread of participatory approaches to environmental governance (Fischer 2009; Fischer and Boossabong 2016).

The result of these participatory activities has also given rise to a new kind of professional orientation, one that challenges the standard techno-bureaucratic approaches of the modern state (Fischer 2000; 2009). These professionals, along with their respective theoreticians, have sought to reconceptualize the role of the public servant as a facilitator of public engagement. Feldman and Khadermian (2007), for example, have reconceptualized the role of the public manager as that of creating "communities of participation." In their view, the challenge confronting those working in the public sector is to interactively combine knowledge and perspectives from three separate domains of knowing—the technical, political, and local/experiential domains. As the three cases outlined above make clear, bringing about more inclusive practices of environmental governance involves inventing participatory contexts in which the representatives of these forms of knowing can discursively share their perspectives in the common pursuit of problem-solving. Beyond merely identifying and disseminating information from these various modes of understanding and analyzing policy problems, such work involves translating ideas in ways that facilitate mutual understanding and environmental deliberation among the participants and discursively promotes a synthesis of perspectives that helps to simulate different ways of knowing relevant to the problem at hand. Most of this work, however, has focused more on the participatory inputs than the specific environmental outcomes. More research is needed to better understand the relationship between the two (Koontz and Thomas 2009).

In many cases participatory expertise involves the development of citizen/expert alliances and the use of practices such as community-based participatory research and participatory action research, as was the case in Kerala and Nepal (Fischer 2000). These methods involve professional experts helping lay participants conduct their own environmental inquiries into problems of concern to local residences.

While there have been important efforts to facilitate deliberation between citizens and environmental experts, there are a number of problems that still need to be dealt with (Fischer 2009). Perhaps most importantly, professionals are not trained to facilitate participation and many—maybe most—don't believe there is any point in engaging citizens in such issues. The successful efforts, more often than not, are the result of activities engaged in by professionals involved in progressive social movements of one sort or another (Fischer 2009). In addition, they raise difficult but important epistemological questions related to the nature of such knowledge: Does it just involve a division of labor organized around the traditional separation of empirical and normative issues? Or does it require a new hybrid form of knowledge, involving a fusion of the empirical and the normative and perhaps a special role for local lay knowledge? Included in this question is the need to explore the relationship of reason to emotion. Although everybody in politics knows that emotion and passion are basic to the politics of governance, this topic has yet to receive the attention it deserves in the literature on democratic governance and policy (Durnova 2015; Fischer 2009: 272–9).

Moving Up the Levels: Local Influence on the National State

But what about the higher levels of government? An important and challenging question raised against participatory governance is the contention that it only works at the local level. There is a growing literature on this subject. Various theorists have argued that it is an important contribution to governance at the local level, but can contribute little to higher levels of politics, national politics in particular. Basic here is the contention that such participation is "unrealistic." The issues of government at the higher levels are simply too complicated for citizen participation to be meaningful. Democracy at this level has to remain representative at best (Sartori 1987; Przeworski

2010). Such participation, it is argued, can only be realistic in small groups. Writing from the perspective of social choice theory, Przeworski (2010) maintains that the "causal efficacy" of democratic participation is a possibility for only a few in elite circles. Insofar as effectiveness and equality generally conflict, participatory democracy "is not feasible at the national scale."

This issue, however, appears not to be so straightforward (Dahl 1989). There is a growing empirical literature that examines the tradeoff between citizen participation and environmental effectiveness as it relates to different levels of governance (Moss and Newig 2010; Reed et al. 2010; Bache and Chapman 2008; Dahl 1994). This suggests that the answer depends on a number of factors related to specific policy issues, the presence of competing goals, the quality of the representatives, and the nature of interest group politics at different levels. For example, Koontz finds that citizen participation on economic issues tends to be most effective at the level of state government, whereas citizens tend to be more active and effective at the national level when it comes to non-economic issues such as the environment (Koontz 1999). Rochloff and Moore (2006) provide evidence that shows effective environmental governance depends on a multi-scalar approach that integrates national, regional, and local levels. With regard to outcomes for environmental stakeholders, participation depends on the nature of the policy processes at the different levels (Reed and Bruyneel 2010).

Especially interesting is the contribution of Pogrebinschi and Samuels (2014). They challenge the assumption that participation only works at the lower level by pointing to the role of the National Public Policy Conferences in Brazil. Convened by the national executive branch with strong civil society participation, these policy conferences start with deliberative participation among diverse groups at the local level, the results of which are carried forward by local representatives to the state and regional levels, before being taken up officially by organized public deliberations at the national level. Covering a range of topics, including environmental protection, such conferences go on for about a year, with outcomes generally judged positively. The research demonstrates a high degree of correspondence between the recommendations in the final reports and the resultant public policies that make their way through the legislative branch. It is presented as convincing evidence that participatory democratic deliberations at the municipal level can have a national impact.

The National Public Policy Conference is not participatory governance per se. But the kinds of deliberations involved are basic to such governance (Avritzer 2012). And it is just here that we can add the experiences of Kerala and Nepal, which in fact serve as such examples. Both illustrate the possibility of local deliberations influencing higher levels of decision-making. As we saw, through a set of hierarchically interrelated deliberative forums Kerala has formulated its state five-year plans on resources planning from the bottom up. The plans have been sent forward by the state planning office to New Delhi to be included in the larger five-year plan for the country as a whole. In Nepal a network of locally based forest user groups, through a federated network (FECOFUN) that extends upward to regional and national levels, significantly changed the policies and practices of the central Ministry of Forestry. The deliberative politics of the network, which we shall examine in more detail in Chapter 7, also had a well-documented impact on the political culture of a newly democratizing nation; indeed, it has been one of the primary democratizing forces. Both of these examples are clear cases of participatory environmental resource planning that, albeit in somewhat different ways, demonstrate that strong participatory governance can have significant impacts beyond the local level.

Interesting here is the fact that in all three cases—the National Public Policy Conference, Kerala's people's planning, and the community forestry movement in Nepal—these deliberative processes emerged and succeeded because political groups at the top joined together with grassroots movements from below. That is, the top and bottom of the power structure must work together for such projects to have an influence at the higher level (Fischer 2009). Activists and reformers must emerge at both levels.

Also important to note is the fact that all of these come from the less developed world. To some degree Brazil is an exception here, but not entirely, especially given that the National Public Policy Conference was developed and introduced long before the country began to join the economic ranks of more developed nations. Moreover, whereas they are often discussed in public administration as measures of "good governance," the experiences presented here show them to emerge from political struggles against unjust and inequitable social systems. As such, they took root in civil society, particularly thanks to social movements, but then moved forward with the assistance of particular political parties and public servants willing to help make space for them.

To be sure, we do not know enough about the social and political factors that have contributed to the successes of these processes. These cases show, however, that such practices can have an impact both at the local level of governance and beyond. Although the clear successes of these cases are exceptions to the rule, their experiences justify the need for more empirical research into these democratic innovation processes.

Conclusion

Participatory governance activities have offered significant new insights into questions that have not received the attention they require in traditional political analysis and democratic theory, including environmental political theory. If a strong system of democracy—representative democracy in particular—depends on vigorous participation from below, then participatory governance has the potential to fill the "institutional void" that much of political theory and contemporary practices of representation have failed to address. Given that this participatory void extends to the environmental or green state, participatory environmental governance offers an alternative to further technocratization of environmental policymaking.

Citizens, as the cases of participatory governance presented show, are able to participate meaningfully in complex environmental decision processes. The supporting findings are anything but mundane, given the fact that citizens are regularly said to be unable to understand the problems and thus meaningfully participate. Citizen-oriented participatory governance can also confront head-on the argument that participation impedes managerial efficiency and political accountability. Indeed, participatory environmental governance can not only improve the delivery of environmental protection, but can also increase the social equity of programmatic outcomes. Further, it has the potential to facilitate the creation of new professional forms of environmental expertise, especially knowledge practices that recognize the value of local environmental knowledge.

Although the practical work on governance generally involves a collection of separate experiments and projects that have common threads, they often involve outcomes that are difficult to interpret. It is essential to acknowledge that not all the experiences with many participatory efforts have been positive. As Newig (2007: 52) sums it up, there are still "empirical questions as to the extent to which participative processes actually contribute to an improved implementation of environmental policy and thus to a more sustainable usage of the environment." An important key to success depends on the contextual factors associated with the particular environmental decision processes. Of special importance, as we have argued here, is the political context. It is the political setting that differentiates participatory governance projects in Porto Alegre, Kerala, and Nepal from governance projects more generally. Where governance initiatives generally are situated in the existing systems of political and social inequality—that is, they try to work within them—participatory governance typically involves explicit struggles against such inequalities, and as such is rooted in and supported by civil society-based social movements independent of the dominant bureaucratic power and influence. Such projects may have one foot in the existing arrangement, but they have the other foot solidly planted outside in civil society and thus can more easily avoid the kinds of cooptation that befall many mainstream governance efforts. Fung and Wright (2003: 263) have referred to the goal of such a strategy as the pursuit of "collaborative countervailing power."

The task of sorting out the positive and negative elements contributing to the outcomes of such participatory environmental projects thus takes on particular importance. We need to learn more about the contextual circumstances—power relationship, degrees of inequality, levels of citizen competence, and more—that promote or hinder such projects, especially those factors associated with the strong models of participatory environmental governance. Given that there is no shortage of elements that come into play, such an assessment is challenging.

What, then, can we conclude? Independent of a good deal of the rhetoric associated with discussions about participation, the evidence from experiences with governance, including participatory governance, illustrates that while participation can work, it can also pose difficult issues with no simple solutions. Given the difficulties involved in designing and managing participatory processes, it comes as no surprise to learn that citizen participation schemes rarely follow smooth pathways. Despite its promise, participatory governance by citizens is a complicated and uncertain business that needs to be carefully thought out in advance (Fischer 2000). In the absence of serious attention to the quality and viability of citizen participation, it is usually better to forgo such projects.

In sum, participatory governance can supply a theoretical political foundation for the turn to eco-localism that follows. But we need to learn more about how to organize, cultivate, and facilitate participatory environmental governance. This should be a first priority of those engaged in both the theory and methods of the practice. We can thus benefit by turning to a more detailed assessment of a well-organized and successfully executed illustration of participatory environmental governance, the community forest movement in Nepal.

Notes

1. Much of what follows is an extended version of a chapter on "Participatory Governance" that appeared in the *Oxford Handbook of Governance*, edited by David Levi-Faur (2012). In particular, this chapter applies the ideas in that essay to environmental governance.
2. Many of these initiatives have drawn their inspiration from the progressive projects of political parties in India, Brazil, Spain, Mexico, and the U.K.
3. Just to cite a couple of examples, governments such as the Democratic Republic of Congo, Botswana, and Mali have set up projects to strengthen national capacities to implement Principle 10 of the Rio Declaration, calling for forms of participatory environmental decision-making (see Kimani 2010).
4. The issue is crucial for participatory governance. It has no special significance if citizens are neither capable nor empowered to engage in decision processes that

affect their own lives. Studies show that many people in the middle rungs of society, and in some cases people closer to the bottom, can competently deal with policy discussions (Fishkin 2009; Delli et al. 2004). Research, much of it related to environmental and technological policy issues, finds that lay panelists on citizen juries increase their knowledge of the subject under discussion and often gain new confidence in their ability to deal with complex policy issues generally (Joss 1995). Many participants tend to describe such participatory experiences as having had a stimulating impact on their personal lives, often leading to further involvement in public affairs, environmental issues in particular (Lawrence 2005). Much more challenging, however, is the situation for marginalized members of society, those who might benefit from it the most. But here too there are positive signs. Participatory projects show that citizens with less formal education can also participate under the right conditions with surprisingly high levels of competence. In many cases, for example, the participants in local deliberative councils concerned with resource planning would be described as simple farmers. Nonetheless, they impressively participated in environmental and resource management planning projects. In particular, local environmental knowledge has often taken its place alongside technical expertise.

5. With regard to citizen competence and empowerment, the practices of participatory governance are put forth as a specific case of the broader view that participation contributes to human development generally, both intellectual and emotional. Empowerment through participation has in general long been part of the progressive educational curriculum, and numerous citizen-based deliberative projects bear out its influence on personal development (Joss 1995; Dryzek 2008). Beyond institutionalizing new bodies of client or user groups, they have created new opportunities for dialogue and the kinds of citizen education that it can facilitate, especially communicative skills, citizenship, and learning (Lawrence 2005). Many progressive NGOs also speak of "people's self-development" and "empowerment" as primary goals, emphasizing political rights, social recognition, and economic redistribution in the development of participatory approaches (Rahman 1995). Rather than merely speaking for the poor or marginalized citizens' interests and issues, they have labored to assist people to develop their own abilities to negotiate with public policymakers. Many of these projects have been concerned with environmental issues.
6. In fields such as education, health care, environmental protection, forestry, and irrigation, participation is seen to lead to better service delivery, quicker responses to emerging issues and problems, more effective development and design of solutions appropriate to local resources, higher levels of commitment and motivation in program implementation, and greater overall satisfaction with policies and programs (Ojha 2006). With regard to efficiency, participation can lead to improved monitoring processes and verification of results. Some, however, argue that by diffusing authority and control over management, decentralized participation can also weaken efficiency (Khwaja 2004). And others argue that it can lead to resource allocations that violate the true preferences of community members, as some may withhold or distort information about their preferences and choices. This problem is perhaps most acute in developing countries, in which community participation is

frequently related to external donor-funded projects. All too often in these cases such participation can intentionally advance preferences that are seen to be more in line with the interests of the donors than local interests. It is a point that can be confirmed in the case of the community forest movement in Nepal. The participants simply try to increase their chances of obtaining available resources by telling the donors what they want to hear (Platteau 2007). In addition, the determination of efficiency encounters methodological problems (Osmani 2007), as firmly establishing the cause–effect relationships can be problematic (Newig 2007). It is always possible that a positive association between efficiency and participation may only reflect a process of reverse causation—that is, community members had already chosen to participate in those projects which promised to be efficient. To know if participation has in fact contributed to efficient outcomes, investigators have to discern if such extraneous factors are at work. Although this is theoretically possible, it is a difficult technical requirement. Such information is often unavailable or difficult to come by.

7. Empirical investigation on social equity, however, requires more research. Many studies suggest that participatory approaches in local arenas can be of assistance to the poor and disadvantaged members of the community, but other research fails to clearly confirm this (Papadopoulus and Warin 2007). Overall, investigation shows that community participation can lead to more equitable outcomes, but it is particularly difficult to achieve such results in inequitable social contexts. Equitable outcomes more commonly occur in combination with other factors, such as those related to the distribution of power, motivation levels of the participants, and the presence of groups that can facilitate the process. One of the difficulties in assessing the impact of such participation is that there is often no reliable information about the distribution of benefits and costs to households, thus making it difficult to render comparative assessments (Osmani 2007).
8. The questions of representation and power related to governance have been widely discussed in political theory and deliberative democracy (Smith 2003). Through critical reflection in participatory processes, disadvantaged citizens have improved chances of expressing their preferences in ways that can make them count. It is a point basic to the environmental justice movement (Schlosberg 1999). Basic also here is the question of how small network groups can represent a larger public. Just as important is how meaningful deliberation can take place against the backdrop of a skewed distribution of power. These questions have also received considerable attention in environmental political theory (Eckersley 2004; Baber and Bartlett 2005; Dryzek 2000; Bäckstrand et al. 2010).
9. Participatory environmental governance seeks to give voice to those without power. But one has to be careful in assessing the degree to which it can generate unmanipulated participation. At the current state of development, participatory environmental governance itself often exists as much or more than as a strategy for struggling against the political imbalances as for counterbalancing them outright.
10. The World Bank, for example, has deftly co-opted various participatory projects and their methods to generate support for their own agendas. Having discovered the relevance of local involvement and participation from many of its Third World

investment failures, the Bank took an interest in the advantages and institutionalized a participatory program designed to facilitate direct local contact with the communities it seeks to assist (World Bank 1994). Not only have senior bank staff members been directed to get to know a particular region better through personal participation in programs and projects in its villages or slums, the Bank has pioneered a technique called participatory poverty assessment designed "to enable the poor people to express their realities themselves" (Chambers 1997: xvi). It has been adapted from participatory research experiences in many countries around the world (Norton and Stephens 1995).

11. Because participatory governance is largely designed to compensate for the failures of representative government to adequately connect citizens to their elected representatives, the ability to bring these two political models together is important (Wampler 2009), a topic to which we return. Closely related to representation is the question of power, or what Osmani (2007) calls the "power gap." A function of the asymmetrical power relations inherent to modern societies, especially those created by the inequalities of rich and poor, this gap poses a difficult barrier to meaningful participation. As environmental justice studies show, when inequalities are embedded in powerful patriarchies such projects are prone to be captured and manipulated by elites, whether political leaders and their patronage networks or those providing development assistance from the outside.

7

The Community Forest Movement in Nepal as Participatory Governance: Civil Society, Deliberative Politics, and Participatory Expertise

In Chapter 6 we pointed to the importance of expertise in facilitating participatory governance, noting that it is possible in many cases to identify a new breed of civil society experts who have worked with such participatory movements. One such case is the story of the community forest movement in Nepal. The experience there with participatory governance through civil society organizations offers important insights into the broader tensions between democratic participation and technical expertise in climate change policy, especially the role of facilitating public engagement. Indeed, the efforts of this community forest movement, given the crucial role of forests in reducing carbon emissions, has come to be considered a model for global protection of forests in countries around the globe. The purpose of this chapter is to present this development and to interrogate its lessons about participatory governance. To properly understand this experience, it is necessary to begin with the political context of which it is a part, namely political revolution and the struggle for democratic governance.

Democratic Struggles in Nepal: Political Backdrop

The emergence and development of the contemporary community forest movement in Nepal was spawned directly and indirectly as a result of the political revolutions that have taken place in the country since 1990 (Sugden 2011; Shaha 2001). After centuries of monarchical, feudalist rule, Nepal's top political parties forced the monarch to accept a multiparty representative political system (Gellner 2007). The result was a constitutional monarchy that introduced significant measures of democracy for the first time. One of the important results to follow was the opening of a vibrant civil society movement across the country (Whelpton 2007; Pandey 2010; Jha 2015).

As was the case in many countries, the fall of the authoritarian regimes in Eastern Europe and elsewhere led to a dramatic expansion of interest in the revitalization of civil society, influenced in part by ideas in political science and sociology from Western Europe and North America and in part from street-level struggles in Eastern Europe and other parts of the world (Pandey 2010; Kaldor 2005). In Nepal, the concept of civil society and the public sphere moved to the center of the society and blossomed (Dahal 2001). And in the realm of forestry, this was especially facilitated by the passage of the Forest Act of 1993 which provided local groups with the right to have a say in the governance of the forests from which they derived their livelihoods. Both are of critical importance for the story that follows. Before looking more specifically at the civil society organizations, however, it is important to introduce the underlying political currents that spurred this development.

Especially significant here is the fact that the political revolution did not stop with new parliamentary arrangements. As these new arrangements were being introduced, a revolutionary Maoist movement was forming in the countryside, representing an even more fundamental attack on the political system, including the new constitutional monarchy (Thapa 2003; Hutt 2004). At first, this movement was largely ignored, treated as an uprising for the Royal Army to put down. But by the mid-1990s, it was becoming clear that this Maoist-led insurgency was gaining support and in the early years of the next decade it was no longer obvious that the army would defeat the Maoist guerrillas. This was especially the case after the movement managed to assassinate a member of the royal family in the palace itself. By 2003 it became clear to most that the movement was not going to be swept aside. And as the decade progressed, there seemed to be a hopeless deadlock between the Royal Army and the revolutionary forces in the countryside. The army could defend Kathmandu, but it could not defeat the rebels outside of the capital.

This led the king to attempt to reassert his monarchical authority, which proved to be extremely unpopular in the country as a whole. It was a miscalculation that played into the hands of the Maoists. Taking advantage of a general public opposition to this development, including that of leading party politicians, the Maoists shifted their strategy. Recognizing that the civil war could go on for a long time, without substantial gains, and that there was major political unhappiness in Kathmandu, they proposed to lay down their arms and enter party politics. Since then there has been an ongoing, and thus far largely unsuccessful, effort to write and adopt a new constitution. Indeed, in 2012, unable to reach basic agreements about the constitution, the government was forced to resign, leaving the country in the throes of political uncertainty, if not turmoil.

Civil Society and NGOs

Essential to this story is the role civil society organizations have played in the political transition. The rise and role of these organizations can be interpreted in the context of a larger set of external developments. After the fall of the Soviet Union and the eastern bloc countries under their influence, the post-Cold War era witnessed a worldwide turn to the role of civil society and NGOs as promoters of democracy in the newly emerging or reorganizing political systems around the world, both West and East. The scurry to build new democratic regimes turned attention to the promotion of civil society institutions and the encouragement of NGOs to serve as primary vehicles for such civil society development. Indeed, the new civil society institutions were widely seen as the alternative to government, in particular the rigid bureaucratic political structures of the former authoritarian systems.

Civil society has never been a concept without theoretical difficulties. As Shah (2008: 3) has pointed out, "civil society is not a conceptual abstraction developed from a range of comparable empirical phenomena but a normative project for a particular philosophical and political standpoint." Moreover, there is no single concept or model to be found. There are instead competing theories and different empirical practices. Despite the differences, the models are united by a set of actors and beliefs which validate their recognition, legitimacy, and claims upon moral and material resources. The network of civil society organizations is made up largely of academic centers, donor groups (in particular foreign foundations and states), activist organizations, and local oppositional groups. It represents a shared ideology and program which has been described by Sheper-Hughes and Courgois (2004: 18) as organized around a desire for a "global economic system based on principles of democracy, human rights, and free markets." The development has, as such, been basic to the spread of globalization in the post-Cold War period in search of a new liberal hegemonic world order. The realities associated with nongovernmental organizations, however, have not turned out to be straightforward and self-evident. Indeed, what we have learned over the past two decades is that NGOs can take many forms, socially and politically (Wagle 2006).

For many, they have been seen as instruments for promoting democracy-building and improving governance capacities. Others, at the same time, have emphasized economic development, the creation of social capital, and welfare services delivery, and still others focus on geopolitical security goals. As an organizational instrument for the latter goals, civil society organizations are seen to be flexible and cost-effective. As Shah (2008: 36) puts it, "Unlike political parties and trade unions, civil society is less tied to the local constituencies or electorates, and therefore more responsive to its external donors," such as the World Bank or the U.S. State Department. It is a development that, as we shall see, can be as much or more of a problem than a positive contribution to a democratic transition.

Some have worked for progressive purposes, others have sought to promote programs of questionable value, and still others have served as political conduits for the money and interests of external forces, ranging from foreign governments to the World Bank. Moreover, for many activists NGOs became another form of employment, sometimes the only available opportunity. In short, they came to mirror the pluralist interest group politics that emerged in these countries after the new waves of democratization in the 1990s. But it is nonetheless possible to find some who do reflect the kinds of hopes and commitments associated with early civil society theorists. The purpose here is to report on two such NGOs in Nepal, FECOFUN and ForestAction.

Nepal is, in fact, a classic example of the emergence of an NGO-dominated civil society (Bhatta 2007). While the concept is not altogether foreign to Nepali society, it is in literal translation more closely related to an older concept of "citizens' society." But this understanding is related to a different set of arrangements which encouraged citizens to assist one another. As "customary associational forms," they traditionally helped to mobilize people and resources beyond the family to work for self-help, community improvement, service delivery, and charity (Shah 2008: 9). Such organizations are grounded in both traditional beliefs and practical goals. Much of this history, however, has been lost in the shift to new-style civic society organizations. The new forms have largely been neglected or replaced the earlier more spiritual and ethnic roots of social engagement of the citizens' society organizations.

The organizations that were to emerge in the contemporary era grounded their existence and justification in modernist ideas related to knowledge claims about policy and organization, normative ideals, educational credentials, and political philosophy. These orientations, taken together, have given rise to an unmitigated belief in the inevitability of democracy. Or at least this is the ideology advanced by most civil society organizations. And it has, in fact, introduced a good measure of pluralism. But this is not to be understood as a template for harmonious progress in matters related to social and economic justices. Like pluralism in general, these civil society organizations also produce a significant amount of conflictual politics.

The proliferation of NGOs in Nepal is very much tied to the larger political turmoil that resulted from the struggles since 1990. Indeed, civil society organizations played unprecedented roles in the political struggles leading to regime change. Even though the struggles were initiated and advanced by a mixed group of parliamentary parties and the insurgent rebel Maoist forces, civil society organizations took on a critical role in mobilizing the mass citizenry and establishing a public discursive context that was critical of the king and his governmental regime. Especially important for this role was the mass media; via newspapers, radio programs, and televisions stations they succeeded in turning public opinion against the regime, leading it to retreat to a defensive posture. These media, in particular radio, were seen to trigger an avalanche of mass political demonstrations. As one observer put it, the interactions among the media, civil society, and political parties meant that radio and television stations filled the air with the challengers' critical views, while the newspapers dealt with interpretations of the contest and sought to set the understandings of what was fair and politically acceptable.

Of equal importance were the substantive efforts of these organizations in areas such as democracy-building, human rights, development, conflict resolution, and advocacy. Over fifteen to twenty years these organizations, often funded by outside sources, increased from around 200 to well over 30,000. According to some important observers, this work prepared the ground for much of the transition that was to follow the political conflict. Both the number of organizations and their networks were critical resources that could be mobilized in the name of regime change. One writer put it this way:

> Using standard "protest repertoires" employed by other actors involved in the political movement, civil society organizations and activists arranged demonstrations, shut-downs, and sit-ins, issued public statements, took part in protest marches, and ran media campaigns. In the course of their political actions, the civil society actors confronted political barricades, faced baton charges, and were arrested like the party activists. (Shah 2008: 32)

Even though there are important similarities here with the protest movements in general, the thing that most clearly set off the civil society organizations and the other political groups in the struggle were their capabilities to mobilize and organize external supporters and resources that were not necessarily available to the political parties owing to international norms of noninterference (Shah 2008: 32; Bhatta 2007). This took on particular significance after the parliamentary parties established an alliance with the Maoists, described by their opposition as "terrorists." According to Shah (2008: 32), it is "precisely the nonstate status and flexible network architecture connecting the local and global that enables civil society to become an effective intermediary between the external political opportunity structures and the internal power struggle." The NGO connections with global organizations such as Human Rights Watch, Amnesty International, and UNESCO, as well as Western donor governments, played a critical role in putting pressure on the royal regime, contributing to its delegitimation and downfall.

Also important in these struggles was the role of a large number of forest development groups. In many areas during the struggle, rural and mountain communities who use the forests were called upon by national groups to take an open stand and support the political struggle for regime change. Indeed, the community forest movement and its civil society organizations can be seen as both a product and an outcome of the new political system.

The Community Forest Movement and FECOFUN

Of the total population of Nepal, about 35 percent is in one way or another involved in the use or protection of roughly 27 percent of the forest land covering the country. These forest activities generate close to a billion rupees of saleable forest products per year. It is an agriculturally based system of livelihood, particularly in the hill regions, that is in all ways connected with issues related to forest ecosystem protection. For generations traditional institutions have had the responsibility for managing these forest lands. As early as the 1970s there were efforts to develop community forestry in an attempt to supply these traditional forestry agencies with collective local wisdom that could improve both the promotion of community livelihoods and forest sustainability (Pokkharel, Branney, Nurse, and Malla 2007). Conceived as an effort to share authority and power, the goal was to cooperatively improve both the policy and management practices of the regional District Forest Offices, responsible to the central Department of Forestry. Over time, however, the methods and practices of these offices have been distorted by the hierarchical interventions of the forest bureaucracy and the claims of foresters to possess superior forms of forest management knowledge.

The story of the community forest movement is basically a story of a struggle with a centralized, hegemonic forestry and its foresters who sought to govern forest policy and practices with scientific forestry principles that provided little or no place for the forest people themselves.[1] The Department of Forestry in Nepal was established in 1925 and over many decades regularly expanded its scope and capacities to assume better control of the country's forests, Nepal's most important resource and thus primary industry. In the view of the foresters at the time, the people in and around the forests were themselves responsible for damaging and neglecting the forests (Ojha 2009).

Since that time there have been numerous laws passed and justified by the same argument; the local dwellers were responsible for forest destruction.

They were thus designed to enforce national controls by the central forest administration at the expense of the local forest communities. And, in some cases, the community forest people did damage the forests. The destruction was, in fact, often a response to the edicts forbidding them to use it. Many among them took the view that "if we cannot use the forests, then nobody else should use them either."

In 1957 there was a further centralization of the national management and control of the private forests, the effect of which was to consolidate the role of the governmental bureaucracy as the expert manager of forests. With this act it was assumed that removing the availability and uses of the forests for private groups would sustain and improve access to forest resources, which largely meant government access. The primary result, however, was to create a strong technical approach that led to stringent rules and regulations which excluded the citizens' role. The ministry, in short, technocratically served the government and the monarchy. The counter-response was the rise of the community forest movement to reclaim the right to use the forests, as local forest peoples have for centuries.

The community forest struggle is closely tied to the struggle to democratize Nepalese society. Indeed, the struggle for democracy is an important condition leading to the development of the current community forest movement. Without the democratic thrust preceding the movement, it would not have evolved in its present form. Emerging in 1990, the democratic processes opened up a political climate in which the citizens of Nepal could for the first time speak and organize themselves freely; it was a new freedom that gave crucial impetus to the movements that followed. The new political connections between the citizens of the country and the newly elected governments were responsible for legislative initiatives that supported the existence of the people living off the forests. Most important in this case was the enactment of the Forest Act of 1993, which acknowledged the rights of forest communities to use and manage forest resources (Bhattarai and Khanal 2005). Although the World Bank had earlier offered monies and assistance to the central government to encourage local involvement in forest policy, it was the Forest Law that provided the crucial legislative foundation for the developments that were to follow. Promoting an engaged segment of civil society concerned with the uses and protections of the forest areas, it facilitated a new political discourse about the devolution of political power and influence at all levels of forest governance, particularly in the Ministry of Forests and Conservation. Implementing the decentralization of forest governance emerged in civil society as an important topic in both political and intellectual circles.

Moreover, the new Forest Law and the activities to which it gave rise encouraged external support, in particular from donor organizations such as the World Bank and the Ford Foundation. In addition to money for programmatic developments, they offered other forms of assistance as well. These included an emphasis on people's rights over business interests and rules and regulations supporting the forest communities' rights with regard to forest use and resource protection.

The primary vehicle to translate this into progressive change was FECOFUN (Federation of Community Forest Users Nepal), which picked up and carried the community forest movement forward (www.fecofun.org.np). As a civil society organization, FECOFUN emerged to play a crucial role in formulating forest policy. Most basically, it has regularly pressed for citizen engagement in forest policy development and management at both the local and national levels (Ojha 2009: 7–18).

Organized by political activists, many of whom had gained experience and skills in the radical struggles for democratic change, these leaders set up a federated civil society network to link together the "Community Forest User Groups" (CFUGs) that were enabled by the Forest Law. CFUGs were first started by a few local activists working with NGOs (such as Watch and the International Centre for Integrated Mountain Development (ICIMOD)). These projects involved bilateral forestry activities that facilitated forms of networking among CFUGs, primarily for sharing and collecting planning information. Although CFUGs had formally developed in response to the new law, they were small, dispersed, and generally poorly organized; they lacked the knowledge and skills needed to advance their own interests. It was here that the FECOFUN activists stepped in. Recognizing the need for strength in numbers, they organized a national network linking these local units in ways designed to progressively support and advance the interests of the community forest people.

In the process of the CFUGs' formation, a number of left-wing leaders who possessed strong mass leadership experience stepped in and took the networking aspects beyond the imaginations of those who initially supported them. These leaders were not intellectuals per se; they were primarily activists and organizers. But they did have intellectual support. As part of the process, they continually engaged with critical intellectuals, inviting them to support, train, and advise the movement. These activists were also given a helpful sum of money by the Ford Foundation, thanks to sympathetic people in the Delhi Office of the Foundation. This money also expanded the reach of the network and paved the way for more money.

Basic to this effort was the development of a FECOFUN constitution. The document set out the primary goal of the CFUGs network to be the facilitation of cooperative collaboration among CFUGs, with an emphasis on promoting learning based on common, shared experiences. The organizing efforts focused on making the members of these groups aware of their rights to forest access, as well as responsibilities and obligations in the management of resources as spelled out in the government's forestry rules and regulations. Specifically, the federation is politically structured through an elaborate set of relationships, beginning with a general assembly, a national council, executive committee, steering committee, regional coordination committees, districts branch committees, range post committees, and village-level CFUGs.

After FECOFUN began to take root, a number of supporting events occurred. One was the development of a socially committed and politically adept group of people who emerged to lead FECOFUN. In addition, there was an extensive mobilization and supply of advisors, both technical and administrative, from an extensive spectrum of organizations, national and international (Ojha, Timsina, and Khanal 2007). Both of these developments served to help forge a "critical mass" of local FECOFUN activists. In addition, they established a multi-layered form of governance and organized periodic elections for the various governance positions. Through all of these efforts the organization maintained a critical stance against the technocratic approaches of government, embracing a form of self-governance.

Most important has been FECOFUN's successful efforts to offset the long-dominating techno-scientific approach to forestry. It did this by working to raise the consciousness of forest communities and thus their interest in and commitment to forest governance. Prior to the establishment of the FECOFUN network, forestry services were administered through either the central forestry bureaucracy in Kathmandu or various bilateral projects. Launching a direct challenge to these centralized practices, common to forestry in general, FECOFUN started providing information about services and often in ways different from the government. Indeed, the organization established itself as something of a shadow forestry agency, monitoring and discussing the activities of the forestry ministry.

Over time the kinds of issues that the organization has taken up have continued to expand. These have included efforts against monopolistic practices supported by the Timber Corporation of the country, opposition to attempts to give the District Forest Offices more power, a ban on green felling that curtailed the rights of CFUGs, struggles against new management practices designed to restrict particular types of production practices, challenges to the addition of unnecessary bureaucratic procedures imposed on CFUGs operations, blocking government efforts to undermine local approaches to the protection of biodiversity, the levying of taxes without appropriate consultations, and pushing for collaborative forest management more generally (Timsina, Luintel, Bhandari, and Thapaliyam 2004).

All of these struggles served to expand the citizens' role in policy decision-making and implementation. To further this end, FECOFUN has employed a series of strategies to influence and extend these activities. Along the way, it has shown itself to be able to learn, adjusting these strategies based on success and failure. For example, instead of focusing primarily on the efficient management of services for the logging industry and other stakeholders, as the agencies and offices of the Forest Ministry and other NGOs have done, the federation has provided civic, citizen-oriented understandings of the same information, especially information about who gains and loses from particular practices. Extremely important to this civic perspective has been the framing of "national forests" as opposed to understanding resources to belong to specific groups or beneficiaries. Approached in this way, community forest groups began to better grasp their own relationship and rights to the forests and their resources.

On a more practical level, the federation has often focused on particular government programs or plans seen to be detrimental to either community interests or the future sustainability of the forest ecosystem, or both. Toward this end, FECOFUN has made a practice of distributing a full range of information about the relevant laws pertaining to forest governance (Ojha 2011). Of special importance, FECOFUN made available information about the legal provisions pertaining to community forestry to groups in areas without community projects, especially to areas in which the District Forest Offices of the Forest Ministry found little motivation to make this information available, most importantly information about rights. To facilitate the spread of this activity, FECOFUN established its own radio program with regular broadcasts, programs that provided information and discussion that contrasted markedly with the official views of the central government. As observers close at hand describe it, such efforts are geared to assisting the political capabilities of the CFUGs beyond the more traditional clientalist connections to the bureaucracy of the Department of Forestry (Ojha, Timsina, Chhetri, and Paudel 2008).

In addition to assisting in networking activities, FECOFUN has emphasized the goal of enhancing the institutional structures and processes of the CFUGs. Important among these capacity-building activities is assistance in developing and preparing the CFUGs' operational plans, as well as managing their financial accounts. Another contribution is the training of personnel, having offered a dozen or more courses for its CFUGs chapters (with topics ranging from how to do forest surveys to learning CFUGs networking skills, grasping technical issues related to agro-forestry, the development of deliberative competencies, and the training of trainers).

FECOFUN has also involved itself in practical issues related to service delivery. Clarifying and supporting management services at the local level, it has established collaborative relationships with a diverse range of NGOs and other organizations to help facilitate the delivery of crucial services at the community level. In some cases this has involved coordination issues, but in others it has concerned strengthening technical capacities.

The staff also teaches skills in conflict resolution, particularly as they have pertained to the harvesting of forest products, the distribution of products, boundary disputes among community forest users, and the role of District Forest Offices in community forest governance. It has also placed a strong emphasis on the role of women and women's issues in community forestry. FECOFUN and CFUGs are required to have equal divisions among the sexes. FECOFUN has also supplied legal consultation to member chapters confronting law-related problems. This has included the filing of numerous cases against the Department of Forestry pertaining to the curtailment of community rights, operational plans, and the transfer of land, among other issues.

In the case of legal rights, FECOFUN has also used various tactics in struggles against attempts to restrict the rights and duties of community forest groups. In some cases, it has organized large-scale public protests in both Kathmandu and at the local level, demanding the turning over of forest lands as specified in the 1993 Forest Act. In addition, it has arranged meetings on questions of local community rights with the Parliamentary Committee for Natural Resource Management and other members of Parliament, as well as submitting letters of protest to the Prime Minister and his Minister of Forest and Soil Conservation. In other cases, it has lobbied political parties and the media to support community forestry rights. Beyond specific forest issues, these activities have been seen as a model for the democratization of Nepal (Ojha 2013; Paudel, Banjade, and Dahal 2008). Not only have these activities strengthened the connections between elected politicians and the broader citizenry, they have been especially significant in the context of national deliberations related to local governance and decentralization, as well as the establishment of relevant laws that empower local governing groups in matters related to the regulation and control of forest resources. One of the many reflections of this is the regular assemblies of CFUGs to elect their executive committees. These assemblies, which include the voices of oppressed minority groups typically left out of the political process, are judged by activists and participants to be more deliberative than national parliamentary debates. It is the practice of many CFUGs to organize discussion groups to learn about and deliberate the issues in advance of the formal assemblies.

It should also be noted that FECOFUN has actively advanced the community forestry movement through numerous global networks. Its members have contributed to forums ranging from Europe to the United States, Latin America, Asia, and Africa. In addition to promoting the movement and the ideas upon which it is founded, it has added new lessons and perspectives for FECOFUN itself. Through such engagements FECOFUN has worked to establish worldwide the view that not only is community forestry one of the most successful development projects in Nepal, it is also a model for forest protection around the globe, a view acknowledged by the United Nations.

There is no question that the federation of CFUGs has helped to solidify the power of community forest people and, in turn, contributed significantly to the protection of the forests. Further, the strengthening of the local power of these organizations has also changed their political relationships with regard to the central government. As Ojha, Timsina, Chhetri, and Paudel (2008) explain, these altered power relations undermined unilateral government decisions, especially with regard to controversial and unenforceable decisions, and in the process made clear the need for pluralistic deliberations. As such, FECOFUN has emerged as a constructive opponent of the governance system, particularly with the forestry agencies. Not only has it supplied a counterweight in forestry governance, it has also promoted greater policy learning in the political system as a whole.

But this is not to say that there are no issues that need attention. In general, the primary issue seems to follow the general logic of a social movement-turned-established organization. Such movements typically begin to falter, even decline, when they start to work closely with the established powers; they lose their progressive ideologies, their organizational structures become hierarchical, and their administrative procedures start to push out creative political activities. All of these potential problems can be described as part of the political cycle of such organizations. While FECOFUN has clearly not lost its political edge, all of these tendencies can be found at one stage or another.

Some have worried that FECOFUN is becoming a little too staid in its ways and thus at times too close to those it otherwise has critically monitored. As the central players of the federation have increasingly become significant figures in the political process, both national and local, their successes have brought them many opportunities for personal advance and gain. The expansion of alliances and networks with other groups, often other NGOs looking for collaborative projects, has meant new sources of activities that can sometimes dilute the focus on CFUGs and local forests. Rather than emphasizing internal network achievements, some have appeared to be influenced by opportunities from outside. Moreover, some of the organization's leaders have themselves reached out to others for assistance in implementing projects suggested by donor organizations. This becomes all the more important given that FECOFUN still largely depends on others for financial resources. Because it is so heavily dependent on donor-funded projects for the large share of its activities, it wittingly and unwittingly finds itself dependent on the advice and interests of outsiders, often with quite different orientations. To some degree FECOFUN has managed to navigate through this process, but there are limits to how much maneuverability this situation offers.

One subtle and sometimes not so subtle shift resulting from this financial dependence is the fact that donors typically focus on programmatic issues, including technical services, rather than critical deliberative issues. Service delivery is, to be sure, an important issue in forest protection; but from an instrumental or technical perspective, important social issues are not usually part of the concern. Advancing women's issues, civil rights, or political development often end up getting short shrift in the process. It is important from both a political and a learning perspective to keep a careful eye on the underlying social issues related to power and inequalities. Short of that, FECOFUN runs the risk of becoming just another development NGO among many. Even where the organization undertakes technical activities, it needs to also focus on the political and institutional conditions which both shape and limit those services, rather than on straightforward technical dimensions of forestry. For example, as Ojha, Timsina, Chhetri, and Paudel (2008: 81) write, "instead of doing technical research on some aspects of forestry, FECOFUN [should] seek to ask . . . why the Department of Forest Resources and Survey, which has a mandate to lead forestry related research in Nepal, has actually very limited research engagement."

In just this regard, the organization needs to guard against an instrumental mindset based on routine practices, common to an organizational growth cycle. Instead of risking the loss of its creativity, FECOFUN leaders need to enhance the federation's capacity to work with the many different cultural orientations, political ideologies, and knowledge perspectives, no small challenge for a country with over a hundred ethnic groups and nearly as many languages. Some of their leaders, according to various observers, have already adopted attitudes and practices that appear to reflect the influence of dominant mainstream development thought, including its limited interest in policy deliberation as a tool for learning. A critical change-oriented perspective was FECOFUN's initial source of strength; thus some worry that it will fall by the wayside as the organization seeks to expand its revenue-generating activities.

In terms of organizational structure, another challenge has to do with FECOFUN's centralized management. One ongoing issue is the need for better representation of marginalized groups. Many members are still not sufficiently represented in FECOFUN committees, leading some to suggest that the federation's members do not sufficiently own the organization. While the executive committee has clearly sought to raise the voices of the marginalized members, they have at times had difficulties due more to hierarchical social patterns in the society as a whole. Aspects of the caste system, for example, mitigate against these groups more generally, which also gets played out in the organization itself, despite its efforts to remove the barriers.

But there are other factors as well. Even though FECOFUN is formally a federation of its member groups, it tends to use language and some practices that can suggest that the CFUGs are subordinate units. Independent investigations show that this tendency has begun to have an impact on internal deliberative practices, which start to reflect the top speaking to the bottom. As Ojha, Timsina, Chhetri, and Paudel (2008) have explained, if the organizational structure were to more openly treat district and community levels of the federation as collaborative partners, the representatives of the various levels would be able to better identify and discuss new issues and the lessons that might be drawn from them.

ForestAction

The development and management of this deliberative governance network is a challenging and even at times intense activity, constantly encountering both new opportunities and hurdles. Not only is burnout a threat, but also time to reflect is often scarce (as those on the line must constantly respond to the challenges, often unexpected). Which brings us to an extremely important component of the community forest network, the role of ForestAction, a think-tank NGO established in 2000 (www.forestaction.org).

ForestAction is devoted to forest resource studies and the development of action programs to assist the community forest movement (Lama 2011). It has served, formally and informally, to ease many of the pressures on FECOFUN and the community forest movement by adding an innovative, reflexive dimension to the deliberative processes, both at the national and community levels. Basic to the theory and practice of ForestAction has been a recognition that development efforts have had little payoff, despite both the resources and rhetoric devoted to them. Toward this end, the organization initiated an upsurge of critical research and reflections on the failure of development practice in Nepal, as well as internationally.

Described as a civil society organization that is non-profit, politically non-aligned, and self-governed, ForestAction blends professional knowledge with citizen power to build pressure for change. Constituting a multidisciplinary team of some forty professionals who combine their knowledge of natural and social sciences together in critical action research and policy dialogue, the organization lists its main areas of expertise as policy process, decentralization, community and local governance, social learning, institutional processes, and gender and social inclusion. It states its vision as that of an "environmentally sustainable society," free from injustice and poverty, and its more specific mission focus is on institutional, policy, and technical innovations that advance an effective and sustainable form of natural resource management. The goal is to inform policymaking processes and to empower forest and other natural resource-dependent communities. In this context, it emphasizes in particular issues that help "the poor, marginalized and vulnerable groups" to take control and play an active part in natural resources and forest management, emphasizing in particular "livelihood benefits."

Although ForestAction emerged formally independent of FECOFUN activities, it was the same community forest problems—particularly those related to community rights and democratic governance—that spurred its appearance and development. Part of this was a recognition of FECOFUN's need for a critical intellectual partner. Developed around a mix of independent and dependent relationships with FECOFUN, ForestAction has organized a variety of conferences, workshops, and information sessions to put forward their work on participatory governance, and has provided useful advice to the leaders and membership of FECOFUN. One such event was what they called "Banchautari I" in which both of these organizations worked with the technical forestry community and other stakeholders to organize deliberative platforms on forestry policy issues, drawing on both evidence and the voices of the community (Ojha, Paudel, Khatri, and Bk 2012).

ForestAction's activities have included participatory field research to study and facilitate local deliberative practices, the examination of forestry practices with a view to their implications for local communities, the creation of methods for developing forest inventories, the identification of sources of local knowledge held by the community, and a focus on related social issues, in particular the role of women, identity issues, and the problems of environment and poverty.

Basic to the strategy is the facilitation of democratic deliberative governance, in which citizens in their local communities are allowed to participate in dialogue without coercion and manipulation. The process provides public spaces to all of the civil society participants involved in creating an organizational vision, mission, and specific objectives. The goal of the organization's researchers has been to make development processes more deliberative, with the assistance of intellectuals, development activists, human rights workers, and civic society network activists who challenge the mainstream discourses and practices of development and, in the process, advocate for devolution, decentralization, participatory governance, and the protection of local people's rights over the use of natural resources. Underlying this effort is an emphasis on social learning, transformative learning in particular (defined as involving a shift in the structure of the basic premises of thoughts, feelings, and actions at the organizational level). Transformative learning recognizes the role of consciousness-raising, defined as permanently altering one's way of being in social and organizational life, in particular a sense of justice and peace. The strategy also involved acting as a critical bridge between local action and global discourses on environmental governance, partly made possible through the *Journal of Forest and Livelihoods* which was set up by ForestAction Nepal in 2000 after the founding of the organization.

Underlying both of these activities is the facilitation of collaborative engagements. ForestAction interacts with partner organizations, FECOFUN in particular, to nurture a collaborative culture that helps in planning and implementing its programs for positive and productive outcomes. Toward this end, it promotes an emphasis on a mutual respect of the environment which acknowledges the contributions made by their partner organizations. In terms of social-scientific action inquiry, ForestAction has developed and practiced a unique participatory model of "critical action research" which might well be taken as a model practice for policy-oriented civil society organizations generally and civic society experts specifically (Ojha 2013). It is thus useful to examine the main features of the model.

ForestAction and Critical Action Research: Developing Theory and Practices through Participatory Deliberation

ForestAction, as already noted, emerged to address a particular shortcoming of both conventional service delivery focused NGOs on the one hand, and radical rights organizations like FECOFUN on the other. ForestAction's work indeed strongly complemented and at times challenged the strategies of FECOFUN by bringing attention to deeper intellectual issues related to its activities. ForestAction also brought evidence from the field and insights from local innovations to the decision processes, which strongly put forward the voices of the communities networked around FECOFUN. In the case of FECOFUN, the organization tended in its early years to take advice from experts employed by international conservation and development projects. This relationship worked in that it helped FECOFUN develop various important organizational skills. But once FECOFUN leaders started to make more assertive claims about community rights, they needed more critical knowledge partnerships, which they could obtain from groups like ForestAction. Yet not all of the leaders of FECOFUN valued a critical knowledge partnership, so the strategy ForestAction took was more proactive than demand-driven in its effort to demonstrate that alternative knowledge and ideas could make FECOFUN stronger. Recognizing that FECOFUN is a large and heterogeneous organization that still had some leaders who take a somewhat skeptical "show us" approach to the issues at hand—that is, if you have useful advice, demonstrate it in action cases or practices. And this is just what ForestAction set out to do.

ForestAction was founded by Hemant Ojha together with Krishna Paudel and Shambhu Dangal in 2000 to add this critical action-oriented knowledge into forestry debates and governance. Two of the founders—Ojha and Paudel—spent over a decade nurturing and developing the organization from ground zero to an international reputation. At a time when Nepal was moving through widespread dissatisfaction with the Westminster-style democratic system, and civil society groups were taking up the critical policy discourses of participatory governance, the founders of ForestAction articulated a vision of environmental democracy working for the poor and local communities across the country. They brought this vision from a strong professional and practice experience: Ojha came from a forestry background with a noted record of left student activism and NGO management experience; Paudel possessed an agricultural background with left-oriented political activism and community-based organizational management experience; and Dangal had a technical forestry background coupled with a close collaborative working relationship with Ojha. Their progressive aspirations, ecological and democratic, were particularly relevant in the context of the forest sector in which they worked.

Although the Forest Act of 1993 had handed over control to the local communities, the decision process was still controlled by the forest administrators, particularly in the District Forest Offices. Ojha and his colleagues were concerned about the fact that FECOFUN was getting increasingly obsessed with the national level and not taking the local communities and their local knowledge sufficiently into account. To ForestAction, the ways in which FECOFUN was advocating for local rights were at times counterproductive due to a lack of evidence as well as intellectual rigor in the articulation of local governance and community rights. Local, everyday concerns did not initially emerge as a priority topic at the national level of FECOFUN, who were busy advancing other pressing policy issues in other spheres, national and international. Moreover, the government agencies and FECOFUN often engaged in confrontations that created more heat than light. One prominent case was a policy issue related to Terai Forest Management (Ojha 2008; Nightingale and Ojha 2013).

To alleviate such tensions, ForestAction investigators began to regularly visit particular cities as part of their critical action research efforts. As Ojha explains, researchers spent months "questioning and challenging the villagers, making them more conscious about inquiring, probing, reflecting and discussing issues at the hamlet and village levels as well as with other stakeholders."[2] These deliberations "helped to improve the confidence of users as legitimate managers of forests to make sense of the legal documents, better understand the technical calculations and terms used in the forest management plan document (through which forest officials were manipulating the harvesting practices)." It also permitted ForestAction researchers to collect and organize evidence in relation to the problems facing local institutions, and thus enabled them to challenge techno-bureaucratic domination, in the language of science itself when necessary.

Most fundamentally, its intellectual leadership established the organization around two sets of interrelated issues. The first concerned an understanding that social and political injustice in Nepal, especially social exclusion, are deeply anchored to centralized control over natural resources and unequal social and political access, in significant part rooted in the caste system. Of special importance were the issues of land tenure, market relationships, and political governance in the natural resources sector, a sector still largely agrarian and pre-capitalistic in character. Although these problems were not unknown, the deeper dynamics of this social reality were not part of the thinking of the parties, ministries, or NGOs.

The second concern was about the course of democratic governance. After the introduction of more democratic political practices during the 1990s, the leaders of ForestAction wanted to make sure that democratization did not get stuck in the top-down processes of either multiparty liberal party politics or civil society organizations, in the forest sector where they worked. Moreover, they sought to ensure that the dramatic expansion of civil society activities in the forest sector did not get caught up in a functional service mentality. Both tendencies were clearly visible, which the founders of ForestAction found worrisome.

Toward this end, ForestAction set out to develop a community-oriented and internally democratic organizational model that combined pressing concerns relevant to the forest community with the collection and wider dissemination of critical knowledge useful for innovative undertakings. It involved, as such, an interconnecting of research related to the activities of both social movements and citizen oriented community deliberation. Toward this end, the organization developed a critical action research model that took its inspiration from the literature on critical social science and the methods of participatory research (Paudel, Green, Ojha, and Barnes 2007; Ojha 2013). Theorized in large part by Hemant Ojha, the model begins with the recognition that social inquiry can never be politically neutral and that useful knowledge must emerge from an attempt to confront challenges on the ground. As such, ForestAction has clearly specified its research as an effort that can generate "counter knowledge," understood as knowledge that helps to uncover social and political hegemony through critical discourse, theoretical and political. It is based on the belief that the "hegemonic formations and practice of forest governance can be challenged and questioned in the domain of civil society and in the realms of knowledge, discourses, and practices." Fully engaging the writings of Freire, Habermas, and Bourdieu, the effort employs participatory research to unravel "systematically distorted communication" in ways that can create an "epistemological break" (Bourdieu 1977; Ojha 2006) in the thinking of both political actors and ordinary citizens.

Basic to the orientation is a commitment to the idea that one can only adequately understand the social world through efforts to change it. Given the barriers and uncertainties associated with change, the approach requires trans-disciplinary methods that step outside of the confines of institutional boundaries to permit framing research questions in ways that capture the practical interests and concerns of ordinary citizens, interests and concerns not captured by standard disciplinary perspectives.

For participatory-oriented critical action research knowledge emerges from the concrete experiences and experimentation with the real world (Ojha 2013). Moreover, it is not just evidential data resulting from the research, but also the human empowerment of the actors who can politically put forward the findings. Toward this end, such participatory research is "learning-oriented." Instead of adopting the standard development orientation, as Ojha explains, the focus is on generating new knowledge that can be experientially learned by the various stakeholders, in particular through discursive interactions. Insofar as the research demonstrates that effective interventions require coordinated multi-level efforts, the dissemination of the research findings is to the wider public as well, including politicians and governmental officials. Indeed, many of the problems confronted at the local level are the manifestations of policies and practices at the national level. To increase the uptake of their research findings, the staff engages a varied spectrum of stakeholders in deliberative processes moving from research design to the later dissemination and deliberation of their empirical and normative results.

In terms of normative investigation per se, ForestAction pays special attention to the mental models that underlie and direct forest policy. The basic premise from the outset is the recognition that forest policy and practices are governed by various established and widely accepted belief frameworks that have been carried over—often tacitly so—from earlier times; there is thus a need for counter knowledge. Indeed, the uncovering and questioning of such mental constructs embedded in these traditional policies and practices is what makes the work "critical," in the sense of regular reflexive interactions between the social and political assumptions and empirical inquiry. This applies for ForestAction's own work as well as for the larger forest policy network.

All of this, not surprisingly, involved a struggle with the forest bureaucracy, as nothing has been more problematic than their mental constructs. The idea of critical action research and its participatory orientation was—and for some still is—foreign to these administrators and was only surrounded by question marks. This made it necessary for ForestAction to demonstrate concrete results. Toward this end, one of the organization's most important efforts was the development of an alternative forest inventory. This helped

ForestAction to win over more forward-thinking bureaucrats and officials within the government system, and they have thus been able to expand their knowledge networks inside the administrative and bureaucratic system.

One of the examples in which the work of ForestAction has nurtured alternative thinking in forest governance has been the technical assessment of forests in the context of community forestry. Although the Parliament passed a progressively oriented forest reform act in 1993, acknowledging local users as the managers of the forests, the established forestry bureaucracy succeeded in manipulating the goals of the new law to suit its own purposes. Through new guidelines, regulations, and by-laws they managed to impose their own conventional scientific approach to management and measurement on the local communities. In effect, this ensured that professional foresters would remain in charge. One of the most important of these directives was the "Forestry Inventory Guidelines of 2000" (Hull, Ojha, and Paudel 2010).

These guidelines made mandatory the acceptance and use of a sophisticated scientific method for forest inventory measurement, which had to be signed by local forest users as a precondition for using the forests. Issued without consultations with the CFUGs, or any other civil society forest groups, the ruling led to extensive confusion and chaos in forest communities around the country; the central ministry's District Forest Offices began terminating users' agreements for those without the kind of properly measured inventory data specified by the new forestry guidelines.

Just as problematic, the scientific inventory was of little use to the local community and came to be seen as little more than a bureaucratic inconvenience rather than a management tool. A big part of the problem was that it attempted to measure forest production and consumption with a degree of precision that was of little help to anyone. Not only did it take a long time to complete, the levels of information were unnecessary (Ojha 2002).

ForestAction used this issue to enter into the debate from the perspective of local community knowledge. Working with FECOFUN, as Ojha explains, they sought "to understand the scale and complexity of the problems at the ground level" (Ojha 2013). As he put it, "we recognized that it was necessary to further probe the issue and to propose alternative ideas on how science and local knowledge can be combined to inform forest management decision making, and how that can be reflected in the national policy." This led to the writing of a discussion paper on the science and politics of forest inventory in Nepal,[3] followed by a survey of the problems facing user groups around the country and then field visits to better grasp community views on the guidelines. With this information in hand, meetings were organized with forestry officials to gather official perspectives before launching an inventory issue case study.

The resultant research clearly revealed the ways in which the foresters' scientific approach functioned—intentionally and unintentionally—to restrict the role of the community forest users. For one thing, these methods were too complex for local people to understand, not to mention carry out. This meant that they had to turn to local foresters for assistance with this task, who in effect translated the discussion into a language that disadvantaged the local community members. In addition, there were not enough foresters available at the local level capable of conducting this scientific inventory, which led to delays. To engage such foresters was also often prohibitively costly to community forest members. At the same time, however, without the official inventory they could not get permission to use the forest.

The empirical and analytical outcomes of the case study led to a range of options for integrating scientific forestry and the local knowledge of the community. In particular, the evidence revealed that the Forestry Guidelines were overly burdensome to the local citizens and were in many ways problematic in terms of the overall purpose of the Guidelines. The findings clearly indicated the ways in which such concrete local problems are not available to centralized policy experts, a point grasped by the Director General of the Forest Department. As a result, the Department initiated a process of revisions to the Guidelines through a participatory multi-stakeholder working group that carried out a range of stakeholder consultations. ForestAction was directly involved, contributing to the deliberations, evaluating the policy negotiation processes, and supplying critical advice on how local and scientific knowledge can be integrated in the forest governance decision-making.

Such an inventory was one of the important activities that assisted ForestAction in transforming the association between local forest communities and the official forest departments and agencies, both local and national. Even though the forest legislation turned control over to the local communities, the decision-making processes, as we have already seen, were still in the hands of the forest managers, especially those in the District Forest Offices. What is more, the everyday local issues did not at the outset necessarily emerge as critical topics at the upper levels of FECOFUN, as the people at this level were preoccupied with putting forward policy issues related to forests in other spheres, both in Nepal and internationally. ForestAction therefore made a careful study of the connections between the official inventory and the actual realities in the forest communities.

The result of the study was to show the exclusion of the community forest movement, despite the fact that the new law called for their involvement. The technical procedures, which were complex and rigid, effectively excluded the forest people on the ground of incompetence. It also led to more substantial contradictions. By disempowering the community users, the very people who manage the resources in the forests themselves undercut the implementation of the plan. For one thing, forest users ended up, practically out of necessity, ignoring the inventory, which contrary to the intentions of the Department of Forestry meant that the harvest of forest products continued to be informally carried out in an unsystematic, unplanned manner. Thus, as Paudel and Ojha (2007: 49) have put it, the "situation challenges the assumption regarding the need of an [Operating Plan] and the whole idea of planned forest management for sustainability and equity."

The underlying issue here is the problem of knowledge, in particular a politics of knowledge. First, it is a question of what kind of knowledge. Missing from the forestry perspective, as ForestAction has shown, is a recognition that the supply and demand for forest products at the local level are as much or more socio-political problems as they are technical in nature. While this does not deny the importance of science, it does raise critical questions about its relation to the contextual realities on the ground. For one thing, the emphasis on empirical measurement ignores the more qualitative forms of knowledge and insights that are available in the community itself. Indeed, this involves forms of local knowledge—rich forms of environmental knowledge and indigenous practices for resource management—that have in effect been tested through applications over centuries. As Ojha explains, the procedurally mechanistic, empirical approach spelled out in the 2000 inventory guidelines did not permit the forest villagers to incorporate their common-sense experiences into the assessment process. As a result, the forest users ended up with a poor understanding of the processes and thus a limited sense of ownership. Often the local people simply do not believe the findings and ignore them.

The research of ForestAction has led to a range of options for integrating scientific forestry and the local knowledge of the community. In particular, the evidence revealed that the Forestry Guidelines were overly burdensome to the local citizens and were in many ways a problem in terms of the overall purpose of the measures. ForestAction's findings clearly indicated the ways in which such concrete issues are not available to centralized policy experts, a point grasped by the Director General of the Forestry Department. As a result, the Department initiated a process of revisions to the Guidelines through a multi-stakeholder working group carried out by a range of participatory consultations. ForestAction was directly involved, contributing to the deliberations, evaluating the policy negotiation processes, and supplying critical advice on local and scientific knowledge that could be integrated in forest governance decision-making.

Based on this and other research projects, ForestAction researchers developed a model of deliberation for bringing their critical research findings to bear on policy decision processes. As the denial of participant involvement had led to a dearth of deliberation, with the Department of Forestry failing to benefit from a discussion about both the local knowledge of the community and the institutional linkages through which the inventory findings must pass, ForestAction created a "deliberative knowledge interface" for natural resource governance (Ojha, Paudel, Banjade, McDougall, and Cameron 2010). Constructed around diverse types of knowledge systems, this work has sought to understand how questions of equity and efficiency and sustainability in natural resources management are shaped, influenced, and determined by deliberative interfaces among knowledge systems associated with various groups of social agents in the practice of natural resource governance.

While the policy and practice of natural resource governance are enacted through knowledge and political interfaces among the competing technical, social, and cultural categories of agency to varying degrees in various ways, there is a problem of persistent divergence of perspectives which has created difficulties in achieving negotiated outcomes. From the perspective of deliberative participatory governance (as a strategy of resolving the conflicts among different knowledge systems), the model of critical participatory inquiry can be used to identify problems of dominance in decision processes (Timsina, Ojha, and Paudel 2004).

Conclusion

What can we learn from this story of participatory governance in Nepal? Crucial to the story is the energetic activism associated with these political transformations, in particular the way in which they gave rise to an experienced cadre of activists who sought to give specific meaning to the unfolding transition of the struggle to democratize the Forest Act, especially its implementation. The way they sought to establish political structures that would embody and facilitate the evolution of democracy and participatory governance was critical to the successes in Nepal.

The experience in Nepal also confirms the view that the success of local participatory processes depends on more than activities at the local community level. Success appears to depend on the bottom reaching upward and the top reaching downward. That is, participatory governance depends on support from the top, in this case the establishment of the FECOFUN network that organized and linked the local CFUGs together. Without the cadre at the top the effort would not have had the support or skills necessary to have taken root and evolved. Moreover, FECOFUN has played an important role in the democratization of national politics in Nepal. Its contributions to participatory engagement have been described as a model for democratization. A form of participatory governance seldom found in the advanced industrial countries, the model serves well to stimulate and encourage thinking about the democratic challenge in general.

Basic to the story is the fact that participatory developments do not just happen. They are promoted and facilitated by external actors and depend in significant part on the learning and skills these actors bring to the challenge. Indeed, the importance of learning and skill should not be underplayed. Many experiences show that there is a need for people who can facilitate the process of democratic participation. In this case, we find the role of experienced activists in FECOFUN who brought their political and organizational skills to the development of the community forest movement. But we also saw that this was facilitated by the critical action researchers at ForestAction, who labored in particular at the community level to uncover and work on the issues from the everyday perspective of ordinary community life, far removed from Kathmandu and the leaders of FECOFUN. It is not that FECOFUN was out of touch with the local level, but rather that it had to busy itself with a wide range of activities at the level of the national federation which were simply different. Given the pressures of time, and the challenges of energy and resources, these leaders could not remain close enough to the local level to facilitate the full development of opportunities and social capital.

But ForestAction could. Working as something of a think tank for FECOFUN, ForestAction has helped to keep the movement fresh. By focusing reflexively on the assumptions that have governed the theory and practices of forest management, ForestAction has served to assist the community forest movement further develop in progressive directions, with emphasis on social equality, participation, and inclusion. It has also supplied a useful model of critical action research that can be adapted to different contexts, particularly as they pertain to the democratization of sustainability struggles in the future. As such, the critical intellectual activities of ForestAction have helped FECOFUN avoid the pitfalls long associated with the general logic of a social movement-turned-established political organization, in particular the typical centralization leading to nondemocratic forms of elite decision-making and an increasing reliance on professional expertise.

Essential here has been a demonstration of the role of civil society organizations to address power imbalances that emerge in the course of a social and political change, the recognition of the participatory role of multiple actors in the production of knowledge and thus a need for "knowledge partnerships," the importance of networking that includes marginalized citizens, and the promotion of investigation that identifies and explores the elements that both constrain and facilitate deliberation for participatory governance.

Given that forest protection is a critically important issue for global warming, and will only continue to become more so in the coming decades, the developments in Nepal offer lessons for forest communities around the world (Chaudhary and Arysal 2009; Adhikari 2009). But equally importantly, the experience provides important experiential and theoretical knowledge that can help us think about the role of revitalizing democratic practices, participatory governance and participatory expertise in particular, in a world that will increasingly face a growing technocratic eco-authoritarian challenge.

Notes

1. Most of what follows is based on interviews conducted in Nepal in 2011. I owe a great deal of thanks to Hemant Ojha for making the trip possible and organizing the visits to forest communities and interviews with members of FECOFUN, CFUGs, and ForestAction.
2. Personal communication with Hemant Ojha, September 2011.
3. Personal communication with Hemant Ojha and Dil Khatri, September 2011.

8

Practicing Participatory Environmental Governance: Ecovillages and the Global Ecovillage Movement

A great deal of environmental talk does not meet up to real-world practices. A major exception to this regrettable reality is the activities of the ecovillages. These communities are essentially made up of people who have decided to rethink the way they live and to reduce their ecological footprint. Most of these communities involve people who have, in one way or another, returned to the land to lead a less complicated way of life. But this is only partly the case; ecovillages and their variants are also found in major cities, although far fewer in number. They are, in addition, more than a local phenomenon; there is a global network of ecovillages. Linked together through the Global Ecovillage Network (GEN), they are found in countries as different and diverse as the Germany, the United States, Thailand, Brazil, India, and Senegal. Some of their members live simple lives that in various ways involve practices that remind one of earlier centuries, but others accept high-tech features such as cell phones, internet connections, solar panels, and even research laboratories.[1]

Perhaps the first thing that one can say about these people is that they "put their money where their mouths are." As concerned environmentalists, they are engaged in a social experiment designed to pioneer new ways of living together sustainably. Indeed, their practices without doubt represent the closest form of sustainable living that can be identified. In this regard, it is astonishing that they have largely been neglected or overlooked, even by ecologists and environmental social scientists. Many have largely written them off as small groups of eccentric idealists who offer no particularly relevant solution for the planetary crisis, an argument that is grounded in significant part in the globalization of environmental problems and politics. Often they are thought of and described as fringe groups, dropouts who have removed themselves from society. But this is scarcely an accurate portrayal. A closer look shows this to be overly simplistic, even a misrepresentation.

First, it is important to concede the point that invariably comes up early in discussions about ecovillages. Ecovillage life, in short, is not for everyone. Some people are not well suited for such communal life. Nor is it a community form that could be relevant to the contemporary lives of many millions of people, given the structures and processes of the dominant urban-industrial society. For people living in ecovillages this poses no problem. They concede that the ecovillage is not a model for everyone. But what they do maintain is that they are developing practices that can be translated, adopted, and used in other contexts, including cities. What is more, and very much to the point here, most of them practice well-developed forms of participatory governance, a reality that has drawn very little attention from political scientists and the scholars of governance.

Litfin (2013) is a prominent exception to the rule in political science. Although she made her mark studying global environmental politics, she has in recent years devoted a great deal of time to visiting and studying ecovillages. As she tells the story, she suffered from the lack of an ability to tell her students what the workable options or alternatives might be, a problem that I have experienced myself. My own course on environmental politics at Rutgers has been called the "bad news course." Although the topic is extremely relevant to the lives of my students, involving important—even at times fascinating—stories about events around the world, it is difficult to tell them what they can do; that is, other than the obvious activities—such as putting on rubber boots to clean up the river, taking mass transit instead of driving a car, eating less meat, taking vacations locally, and so on—none of which strikes most of them as very workable given the structures of the environments in which they live. How, they ask, could things be done differently? And just here the ecovillage experience becomes useful. It is, in fact, about people who have decided to live differently. Not only can they offer interesting and important insights, they are also inspirational, even if seemingly utopian.

Ecovillages: What Are They, What Do Ecovillagers Do?

The ecovillage movement, as Litfin (2009: 4) explains it, "may be understood as a conscious and pragmatic response to the material and ideational crisis of modernity, a response that is grounded in a holistic ontology." Pointing to the ecologically destructive consequences of modernity—"the ability of human reason to control nature, materialist ideas about progress, the dominance of possessive individualism, and more"—environmentalism, as she points out, has long called for a more holistic reorganization of life on the planet, or what she refers to as "new ways of living premised on a sense of deep connection to the human and biotic community." For Kunze (2009; 2015) ecovillages have created a way of reclaiming or decolonizing the lifeworld.

Such language is scarcely new to environmentalism, radical environmentalism in any case, but there is little in the contemporary world that embodies this connection. Ecovillages, however, are a very important exception. And rather than an isolated phenomenon, we find hundreds of ecovillage communities, taking different forms, spread all over the world. Indeed, the Global Ecovillage Network (GEN) shares information and experiences through a wide variety of conferences and meetings of various sorts. In fact, what first seems to be a group of isolated rural dwellers turns out to be part of a global movement.

Attempting to put their beliefs into practice, the people in these communities have dedicated themselves to experimenting and innovating with new ways of living together sustainably (Bang 2005). An ecovillage can be defined as "a semi-self-sufficient, human-scale, cooperative sustainable settlement that integrates all the primary facets of life—sociality, alternative economics, food production, energy, shelter, recreation and manufacturing—with sensitivities toward the environmental and its natural cycles" (Parr 2009: 62).

Basically, their members are rather regular sorts of people, many of whom strike one as the kind of person that might be encountered on a university campus. Indeed, many of them have earned college degrees. In Litfin's (2009: 125) words, they are "small groups of people the world over [who] are coming together to create modes of living in harmony with each other, with other living beings, and the Earth," in response to both the steadily deepening global environmental crisis and the disintegration of socio-cultural institutions and structures that define modern life in advanced industrial countries. The movement represents a remarkable array of socio-ecological practices and diverse cultures. Although ecovillagers join for different reasons, it is their commitment to environmental sustainability that binds them together.

While ecovillages have their origins in various philosophical traditions, Dawson (2006), a former global ecovillage official, points to an emphasis on self-sufficiency and sustainable agriculture. Some, but not all, emphasize spiritual renewal that has long been common to ashrams and monasteries, as well as ideas and principles found in movements based on the thinking of Gandhi. One also finds strong influences from environmental, feminist, and peace movements, as well as the appropriate technology movement of E. F. Schumacher (1973). Typically underlying these principles is a return-to-the-land movement, emphasizing biotic harmony. Litfin, it should be noted, has found very few tendencies toward sectarianism or cultism and no indoctrination in the ecovillages she studied.

Even more specifically, these groups generally devote themselves to the ideas of permaculture and Gaia. Permaculture is a philosophy focused on working with the natural ecology, in contrast to working against it. Although the term originally related to "permanent agriculture," it was later extended to mean "permanent culture." It has evolved, as such, as an approach to socio-ecological design and sustainable agriculture. Basic to the design principles is a holistic method for integrating societal institutions and practices with ecological requirements in ways that minimize the use of energy, reduce human waste, produce ecological building materials, and supply sustainable foods. For the practitioners of permaculture, the whole is considered to be greater than the sum of the parts.

Gaia theory, or the Gaia principle, maintains that living organisms interact with their inorganic surroundings to establish a complex and self-regulating system that plays a central role in maintaining the conditions for human life on Earth. The basic hypothesis is that organisms and the physical environment co-evolve. Put forward by Lovelock (1979), it focuses on the ways that the evolution of the biosphere has affected the stability of oxygen in the air, the salinity of the oceans, and the worldwide temperature of the planet. Insofar as the general processes of evolution—both organic and inorganic—affect one another, the planet itself can be considered something like a living organism regulating the conditions that make life on it possible.

It is often the return to the land that leads people to find ecovillages problematic. Many argue that while this might be nice, it is irrelevant to modern life. But in the context of the questions posed here, one can argue this quite differently. As we move toward a serious confrontation with the implications of climate change, the concerns of survivalism—to one degree or another—will become more crucial. Under these conditions, certainly in severe form, large numbers of people will be forced to flee the cities and head for the land. Whereas this was once considered improbable, it will most likely no longer be a matter of choice but rather a hard necessity for many. Desperate people will be on a search for ways to survive. Already we can see signs of such a dynamic in European countries such as Greece which have been facing extremely severe financial hardships and high levels of unemployment.

Significant numbers have left the cities to return to the land where they can grow their own food, often communally with other family members. Most urban dwellers know little or nothing about living on the land, including the knowledge necessary to be able to grow their own food. But hungry people will learn, even if not overnight. And this is just where the ecovillage concept can be significant, as their members have already been experimenting with ways to do this, similar to what Hopkins (2008) has referred to as the "Great Reskilling." They have developed a substantial wealth of local knowledge that is readily transferrable. The members of ecovillages are fully aware of this and discuss it quite often. At Sieben Linden, for example, people wonder how they will manage if throngs of people arrive at their gate looking for help, as they would not be able to accommodate many of them.

It is significant to note, in this regard, that the government of Senegal has turned to the ecovillage model for just such help. Given the country's high level of rural poverty, the government of Senegal has asked GEN to help transform its many rural villages into thriving ecovillages to fight both environmental degradation and poverty. As a consequence, there are some 350 of these ecovillages of one sort or another in the country, some more developed than others. To facilitate this development, the government has even set up a public agency to assist with this process, called the National Agency for Ecovillages, which has the goal of transforming 14,000 existing villages into ecovillages by 2020.

Ecovillage Illustrated: Sieben Linden

Sieben Linden offers a very good example of an ecovillage that primarily stresses ecological innovation and renewal.[2] Founded in Germany in 1997 in the rural Altmark region of former East Germany near a small village called Poppau, Sieben Linden puts its emphasis on reducing the ecological footprint of the community, developing an alternative model for sustainable living, and educating other people about the possibilities of a different way of life and how to organize it (Kunze 2009; 2015). Spreading across about 200 acres of forest and farmland, it is an "off-grid community" (meaning that it generates its own electricity) with an ecological footprint that is dramatically lower than that of the German average as a whole. Committed to vegetarian agriculture, the work in the fields is labor-intensive.

With a population of about 140, some forty of whom are children, people live in a combination of six multifamily dwellings and some fifty-nine converted "Bauwagen" (or caravan trailers), which are generally designed for use by workers at construction sites (for storing equipment and the like), distributed across the nonagricultural area and divided into six or seven "neighborhoods." The multifamily buildings, constructed with strawbale (including the largest strawbale building in Europe), are largely shared by couples, families, or singles, although most singles live in the converted wagons. There are, in addition, a guest house, a common hall for meetings, offices, a convenience shop, and a goldsmith.

The plan has been to grow to the size of 300 residents, although there are some current discussions about the social feasibility of that goal. Some fear that further expansion will occur at the expense of community life, or

Gemeinschaft. In addition to the residents, there are at any one time a number of visitors, young and old, who stay for varying lengths of time. Some of the visitors are interested in learning about Sieben Linden as an ecovillage; some of the younger people have chosen to work there for a period; others are there for various projects, educational projects in particular. Those who wish to live permanently at Sieben Linden must be accepted by the group as a whole, after a one-year trial period designed to judge their compatibilities with the other members, their lifestyle, and their interests and contributions generally.

Many people there also have a spiritual commitment of one kind or another, but not all. In any case, there is no standard doctrine—such as Buddhism, for example—but this is sometimes the case in others, especially in Thailand where it plays a major role. Many have attended university and there is a wide array of intellectual and cultural activities that reflect that. Among the residents are an architect, an engineer, a medical doctor, a social worker, a nurse, two teachers, a trained landscape ecologist, another formally educated in ecology and media, a carpenter, an editor, a number of artists, and a philosopher, among others.

There are also other activities such as dancing, choir, sports (including boxing), meditation, seminars on Buddhism, holistic ecology, and much more. Each year Sieben Linden organizes an annual festival attended by many outsiders. They also set aside a week every year which they call the "intensive week." During this period they drop most of their other activities and engage intensively in a form of collective introspection, examining their own policies and practices, the evolution of social relations in the community, and the like.

In addition, Sieben Linden maintains a close relationship with the neighboring community and often reaches out to assist their neighbors. At first their neighbors found them strange and kept some distance. But later through interactions they came to better understand them and in some cases even depend on them. Members of Sieben Linden have, for example, used their conflict mediation techniques to help solve an ongoing conflict related to the local fire department in Poppau. A number of them work in the outside community, including a social worker and medical doctor. And the children go to local schools.

With no common ideology or spiritual orientation, most are primarily dedicated to ecological sustainability and reducing the human ecological footprint. This is the main consideration that holds them together, although some people also stress the importance of communal living for them personally. Numerous people are there to escape the anonymity and alienation of modern big city life, as is the case with people in many ecovillages. For those who are practicing Buddhists they have constructed a small building for meditation situated out of the way of the main flow of activities. But others are mainly interested in practical ecological experimentation. There is no particular attire that sets any of these people off, although a few wear colorful Indian-like garments.

Essentially, Sieben Linden is a well-developed sustainable community. It gets enough electrical power from an array of photovoltaic solar panels to take care of the needs of the community as a whole (selling whatever is left over to the local power plant). As an additional supply for heating, it also has plenty of wood from its forests. Although they are strong advocates for public transportation, some have automobiles. But they tend to share travel with other community members. The buildings are outfitted with composting toilets, which provide manure for the field, and plowing for farming is done with draft horses. Almost all of the residents consume only organic food, most being vegetarians or vegans. Their vegetables, maybe 70 percent, are grown in their gardens, irrigated by recycled "graywater" that primarily runs off from washing (bathing, cleaning clothes, and washing dishes). Most of the members eat in the communal dining hall and wash their own dishes before departing. A study by the University of Kassel conducted in 2002 examined the way Sieben Linden creates and transports electrical power, water, heating, and food, showing that it used far less than the typical German home. The overall level of CO_2 output in the ecovillage was only roughly 25 percent of the typical German household. Because of the use of renewable energy, sustainable construction materials such as strawbale, and insulation, the construction and use of their buildings emit only a tenth of what is normally the case in Germany. When it comes to producing food, the ecovillage's emissions are only half of those in the rest of the country, in significant part because most of the food is locally grown coupled with the fact that the residents mainly consume vegetables rather than meat. But in matters related to transportation they do not do nearly as well. CO_2 emission for travel is only 70 percent of the average of the rest of their fellow citizens, despite the fact that they mainly travel by car-sharing or train. One explanation for this is that a number of the members of the community travel on business to other ecovillages in Europe and in some cases to other parts of the world, mainly on matters related to GEN. In this regard, Sieben Linden served at the time of my visit as the European office of the GEN. Some also travel for educational purposes as part of their work at Sieben Linden, informing others about the ecovillage experience or giving seminars on their pioneering efforts related to the construction of strawbale houses and other practical activities. Insofar as experimenting for the future is part of their mission, they eagerly engage in the communication and exchange of such ideas.

Some members of Sieben Linden would be considered ecologically radical, although most are much more practically oriented (unless living in an ecovillage is in and of itself radical). One example of the former was an early group that called itself "Club 99," which sought to see how far its members could go when it came to stringent ecological practices. In fact, they built a neighborhood that existed without the use of electricity. As the most ecologically radical of the neighborhoods, they built their primary residence by hand without the help of machinery and electricity, using only local materials—wood, straw, clay, and various recycled materials, in an effort to use 90 percent less carbon dioxide than the standard housing construction methods employed elsewhere. Further, they also engaged in income-sharing. Club 99, however, no longer exists for a variety of reasons. For some, the way of life was simply too hard and they joined different Sieben Linden neighborhoods. Others left for reasons related to disagreements that emerged between the radical members and those who were more pragmatically oriented in ecological terms; there was, in short, a parting of the ways. Without judging either side of the issue, it counts as one of the learning experiences at Sieben Linden.

Unlike some ecovillages which engage in income-sharing, the residents of Sieben Linden earn their own income and do so in a variety of different ways. Though the average monetary income is about 12,000 euros a year, this figure in no way reflects a lower standard of living, thanks to their shared communal resources. A number of them work as tradespersons of various sorts, such as carpentry and construction, some are artistic performers and teachers, there are gardeners with an organic vegetable shop, there is a small general store, some are cooks, and there is a jeweler with a shop. A couple of them operate a small publishing company, *Eurotopia*, that prints a European directory of ecovillages. A few are employed outside of the ecovillage. Two, for example, are teachers in neighboring schools, one is a social worker for the local government, and another is a medical doctor at a nearby hospital. If there is one main business, however, it is the many educationally oriented seminars that they offer to outsiders. Taking place on a weekly basis, involving quite large numbers of visiting groups during the summer months, these seminars deal with a wide range of topics, but the building of strawbale buildings is the largest and most significant income-earning educational activity. Indeed, they have become one of the world's leading experts on such construction. The strawbale seminars draw large numbers of people from many different parts of the world. Other seminars deal with topics such as meditation, massage training, conflict mediation, transformative social learning (similar in approach to that taught by ForestAction in Nepal), and methods and practices for successful community living. There is also a long-run plan to develop Sieben Linden into something of a village-sized economy featuring craftspeople, artists of all sorts, various medical and health services, a number of small businesses, and social events. For this they hope to construct an extra site to house these activities.

The Global Ecovillage Network (GEN)

The GEN is an organization that represents and connects a large and growing number of ecovillages in more than 100 countries across all continents around the globe (that are associated as well with thousands of eco-oriented traditional villages, intentional communities, various urban ecobarrios/neighborhoods, eco-districts, and co-housing communities).[3] GEN defines an ecovillage as "an intentional or traditional community that is consciously designed through locally owned, participatory processes to regenerate social and natural environments... integrated into an holistic approach" (cited in Kunze and Avelino 2015: 6). The basic task of the network is to facilitate the sharing of information and ideas about ecovillage life, both in terms of cultural exchange and technology transfer.[4] The goal is to showcase "high quality, low impact ways of living that have led to some of the lowest per capita footprints in the industrialised world, and a healthy integration of heritage and innovation in more traditional settings."[5] Moving beyond the early "lifeboat" or "green islands" concept that informed its early decades, GEN today seeks to shift "into becoming a transformative knowledge network that works in close alliance with like-minded organizations and other sectors in an effort to optimise sustainable development strategies for whole societies." Combining with the Gaia Education organization, GEN has established "a set of training programs to facilitate the transition to resilience," described as "learning journeys for change makers and design processes for communities that chart their own pathways into the future."

Emerging initially as the result of activities of the Gaia Trust, a charitable organization designed to promote ideas about Gaia, these were set up to respond to the fact that while there were many disparate ideas and practices about ecovillages in the early 1990s, there was no well-defined concept that captured their purposes and practices. This effort to sort out the features of these communities subsequently led to the first international conference on "Ecovillages and Sustainable Communities for the 21st Century" in 1995, held in Findhorn, Scotland. After that step, the movement began to develop rapidly involving thousands of communities, ranging from large networks such as Sarvodaya in Sri Lanka with 11,000 villages, or the network of 350 ecovillages in Senegal, to smaller groups of fifteen to 150 people living together.

Although GEN has no formal procedures for selecting ecovillages to be members, its goal is to encourage and support the development and evolution of sustainable communities around the world, doing this through communicating information about best practices, networking, and establishing partnerships among its members. This role is of particular importance given the differences between ecovillages with roots in diverse traditions (Dawson 2006). Most are rural (such as Huehuecoyotl in Mexico or Gaia Asociación in

Argentina), but others are urban (such as Christiania in Copenhagen, Los Angeles Ecovillage, Ecovillage Cleveland, or Avalon Eco-Village in Detroit). Some emphasize permaculture (e.g., Crystal Waters in Australia and Barus in Brazil) and ecological design (such as Sarvodaya in Sri Lanka) while others focus on serving as educational centers (e.g., Findhorn in Scotland on alternative technology development or ZEGG in Germany on a range of ecovillage practices). The High Point Ecovillage in Brazil even teaches courses on ecovillages and Gaia at the University of Brasilia. Given this diversity, GEN also offers a "Community Sustainability Assessment" methodology for determining how well such a community is moving toward sustainability. By 2001 GEN was given recognition by the United Nations. Its Economic and Social Council formally gave the network "consultative status."

URBAN ECOVILLAGES

Some ecovillages are located in cities. Contrary to the standard conception of ecovillages being out in the country, there are numerous cases to the contrary. One is the example of the Los Angeles Eco-Village established in 1993 after a decade of planning and preparation. Located in the East Hollywood and Koreantown neighborhoods on a former landfill, in a depressed area that had witnessed riots about twenty-five years earlier, an ethnically diverse group working for a nonprofit organization decided to establish a supportive intentional community. After making friends, organizing neighborhood events, working with children, and planting vegetable gardens and trees, they renovated two tenement buildings and constructed a green area behind them. Today, there are fifty housing units. In their large flourishing gardens some sixty-five types of vegetables, fruits, and chickens can be found, as well as benches and chairs for sitting and socializing. In addition to on-site food growing, there are a community land trust, small-scale green enterprises, pedestrian-friendly streets, bike paths in a car-free community, a community bike shop, gray water systems, a community tool shop, a sewing room, an interactive art studio, and a variety of activities devoted to democratic community-building. The website states that the members "moved to the neighborhood to live more ecologically and more cooperatively. Most of us are demonstrating the processes—ecologically, economically, and socially—that manifest an ecovillage. We are raising the quality of community life while lowering our environmental impacts, and expanding public awareness about more sustainable urban living" (http://laecovillage.org/). It is described as a space where "energy and creativity are not only transforming physical spaces like the garden or the street, but also social spaces in ways that demonstrate the very real challenges and benefits of living in community and making decisions together." As both the website and personal communication show, the people living in the fifty housing units are engaged in a wide range of activities basic to ecovillage life, from workshops related to permaculture to discussions about issues of environmental justice and intentional communities, events related to urban sustainability, public awareness and demonstration project tours for visitors, open dinners with the local community, a bicycle repair service, and a food coop. Some of the members work in nonprofit organizations located within the ecovillage, some have developed small enterprises within the community, and others work outside of the neighborhood in jobs related to environment justice. https://www.youtube.com/watch?v=FdQGozSavz8

Another urban illustration is the UfaFabrik in Berlin Tempelhof, Germany. Located in an area that used to house the UFA film laboratory in the 1920s, an early pioneer in the history of German filmmaking (for example, *The Cabinet of Dr. Caligari*, among others), UfaFabrik was founded in West Berlin by peaceful squatters in a decayed area as "eine ökologische Kulturoase" during the height of the Cold War, with a message of "peace through culture and ecology." Located on the front line of the conflict between West and East, the goal was described by some as a project devoted to finding a middle way between capitalism and socialism. A unique combination of living, social life, working, creativity, and culture, UfaFabrik, covering 18,556 square meters (about 4 acres), houses some thirty to forty residents (almost all of whom work there) and employs another 160 people or so in its welcome center, four performance stages (for music, film, dance, variety, and literature), conference center, organic bakery and cafe (supplied by local farmers), gardens, children's programs, social services, a children's farm, martial arts studios, second-hand market, a circus school (with an emphasis on political satire), sports activities, and more. It serves as an international meeting center for conferences and congresses, holds festivals of different sorts, and features activities and programs related to ecology and future-oriented planning for sustainability. UfaFabrik's communal members share no particular ideological beliefs or spiritual practices. Primarily they are committed to an educational mission, offering a wide range of programs with an emphasis on ecology. Some 200,000 visitors take advantage of these activities each year. With a decentralized energy system, it produces its own electricity and heat, has a rainwater storage system to water the green vegetation and supply the toilets, and has been described as the "green oasis in the big city." In 2004 the UfaFabrik was honored by the UN-Habitat as one of 100 projects worldwide awarded the status "Best Practice to Improve the Living Environment." http://www.ic.org/directory/ufa-fabrik/

The yearly conferences of GEN draw participants from around the world. The 2014 meeting of GEN Europe, for example, was held in July in ZEGG, an ecovillage about an hour outside of Berlin in Germany, with 445 people in attendance. Titled "Connecting Communities for a Sustainable World," the program was organized around a diverse array of activities, ranging from practical exercises such as meditation and yoga to a wide variety of panels dealing with issues related to life and work in ecovillages, including individual as well as collective development. Open to interested parties such as myself, panel discussions included the methods and practices of consensus decision-making, the theory and techniques of sociocracy, the contributions of ecovillages in reacting and adapting to climate change, ecovillages in transition, ecovillages as learning centers, collective intelligence, communities and urban neighborhoods in transition, permaculture, North–South dialogue, and the exchange of experiences of a wide range of ecovillages from around the world (including Italy, France, Germany, Israel, Sweden,

Russia, Hungary, and the United Kingdom, among others). One attendee, Edgar Goell, from the Institute for Future Studies and Technology Assessment in Berlin, found a contrast between the people attending the conference and those more representative of earlier phases of ecovillage development. As he put it, "the current generation which is gathering in GEN meetings is open to discuss and deliberate both local and global society and wants to help transform them by cooperating with external groups and institutions," and further that "neighboring communities and cities are also increasingly open to such 'alternatives'," given that their young citizens increasingly find that traditional settlements offer too few prospects for their own futures.

One such important urban alternative is the eco-neighborhood (or "ecobarrio"). The eco-neighborhood movement, which draws on ecovillage principles, is found in Europe, the U.S., Latin America, and elsewhere around the globe. Often these neighborhoods include co-housing projects as well. In Latin America in particular there are a large number of ecobarrios influenced by ecovillages but also other related environmental approaches such as deep ecology and spirituality. Many, such as those in Brazil, Mexico, Argentina, Chile, and Colombia, represent partial variants of the ecovillage model, and in some cases are associated with GEN. These urban eco-innovations are organized in neighborhoods in which the local residents take charge of or play a significant role not only in the delivery of public service, but also in the larger struggle for sustainability. Indeed, one of the main objectives of the ecobarrio, as Ome explains, "is a change of consciousness in order to redefine our relation with nature and other living beings."[6] Sometimes, as in the case of Bogota, Colombia, these eco-neighborhoods were sponsored and facilitated by progressive city governments. The ecobarrio project there created programs geared to the particular needs of diverse physical environments, as well as economic, cultural, and spiritual needs.

Using participatory strategies and the labor of local residences, they have planted trees and gardens, advanced urban agriculture (including seed exchanges), produced public art, started small enterprises, managed waste disposal, and developed a school of "Forgiveness and Reconciliation" to teach and promote peaceful co-existence and conflict resolution, to name a number of the more important activities. Ome has described the ecobarrio program as seeking "to build sustainable alternatives and individual solutions." She quotes the Bogota Mayor's office as having stated that they seek to "transform communities into ecological neighborhoods, with the aim of promoting social and environmental welfare through changes in behavior, participation and the capacity of individuals to become development agents." The ecobarrios in Bogota and elsewhere thus provide an important practical intermediate step between the urban city neighborhood and the ecovillage more typically located in the countryside.

Eco-neighborhoods are also the basis of larger eco-districts which combine together a number of neighborhoods into a larger area in the city. A fast growing movement, eco-districts includes projects such as Mammerby Sjöstad in Stockholm, BedZed in London, Bonne in Grenoble, and Vaudan in Freiburg, to name just a few (Paris Tech Review Editors 2013). To take an example, "Vaudan" is built on the site of the former French military base of that name. It is based on low-energy lifestyles, featuring a "Solar Settlement" which sells electricity back to the city's grid. It features over 100 housing units with more than 6,000 residents and 600 jobs.[7] Supported by "Eco-City" Freiburg, Vaudan is structured from the grassroots level up to the city's administrative structures. The primary "goal of the project has been to implement a city district in a co-operative, participatory way which meets ecological, social, economical and cultural requirements." Across different neighborhoods the Vaudan district has introduced new ideas about renewable energy, energy-efficient buildings, mobility and traffic, participation, social interaction, and the creation of public spaces.

Finally, before turning to the issue of decision-making, it is important to stress that there is nothing easy about starting and maintaining an eco-village or intentional community in general. These are communal forms that stand outside of mainstream society and thus their structures and processes do not generally correspond to existing legal systems or public administrative practices. There has thus been recognition of the need to establish special definitions and rules to both help govern and facilitate those who intentionally share a way of life with regard to issues related to their nonprofit status, tax considerations related to the nature of their income, and incorporation and zoning regulations that address the needs of these organizational forms, both rural and urban. Other practical issues include the development of an eco-bank, more flexible building permission, the subsidizing of natural ecological materials, certification systems for seeds and plants grown in eco-communities, and the creation at the national level of eco-relocalization advisory boards—if not a national agency such as in Senegal. More generally, the introduction and advance of ecological awareness would further the cause, particularly through teaching ideas and practices for sustainability in the school, and research on issues and concerns related to sustainable eco-communal ways of life. Other considerations might also pertain to participatory decision-making practices regarding responsibilities and rights of community as they relate to joint ownership and the organization of work.[8]

Participatory Governance in Ecovillages: Practicing Consensual Decision-Making

In the typical ecovillage all of these activities are organized and carried out through a mode of participatory democracy that one scarcely finds anywhere else. These communities are mainly committed to forms of consensual decision-making. As Liftin (2014: 117) has put it, they are "experimenting with small-scale postmodern models of governance that are devoted to high degrees of consensus and legitimacy" based on a culture of sharing.

Institutional structures, as Kunze (2009; 2012) explains, are used to rule and organize daily life, but the decisive elements remain the communal and personal communication processes that provide the foundation for empowered governance. The key to success, in her view, is the ability of the ecovillage to keep formalized governance structures responsive and flexible to individuals and particular needs. The task of creating sustainable communities, as her extensive research shows, depends on mutual acknowledgment through community learning, open and constructive forms of communication, and corresponding social competences.

There is, to be sure, nothing simple or easy about these practices and the people in such communities will be the first to say that. For this reason, such communities typically devote a substantial amount of time to education and training in consensus decision-making. Members are typically required to read specific books on these topics and engage in discussions about the role and practices of deliberation, including the emotional aspects that come into play in such interactions, especially as they pertain to conflicts among the members. They learn about ways to deal with conflicts and not infrequently participate in training programs. Some members devote a great deal of attention to these issues, in particular the procedures of conflict management.

These techniques can be quite simple but at the same time rather ingenious. At Sieben Linden, for instance, when discussions get overly heated, one of the members rings a small bell and this brings the group to silence. During this minute or so each member is supposed to ask him or herself how important—really important—is whatever he or she is about to say. Does it involve a constructive point or raise a significant question? Or is it more a matter of the ego calling attention to itself?

The importance and intensity of their participatory practices have not surprisingly led to various ideas about what works best under different conditions, and have given rise to a number of "theorists" on the topic. One, for example, is Christian (2003), author of *Creating a Life Together*. As a resident of Earthaven Ecovillage in North Carolina, she has focused on the "built-in tensions" that arise as people with different concerns—from raising food to bringing up children, building homes, and running an educational center—encounter in the course of getting on. Indeed, basic to her discussion is a distinction between two different types of people, which she calls relational (process-oriented members) and strategic (goal-oriented members). While it is not the purpose here to examine these concepts, it is worth noting the degree to which they sound like serious social psychology (despite the fact that she describes herself as an "informal and anecdotal, not academic" ecovillage researcher). Another example is the book by Buck and Villines (2007), *We the People: Consenting to a Deeper Democracy*, on participatory practices. Or the writings of Bressen (2012) on consensus decision-making.

Especially interesting, moreover, is the fact that many of these communities have discovered and grappled themselves with basic problems that have long plagued participatory democratic theory. At Sieben Linden, for one, the degree of participation involved tends to fall off, sometimes to surprising degrees, which has led to long debates in search of the reasons (Kunze 2003). It comes as no surprise, at least on second reflection, that many members start to say that they are too busy for so many meetings and the long-winded discussions to which they give rise. Some say that there is no need to participate when they are happy with the way things are moving along. Such realities have led to the innovation of new rules about quorums and voting procedures, in some cases accepting a two-thirds majority as the basis for particular issues.

Consensual decision-making, of course, is anything but easy. By all measures, it is time-consuming compared to other methods, especially when the approach is formal rather than informal. The argument for it—that ecovillage members regularly make—is that it leads to better rounded decisions that, after being made, are more easily and effectively implemented. Giving all members the opportunity to think through and voice their concerns surfaces issues and takes care in advance of concerns that can later arise and be problematic.

Where voting processes create winners and losers, and can often lead to polarization, consensus decision-making involves taking the necessary time to find unity before making a decision and implementing it. The process is described as building and extending the connections between the members in ways that enhance community life. At the same time, it is said to combine the knowledge and wisdom—"co-intelligence"—of a larger number of people and thus forge smarter decisions. Moreover, when the considerations of all are included, they are more likely to understand and participate in the implementation of the decision, increasing the chances of its success.

Typically, ecovillages offer their members—old and new—training in consensus building. This involves learning about how to listen carefully to the others, engage in open authentic discussion, and to exercise tolerance along with a fair amount of patience. Toward this end, members learn how to use "gentle" candor, air their concerns and interests out in the open, faithfully attend and participate in group discussions, and be flexible and compromising whenever possible. Participants are taught to never reject a proposal too quickly. They are instructed to pose this question to themselves: "Can it be adapted to meet my needs as well?"

Although these concerns are not new to theorists of participatory democracy, having received considerable attention in earlier years in the new social movements literature, especially that of the feminist movement, these micro-level issues and practices hardly make it onto the agenda of modern-day political theory, despite the emphasis on deliberative democracy. These activities would be a rich source of experience for deliberative theorists, but one would not know from the literature that ecovillages exist. This is all the more surprising given the seriousness with which ecovillagers approach the topic. One of the first things to notice when examining the ecovillage literature or talking to their members about participation is the degree to which they are engaged in very thoughtful, disciplined processes to which a great deal of discussion is devoted. Indeed, there are formal principles involved, analyses of different types of decisions and how they should be treated, discussions of how the rules can be applied in different contexts, proposals for streamlining their processes, methods for dealing with serious conflicts, and more. As mentioned above, it is not unusual for some of the community members to undergo training in conflict resolution techniques.

Consider some examples. One of the main issues confronting consensus decision-making is that the process can be blocked by a small number of people, or even one person for that matter, a situation that can be a potential threat to the future of the group, short- or long-term. To deal with this problem of "blocking," which poses the possibility of the "tyranny of the minority," there is an extensive discussion about what constitutes an appropriate block and how to deal with inappropriate blocks (defined as blocks related only to personal preferences or individual values). One approach, practiced for example at Earthaven, is to require the person doing the blocking to take major responsibility for resolving the problem. Another is to limit the number of blocks that a member can have in his or her lifetime in the community (ranging from three to six blocks), and yet another has been to introduce a policy of "consensus minus one" which denies one person's ability to hold up the group as a whole (Christian 2003). Also, some approaches, such as those at Port Townsend Ecovillage, allow members to "disagree by acknowledging that they have unresolved concerns but consent to the proposal anyway and allow it to be adopted."[9] Thus, "reaching consensus does not assume that everyone must be in complete agreement, a highly unlikely situation in a group of intelligent, creative individuals."

To facilitate the reaching of consensus, White Hawk Ecovillage has established varying procedures for two different types of decisions, what they refer to as "normal" and "major/urgent decisions," depending on their social, financial, and time impacts on the members. Normal decisions involve "policies or small dollar amounts without an urgent deadline" that can be made by the standard method of consensus. This requires a minimum number of people at the meeting. If there are no blocks, the decision is accepted and entered into the minutes of the meeting, which are distributed afterwards to the other members within several days. The other members have two weeks to review the minutes and objections can be raised in writing or at the next general meeting two weeks later. If not blocked at this meeting, the decision becomes final.

Major/urgent decisions involve decisions where time is an issue or an expenditure is larger than $3,000 for a household or the entire community. In these cases a decision can be made without delay, although it requires that all members be present or represented at the general gathering. If not all are available, those absent receive a detailed account of the proposed decision and a deadline for a response. Those who wish to block or further discuss an issue for decision are required to reply to all other community members, explaining their concerns and suggested alternatives. For this the community uses email to ensure that all members receive the response. Afterwards, a major/urgent issue can be decided before or at the time of the deadline if more than 50 percent are favorably disposed.

Some ecovillages have increasingly borrowed from or adopted other models designed to combine harmonious and effective community decision-making. In particular, various ecovillages such as Earthaven have taken an interest in an approach referred to as "sociocracy." Sociocracy, as a method of structuring and governing organizations, has roots in systems engineering (although the concept was first used in the nineteenth century by the founder of sociology, Auguste Comte). In large part, sociocracy has emerged to get around problems associated with consensus decision-making. While it restructures the deliberative process to circumvent the difficult standards set by consensus decision processes, in particular the problem of the time it takes to reach unanimous consensus, it nonetheless maintains the commitment to participation and deliberation.

Emphasizing a mix of effectiveness and simplicity designed to produce high-quality results in a short period of time, it offers a set of work-oriented structures that decentralize community decision-making while offering participatory procedures for dealing with the problem of blocking. As such, the sociocratic organizational structure distributes decision-making power to semi-autonomous, self-organizing teams called "circles," which are defined as the primary governance units. Functioning as equals in determining the policies that will guide their work and the work of the community as a whole, these "circles" focus on policies that are at the same time effective and harmonious with those of the larger organization or community. Although the sociocracy accepts a hierarchy of circles, or what is called a "heterarchy," designed to correspond to different levels of decision-making rather than degrees of control, the higher levels are seen to encompass a broader more inclusive spectrum of the overall work concerned with general coordination and organizational policy. The circles are synergetically linked in ways that permit participatory interaction and exchange of ideas. The community strives for consensus but it is not in all cases mandatory. Instead of attempting to develop proposals that take all conceivable objections into consideration—which is often as impossible as it is time-consuming—leaders seek to find the most effective way to achieve the group's goals as fast as possible.

A decision proposal need not mean that all people are "for" a decision, but at the same time they need not be entirely "against" it. Here the focus is on the "tolerance" of the members rather than their preferences per se, which introduces an important element of flexibility often missing in strict consensus-oriented decision-making.[10] Such decisions are implemented for a fixed period of time agreed to by the members and then reconsidered in terms of their measured impacts. Given that decisions can then be reconsidered based on real outcomes rather than those presumed at the outset, the principle generally works to soften objections. When an objection is put forward, it has to be accompanied by an argument about the merits of the issue. It needs to indicate whether the objection is considered to be "paramount" (a situation of paramount importance) as opposed to "normal" (more closely related to everyday functions), the former being defined as a decision that will help to avoid or avert major trouble ahead leading to a blockage or breakdown. Sociocracy allows for less consensual approaches to various everyday concerns, but always includes procedures for a reconsideration of the decisions.

At the present time, while adopted or explored by an increasing number of ecovillages, the approach is still for most ecovillagers seen to be in experimental stages. It is too soon to say it if will altogether eclipse the traditional emphasis on consensus decision-making, and become the dominant decision framework of the ecovillage movement. This will depend on an assessment further down the road.

Conclusion

In this chapter we continued the exploration of participatory environmental governance, this time turning to the much overlooked role and practices of ecovillages. After examining what ecovillages do, as well as their basic philosophies of permaculture and Gaia, we looked more specifically at the way of life at a particularly prominent ecovillage in the ecovillage movement, Sieben Linden. The emphasis there is typically on reducing the human ecological footprint. Of special interest, however, has been how the members of Sieben Linden and other ecovillages live and work together, especially how they make decisions. We saw that most ecovillages adopt rather pure forms of participatory democratic governance based on methods of consensual decision-making. The discussion revealed that there is nothing easy or simple about this; indeed, in numerous instances they have run into issues that have plagued participatory democracy from its earliest origins, ranging from waning participation on the part of community members to the ability of a small number of participants to block the will of the majority—or "tyranny of the few." But equally interesting, as we saw, many ecovillages have recognized the experimental or trial-and-error nature of these practices and sought to adjust them to their own circumstances or to seek out other participatory models that help them skirt around such problems without forsaking their principles. In this regard, these are genuine experiments in participatory democracy that deserve much more attention by political scientists and sociologists.

Despite the fact that these real-world ecovillage experiments largely operate outside the conventional social and political systems, they are nonetheless still located within them. Given that they are busy learning—and relearning—things that will be useful for the climate challenges ahead, as well as developing or redeveloping rural and urban areas, it would be of benefit to society as a whole, especially future generations, if their existences were made easier. Toward this end, as we noted, there are a large number of policy initiatives that would be of considerable assistance to the ecovillage movement.

In the final chapter we will return to the implications of these observations and the reasons why ecovillages should be taken more seriously. As we have seen, they are generally marginalized because they are local and agrarian. The most typical question concerns the relevance of these small-scale ecovillages to the rest of society. Often this is posed as a question of scale: Can the ecovillage model be "upscaled"? The answer, as we shall see, depends on the meaning and use of the term.

The local, small-scale character of ecovillages, it is also argued, can take on new meaning if or when the conditions of crisis lead many city people facing unemployment and chronic shortages, especially of food, to look for alternatives. Under these circumstances the practices of ecovillages will constitute a viable and much-needed source of practical local knowledge highly relevant for survival, a consideration that has already come into play in Senegal. Eco-neighborhoods and Transition Towns, which we take up in the next chapter, can also play a role, especially in conjunction with ecovillages. This might then lead to a reconsideration of earlier strands of environmental thought which focused on localism and decentralization. Instead of focusing on globalization, we may well find that the contributions of radical social ecologists such as Sale and Bookchin, including the communal theory of libertarian municipalism, might be a better path when it comes to preserving or salvaging democratic practices in the face of a techno-authoritarian threat from the national state and the corporate world. We turn, then, to their ideas in Chapter 10.

Notes

1. This chapter is based on visits to ecovillages, interviews with people in six ecovillages, a discussion with an official of the Global Ecovillage Network, and secondary sources. The ecovillages are Sieben Linden, ZEGG and UfaFabrik in Germany, Wongsanit Ashram in Thailand, High Point in Brazil, Christiana in Denmark, and LA Eco-village in the United States. A short version of the main argument in this chapter titled "Environmental Democracy" appeared in Bäckstrand et al. (2010).
2. I am indebted to Iris Kunze for initially making me aware of Sieben Linden. Special thanks also go to Gabi Bott and Christoph Strünke of Sieben Linden for facilitating and supporting my visit to Sieben Linden.
3. GEN International has estimated that there are more than 1,000 ecovillage networks and projects in over 100 countries around the world, and that GEN reaches out to an estimated 10,000 communities worldwide.
4. See http://www.findhorn.org/programmes/559/#.VMpWIV9ghwY.
5. For a detailed exploration of two other ecovillages, one in Portugal and the other in southern Germany, see the research of Flor Dinis de Aruajo Avelino and Iris Kunze. http://www.transitsocialinnovation.eu/content/original/Book%20covers/Local%20PDFs/192%20Case_study_report_GEN_FINAL.pdf.
6. These remarks are based on personal communication with Tatiana Ome, August 3, 2016. She provide me with information about ecobarrios in Bogota based on her dissertation research at University College London. See http://n-aerus.net/web/sat/workshops/2011/PDF/N-AERUS_XII_Ome_Tatiana_RV.pdf.
7. http://www.vauban.de/en/topics/history/276-an-introduction-to-vauban-district.
8. Although such initiatives are only in their early stages of conceptualization, enough signatures were obtained in Italy to submit a draft law dealing with such issues to the Italian government. While the proposal has not become law, an organization was formed to further pursue the effort.
9. See www.ptecovillage.org.
10. These observations are based on the presentation by James Priest at the GEN conference workshop on sociocracy in ZEGG.

Part IV

Making Theory Matter: From Resilience to Eco-Localism and Participatory Governance

9

Urban Sustainability, Eco-Cities, and Transition Towns: Resilience Planning as Apolitical Politics

In the three previous chapters we explored a number of locally oriented projects in participatory governance, experiences presented as practical alternatives to the more theoretically oriented emphasis on deliberative democracy. The return to the local level, as we have already seen, is in significant part a response to the failure of global environmental agreements (Salih 2013: 2; Esty 2014; Litfin 2014). In this chapter we examine the democratic participatory potentials and realities of other local initiatives. We look first at the participatory activities of cities, including large cities, with a particular focus on the role for citizens in programs designed for adaptive responses to the consequences of climate change. Sponsored by city officials, these participatory initiatives are seen to be largely top-down in nature and not generally democratic per se (Purcell 2008). We then turn to the Transition Town movement, often cited by environmentalists as a progressive ecological alternative founded on citizen engagement. Located within many cities and urban areas generally, although independent of them politically, the primary participatory activities of this movement, while ecologically credible, are not fundamentally geared to the furtherance of democratic practices.

There are many reasons for the lack of democratic governance at the local level, but one of them would be the ecological theory prescribed to inform and guide the environmental decisions of these cities and the Transition Town movement, namely the theory of ecological resilience. Resilience, which has come to dominate the thought and rhetoric of environmental planners, is seen to capture and underscore the tension between the logic of ecological substance and the logic of democratic process, taken up earlier in Chapter 2. One can respect resilience as an ecological theory but, at the same time, recognize this conflict. Resilience, as we see in this chapter, is not a program for local environmental democracy.

Eco-Cities, Sustainability, and the Challenge of Adaptation

The idea of an ecological city is not new. In the 1980s discussion emerged about the "eco-city," and in the 1990s it received new emphasis as part of the sustainability discourse to which the Rio Earth Summit gave rise. Then, in the 2000s the topic expanded in more practical ways with globalization, especially with the growing concerns about climate change (Joss 2011). One of the major reasons for the growth of this newer emphasis on eco-cities and eco-localism generally has been driven by the realization that global agreements have made very little progress and that it is time to take action at home. Coupled with this is a renewed appreciation of the fact that most of the impacts of climate change occur locally.

Indeed, it is estimated that 60–70 percent of all environmental problems are local in nature, either originating locally or having an impact on urban and rural areas. For this reason, even in the global context, it is increasingly easy to find studies of the impacts of global environmental problems on local governments. Cities, for instance, have to expect that their shorelines will decrease as the oceans rise, leading to the need for large numbers of people to move from their coastal homes and businesses (Gillis 2016). Others confronting agricultural droughts will have to search for new homelands with arable lands for agricultural production. And still others will have to move away because of increasing landslides, flash floods, and health-related risks (such as consequences of heat waves, reduced drinkable water, and the migration of particular communicable diseases). All of this will come at enormous cost and effort. In many if not most cases, it will bring increases in civil strife. It means that local actors will necessarily be primary players in the politics of climate change. This is particularly the case when it comes to adapting to these growing pressures and consequences.

A focus on climate change "adaptation" has in fact taken its place along with the global focus on mitigation. Cities, towns, and regions are now faced with the task of adapting their practices in ways that reduce the vulnerabilities of cities and other local areas to the consequences of warming. Adaptation, as such, refers to the "actions to reduce the vulnerability of a system (e.g., a city), population group (e.g., a vulnerable population in a city) or an individual or household to the adverse impact of anticipated climate change due to the emission of greenhouse gases" (Satterswaite et al. 2009: 9). While some local areas have tried to reduce their greenhouse gas emissions, and even continue to try to do so, the effort to mitigate the overall effects is recognized to be relatively futile without a coordinated effort at all levels. The focus has thus turned in significant part from mitigation to the question of how to adapt cities and other local areas to protect them from the dire consequences of such warming. Much of this has occurred with an inadequate understanding of the political implications of adaptation (Nightingale 2005).

In a recent book Barber (2013) describes the ways that cities have been responding, thanks to the creative energies that have long defined many of them. Similarly, Krause (2014: 102) writes that "the past decade has seen unexpected and unprecedented levels of involvement by cities in the global issue of climate change mitigation." Although these efforts vary substantially in nature and scope, coupled with the fact that specific data about their outcomes is relatively scarce, these efforts when taken together account for the lion's share of the local activities dealing with climate change and sustainability, especially in North America. And many European urban areas are even more advanced when it comes to adapting to climate change.

Indeed, numerous cities have developed sustainability plans, generally emphasizing adaptation as much as or more than mitigation. In some cases, city administrations have built sustainability goals into their comprehensive, long-range strategic plans (Portney 2003; Bulkeley and Betsill 2003; Pettibone 2015). Such measures are called for by the Cities Climate Leadership Group (commonly referred to as C40), which was established in London in 2005 as a network of the large cities around the world. Recognizing that big cities are the most vulnerable, the goal of C40 is to encourage and facilitate the exchange of technical environmental expertise and other best practices. Smaller cities have formed the International Council for Local Environmental Initiatives (known as ICLEI), which focuses on climate security, resilient cities, and other local environmental initiatives.[1] In general, local activities, even in these networks, come in different forms (Flint and Raco 2011). The full array of such activities are reflected also in the field of urban and city planning, which have become more and more about environmental planning. Conference after conference these days is about sustainable cities and communities and the need for adaptive strategies, often now discussed as the pursuit of urban resilience.

Such conferences feature a range of concepts such as eco-cities, "smart cities," "slow cities," and "cool cities," all focused on sustainable adaptation in one way or another. The smart cities concept is, for example, an eco-efficiency strategy that emphasizes updating the adaptive capacities of the physical and communicative infrastructures through the use of new technologies, digital information technologies in particular. The focus is on a wide range of concerns that increase community resilience, such as cleaning the air and water, developing renewal sources of energy, dealing with migration, ensuring food supplies, introducing new building materials, minimizing waste, redesigning transportation systems, and more.

Urban Adaptation as Resilience

As this emphasis on urban adaptation has evolved, it has become more and more defined in terms of ecological "resilience." Indeed, resilience is now often spoken of as the end goal of adaptation (O'Brien and O'Keefe 2014). This understanding of resilience draws on several theoretical traditions, most importantly complex systems theory from engineering and ecological resilience from biology. In the science of ecology, resilience focuses on the capacities of ecological systems to deal with major perturbations and disturbances such as earthquakes, flooding, windstorms, forest fires, or insect population plagues. But it also includes human-caused disturbances resulting from loss of biodiversity, climate change, pollution, exploitation of natural resources, and the like. It examines the ways that ecosystems resist destruction or damage and how they recover and restore themselves. Large-scale disturbances can push ecological systems to thresholds that require new structures and processes, or ecosystem "regime shifts." For ecologists such as Walker, Holling, and colleagues (2004), resilience is ultimately concerned with measuring "the probabilities of extinction" (Holling 1973).[2]

Sustainability is not "a utopian end-state," as the Center for Resilience at Ohio State explains. From the perspective of complex systems theory, mainly dominated by engineers, resilience "is an attribute of dynamic, adaptive systems that are able to flourish and grow in the face of uncertainty and constant change." Although we cannot foresee future events, especially those of the sort climate change will bring, "we can equip ourselves to adapt to the turbulence ahead." In complex systems theory resilience is thus "the capacity of a system to survive, adapt, and grow in the face of unforeseen changes, even catastrophic events"[3] (www.resilience.osu.edu). Resilience, as such, involves resisting and responding to social, economic, and political disruption and the disorder that it brings. In their words, "it may be the key to global sustainability."

Similarly, Satterswaite et al. (2009) define resilience as "a product of governments, enterprises, populations and individuals with strong adaptive capacity." It refers to "a capacity to maintain core functions in the face of hazardous threats and impacts, especially for vulnerable populations." As such, "it usually requires a capacity to anticipate climate change and to plan needed adaptations." Thus the resilience of any group, as adaptive capacity, refers to the consequences of climate changes that have to be understood in terms of its impacts and relationships to other pressures such as economic changes, political conflict, and civil violence. Increasingly greater awareness of ecological resilience is seen as necessary for achieving the goal of sustainable development. In environmental policy generally, understanding of ecosystem resilience is now widely taken to be the basis for environmental practices.

Resilience Theory and Planning: Apolitical Politics

This focus on resilience is widely heralded as a step forward. In terms of environmental planning, resilience theory offers a common perspective for thinking about the challenges of adaptation, especially when it concerns ecologists, environmental engineers, and planners. But it is not without its problems from a political perspective, in particular a democratic perspective. Although many of the documents underscoring strategies and plans for resilience speak of the need for citizen engagement, much of the practical focus appears to be instrumental in nature. Often in real-world practices such participation is understood to revolve around pre-established planning goals derived from the principles of resilience based on ecological science. Hajer and Dassen (2014), for instance, worry that the focus on sustainability and resilience through technological change is occurring at the expense of societal reflection about the goals and values that these technologies are to serve. Without meaningful political engagement by citizens across a wider spectrum of society, participation, in this view, will be narrow and often one-sided. It is an issue that becomes all the more important when we observe, as Hajer and Dassen argue, the long road still to travel to achieve resilience and sustainability. For them, much of the discussion about sustainable communities and resilience, as part of the smart city approach, is at root technocratic. In their view:

> Cities are social organisms. You cannot just "pop the hood" and fix them. Thus far, smart city discourse is a-historic and has shown little appreciation for the societal contexts in which our cities will have to be built, rebuilt, and retrofitted. Yet, if we really want to get this right, we must act now and quickly correct the technological orientation in smart city discourses. The fate of 20th century modernism shows the pitfalls of a technocratic approach. (2014: 43)

Such approaches can be seen as an urban variant of the technocentric orientation we presented in Chapter 2. These are scientific planning discourses with little place for citizen participation, beyond the kinds of political rhetoric that accompany them. In this view, at least as tacitly understood, the formal goal is to align economic and societal processes with the scientific requirements of resilience. Insofar as these principles are drawn from nature, they would not, at least ideally, be matters of opinion; there would, as such, be no basic need for democracy and citizen participation. In this view, it could get in the way of the task at hand.

The strategy corresponds, in this respect, to the dominant technocratic—and thus apolitical—thrust of most contemporary political-administrative systems. Given that modern-day cities are not democratic, being run mainly by top-down political elites with the corporate world standing behind most of them, this should come as no surprise. Few of them provide a genuine role for citizens in the planning decision-making processes. Such participation would be a threat to these systems, as it would lead to calls for fundamental societal changes. Thus such efforts to plan for an urban transition are largely focused on ecological change in ways that seek to apolitically hold the existing social and political relations relatively constant.

Ecological Resilience and Public Consultation: Participation from the Top Down

Some cities, to be sure, have included public participation as one of their programmatic criteria in sustainability planning. But much of this is organized from the top down. An important example is the City of Vancouver's "Greenest City Action Plan" for a resilient urban future. As a response to perceived democratic deficits in the city, the plan was designed as a way to bring in citizen participation and consultation through climate policy. Coupled with a sophisticated planning process, supported by a progressive mayor, "resilient" Vancouver has sought to integrate climate change with a range of other social and economic issues. The initiative has become an important model from which other cities can learn. But one needs to be clear about the role of citizen participation in this project, much of which has been organized administratively from the top down as much as from the bottom up. It would thus be difficult to describe these activities as participatory environmental governance in action, let alone environmental democracy.[4]

Another example is the urban eco-economic revitalization of the Copenhagen-Malmo area that spans southern Sweden and Denmark. Both cities have emerged as leaders in sustainability, using the environmental restoration as a planning strategy to revive their economies more generally. The Malmo experience is perhaps most interesting for present purposes. The city government describes its effort as establishing local resilience through holistic sustainable development that integrates economic, ecological, and social environments of the city.[5]

Taking hold of an opportunity to reinvent the city in the mid-1990s, after a period of chronic industrial decline, the city government of Malmo developed a knowledge-based approach to economic and ecological change. Indeed, this "resilient city" has subsequently been able to portray itself as an "environmental pioneer." To achieve the guidelines for this "re-branding" of the city, the mayor established a participatory "visioning process," which resulted in "Vision Malmo 2015." This is described as offering an overarching identity and "*Leitbild* for the city centered on rejuvenation through the creation of Malmo University and the redevelopment of the Vestra Hammen (Western Harbour) area of the city." The visioning process started a more comprehensive planning process in which the city assertively positioned itself as a significant urban actor on the international commercial stage (Fleiming et al. 2006: 187). While the visioning process is heralded as an important example of participatory planning, it has mainly involved a participatory process initiated from the top down with the primary roles played by local political and economic elites.

But there has also been a more bottom-up process focused on experimentation through what are called the "Living Labs Malmo," generally linked to Malmo University. Lab users obtain access to various planning tools, digital technologies, and a range of skills that facilitate experimentation with new ideas, projects, and services. These are generally community-based initiatives which have had connections to a number of social movements. Described as participatory experimentation based on user-developed innovations, the participants constitute an array of actors. Citizen groups have a role, but other major stakeholders are involved as well: researchers, public agencies, business companies, and nongovernmental organizations also have an active part to play. This means that citizens have a voice in the process, but it is only a voice among voices, generally less powerful than the others. While it has been an important environmental effort in the world as it exists, it too would not qualify as participatory governance.

These cities are thus only somewhat different from other cities with environmentally oriented change strategies. The environmental element is typically built into the urban plan but is generally viewed as only one component, left in reality to compete with other economic and social goals. Moreover, although officials tend not to say it in public, participation in the environmental planning process tends to be approached in instrumental terms, with an emphasis as much on legitimation and motivation as goal-setting per se. Many, along with Giddens, see such public involvement as potentially impeding the task of dealing with the ecological crisis at hand. Democratic participation is, in this view, a matter for different spheres of political engagement, but not so much in the environmental realm. In short, there is very little, if anything, in these efforts that would be considered participatory environmental governance.

Ecological Resilience and the Transition Town Movement: Participation from the Bottom Up

When it comes to citizen engagement in the struggle for ecological resilience within cities and towns, a more significant progressive effort on the landscape is that of the "Transition Town" movement. Started by Rob Hopkins (2008) in the United Kingdom, Transition Towns (or sometimes now called "transition initiatives") are designed to get ordinary citizens involved in participatory initiatives that contribute to the sustainability of their local communities. Emphasizing that effective global agreements have failed to materialize, and that most politicians are influenced or even captured by economic interests creating the problem, the founders of the movement have set out to develop a do-it-yourself approach. The Transition Town model as such differs dramatically from the Dutch transition approach we discussed in Chapter 3, seen as organized more from above than below.

Only emerging in 2006, the Transition Town movement has spread rapidly across continents to some forty-four countries with over 1,100 initiatives of one sort or another. The movement began with the recognition that governments have dragged their feet on all levels of government: local, national, and global. But rather than just express disappointment, the founders also recognized there is no need for community members to sit by quietly and wait for their politicians to finally wake up to the huge challenge posed by ecological crisis, the decline of affordable energy, and climate change in particular. Instead, small citizen groups can begin with projects themselves that not only improve their own communities, but also serve as models for others to take similar actions.

Before examining Transition Towns more specifically, it is important to add a note of qualification. Given the large number and variety of Transition Town initiatives, the concerns discussed here—positive and negative—pertain to general patterns of development. In particular, the criticisms that follow apply to most, but not necessarily all projects. Having said that, however, it is easy enough to identify their common commitments. The underlying philosophy of the movement focuses on "taking back control over meeting our basic needs at the local level" (Hopkins 2008: 27). This is typically understood as creating "resilience" for a present and future society of "energy descent." Following resilience theory more generally, emphasis is placed on involving citizens in efforts to restore the capacities of cities and towns to absorb social and physical disruption and disorder. The movement seeks to encourage and support experimentation and innovation that can help people to cope with the unexpected.

Barry (2012: 82) suggests that we can see resilience and permaculture as "locally based and robustly contextualized implementations of sustainability, based on the notion that there is no 'one size fits all' model of sustainability." Whereas theories such as ecological modernization seek the stabilization, if not expansion, of existing energy consumption patterns (by decoupling energy use from growth and advancing green clean tech innovations, along with pursuing sustainable levels of wealth creation), permaculture and the transition movement assume economic and ecological decline. They thus seek to understand how one can engage in an ecological transition under the conditions of marked by diminishing resources (Biel 2012). Whereas the mainstream perspective assumes that there will be resources to manage the shift to sustainability, resilience focuses on how to do it with less.

Essentially, the transition process is a self-organizing approach designed to bring local neighbors together to think about ways they could themselves begin to organize, participate in, and work on particular projects that reduce their energy consumption and thus increase the resilience of their communities generally. A people-led endeavor initiated by activists rather than social scientists and government officials, as was the case in the Dutch approach, it is driven by home-grown ideas and the enthusiasm of citizens. Although Hopkins (2008) offers a twelve-step approach for developing Transition Towns, there is no one fixed model for transition initiatives. In some cases, the groups are quite small, but in others they are fairly large. Moreover, the projects tend to look different just about everywhere one finds them. Typical projects include solar energy panels, community-owned local electricity companies, local currencies, food gardens, food banks, farmers' markets, bakeries, produce markets, breweries, community gardens, planting trees, waste disposal, transportation, building bike lanes, community retail stores, and more.

Hopkins (2013: 60) describes the transition approach as providing "a powerful catalyst, an incubator of new ideas and possibilities" that supplies "support and a structure that benefits projects that were previously run in isolation." As such, "it can underpin the whole process with attention to how the group works and how people relate to and support each other." Rather than offering utopian visions the focus of the group engagements is on "possibilities" rather than "probabilities" (2013: 63). Project emphasis, in this regard, is on the practical and doable rather than long-run plans grounded in social-scientific theory.

Fundamental to the orientation is the development of a self-supporting and diverse local economy that can withstand the shock of the outside economy, the global economy in particular. As Hopkins (2013: 107) has put it, the task "is to build another economy, alongside the current, highly vulnerable, energy-intensive, debt-generating high-carbon economy—one that is more appropriate to our times." The point of departure is typically a bottom-up focus on internal investments, in particular a search for ways to shift savings and pensions from distant, and often unethical, investment funds that promote the very system creating the ecological crisis. The process of internal investment involves a community investing its monies locally. As Hopkins explains, this means developing new organizations and institutions and discovering creative ways to permit money to cycle and recycle through the local economy. As such, this requires generating community spending in a fashion that sustains local economic activity and threads this way of thinking through other local enterprises and institutions. It means developing an alternative economic model for transition to a sustainable future (Hopkins 2013; 2008).

Especially interesting examples in this regard are found in the transition movements in the city of Bristol in the U.K. and the Transition Town Brixton, a large residential area in London. Both have established their own local currencies. In each case, local citizens can turn their British pounds into local money which can only be spent in their own community area. The use of these currencies then facilitates the circulation of local monies in the community itself, as opposed to being drawn out of the area by large corporate chains. The goal is to support the local economy through the use of a form of money that "sticks" to Bristol and Brixton, with the emphasis placed on boosting smaller stores and trading establishments that confront heavy competition from big chain stores.

In Brixton the currency has facilitated the building of community ties and consciousness in a variety of ways. One is that the notes are designed by local artists with heavy consultation on the part of community residents as to who they would like to see on the bills. Typically the notes feature local history, culture, and art. Originally only a paper currency, it also became a mobile cell phone pay-by-text service used by 250 independent businesses and thousands of local residents. There is also a Brixton "app," which allows local users to easily add to their account, and a map showing hundreds of establishments which accept the Brixton pound.

Furthermore, Brixton has established its own community solar energy company, which not only offers local residents affordable energy, but also provides good returns for the local investors, employment for citizens in the area, and various opportunities for job training. As a cooperative social enterprise, the Brixton project has been organized in a way that puts a portion of its earnings back into a community energy fund which is used to improve overall energy efficiency in the area. In the words of its organizers, it serves to ease "fuel poverty" for its poorest citizens. Started by dedicated local residents knocking on doors and asking their neighbors to get involved, this innovative project has spread rapidly, now involving its third major installation locally, and has drawn wide attention across the U.K. and elsewhere (Hopkins 2013).[6]

TRANSITION TOWN FRANKFURT

The Transition Town movement in Frankfurt offers another illustration. This transition project involves a group of core team members and various project groups. The groups have organized numerous projects which can be found at http://www.transition-town-frankfurt.de/projekte.html. Below is an outline of a number of the more prominent activities carried out by the working groups, which meet on a regular basis.

Energy and Mobility. A working group explores and discusses alternative ways of organizing climate-friendly energy and environmentally sustainable modes of transportation. Toward this end, its members work together with other citizens and the city government to develop alternative models. For example, they have organized an electric automobile for the people who transport fruits and vegetables produced by the community-supported agriculture program. In addition, they have sought to improve public transportation, as well as introduce extended bicycle lanes in the city.

Collective Farming/Community-Supported Agriculture. In this project people pay money to a farmer in advance of planting crops and receive a portion of the harvest. Private citizens thus help the farmer bear the costs of farming and receive in return a share of the local produce. This not only supports local farms but is also an alternative to market-driven industrial agriculture. This activity is part of a larger network of community-supported farming in Germany.

Inner Awareness and Reflection/Inner Transition. This working group explores and investigates ideas and visions, as well as hopes and fears related to ecological destruction and the local environment. In particular, it seeks to examine the relationship of Transition Town Frankfurt to the larger picture of events and developments in the city. In addition to attempting to foster cohesion within the Frankfurt group, emphasis is on coordinating activities and networking with other environmentally active groups in the city.

Repair-Café. This offers a place where local people can repair things on their own rather than disposing of them in the garbage. It is also designed to promote neighborhood self-help and reskilling, basic concepts in the Transition Town philosophy generally.

Transition Day and Local Transition. Two groups work here on interrelated goals. "Stadtwandeln" involves guided tours through the city designed to bring different people in contact with one another to learn about various projects that are reinventing conventional city life and are considered places of transition in the city. In the process, people learn about the Transition Town movement and the different projects in which it engages. "Wandeltag" is a particular action day in which TT Frankfurt is introduced to others with different but similar activities and initiatives in Frankfurt. The goal is inspiration and motivation for change.

Urban Gardening and Bees. TT Frankfurt cooperates with and supports a number of urban gardening projects in various neighborhoods throughout the city. They also work to support bee-friendly gardening, as bees are an endangered species essential for plants and trees. The project is designed to raise awareness of this problem by organizing seminars for beekeepers and others who seek to rescue them.

Thanks goes to Stefanie Burkhart of Transition Town Frankfurt for providing me with this information.

There is, however, another aspect of the Transition Town project that has come under criticism, namely its lack of political engagement with the larger political system. In an effort to avoid both governmental delay and political cooptation, Transition Town projects are seen as an end run around the traditional political structures (Kenis 2016: 956–61). Although the movement is fundamentally based on the idea that local participatory action can have significant impacts, Transition Town participants are advised to stay out of the formal political process.

Without suggesting in any way that the efforts are unimportant, the participation in this movement is local and ecological rather than political per se. Indeed, the philosophy spells this out. Hopkins and others argue that the movement should stay out of political campaigns and other issues that will detract attention and energy away from their own local efforts. This nonpolitical orientation is spelled out in the Transition Town literature and can be confirmed as well by looking at the issues of the *Transition Network Newsletter*.

Arguing that the political system, local as well as national, has failed to address these problems, there is seen to be no need to align the movement to them. Such an entry into the political process, it is feared, can potentially corrupt the transition efforts, especially as it would typically involve money with strings attached. There is, to be sure, a long history of experience to document the concern. Second, they argue that identifying with a political party, even a green party, would turn off a large segment of the public, leading to even more general disaffection among the citizens. Indeed, for this reason, the goal is to avoid all campaigns. This would include, for example, taking sides with political groups struggling elsewhere with environmental issues, toxic waste struggles being an illustration. The movement leaders argue for an approach that will bring the largest number of people together while seeking to avoid conflict. The goal is to show that people don't have to wait for the political process to respond; they can "just do stuff" that transcends politics, a perspective borne out by reading Transition Town movement literature. The idea here is to eschew the negative and only take positive stances. In short, the movement seeks to be apolitical.

It is surely not the case that all members of Transition Towns are apolitical. It may even be that the theory is somewhat out of sync with the practices. My own interviews with people offer some evidence for this. While it seems fair to say that most members do in fact refrain from participating in the activities of the established political processes, which is true of most citizens generally, there are clearly exceptions to the rule. Another difficulty here is that transition initiatives take many forms and some would not formally be considered a Transition Town movement project, though their members might engage in a wider range of political activities. And yet another is that it is often difficult to judge what is political and what is not. Running a food co-op can be seen as political, especially if it is understood as a challenge to large food corporations. But this need not involve public political protest.

While taking issue with these apolitical dimensions, one can still consider the democratic potentials of the Transition Town movement. Even though Transition Towns are not democratic projects, the ways they bring people together could lead to democratic engagements. One aspect that could potentially contribute to participatory democracy is the practice of Open Space, designed for initiating participatory events that bring people together to discuss the various issues that emerge related to Transition Town projects. Point number 10 in Hopkins' twelve-step approach describes Open Space meetings as a self-organizing approach that emphasizes individual responsibility for learning and involvement (Hopkins 2008).

But such meetings, as they are run today, appear not yet to have adequately served these goals. Cohen (2010: 44), for example, worries that this stress on "empowered individualism" also influences who shows up to participate. He also contends, with good reason, that some local citizens will not feel "expert enough" to join such a group discussion. Further, it has been noted that admittance to the annual Transition Network Conference in the U.K. costs £100, a sum that many cannot afford. However, while these are concerns that impede democratic participation, they do not amount to insurmountable barriers. There is no reason why the movement could not eliminate them and foster a more participatory democratic culture.

Another barrier to eco-democratic Transition Towns, typical of environmental movements generally, is a middle-class bias of its activists. In this regard, the Transition Town membership movement tends to reflect the demographics of mainstream environmentalism. In the main, the coordinators and local participants are typically white, well-educated, environmentally conscious post-materialists. Some have even cynically referred to Transition Towns as another "pleasurable leisure based community movement" (James 2009: 19). On could here speak of weekend environmentalists. Although Hopkins does stress having fun, such a judgment seems too harsh. Nonetheless, this is not in any case the basis for an inclusive democratic political movement. As we know from the study of environmentalism broadly conceived, the general modes of thinking and lifestyles of the middle-class members often intimidate members from the lower classes of society, even if unintentionally.

There is, as such, insufficient attention to the power differentials and dynamic within communities and the way ethnicity, gender, and socio-economic background play a role in shaping community relationships. In the context of the transition movement, it can involve "assimilating others to our way" of thinking about the world. The challenge, therefore, is to find ways of being open and encouraging diversity without "othering," perpetuating social stratification, denying inequality, or claiming superiority (Cohen 2010: 51). It has also been suggested that the twelve-step model is something of a "managerial approach" that is rather culturally blind. The top-down steering of the Transition Network and "brand management" by the founders in the U.K. (e.g., through the accreditation process to become an official Transition Town) have been perceived as running counter to the bottom-up grassroots creativity and diversity the movement seeks to embody.

Another concern is that many projects tend to be run by a small number of people who organize the Transition Town movement. It is they who seek to involve others and thus determine who plays which roles. This hierarchical dimension is not altogether surprising, but it does not advance the democratic potential of the movement (Cato and Hillier 2010). Further, many projects seem to depend on raising money and when it dries up the people and activities seem to fade away. There is evidence of this in Brixton; the storefront office is no longer to be found (and the salesperson in the novelty store that had taken its place had never heard of a Transition Town). Moreover, it is often the case that staff workers, in any case those who coordinate the larger transition networks, are paid salaries which sets them off from volunteers.

Others raise even deeper concerns. Chatterton and Cutler (2008), for example, argue that a movement that focuses on outcomes but ignores the causes of the ecological deterioration is in jeopardy of being co-opted. For one thing, local efforts to reduce the carbon footprint can exist without problem alongside the economic and political structures creating the problem. An agreement not to rock the boat, in their view, can scarcely challenge the rules of the game. To get to a low-carbon world, the movement needs to have a stronger political engagement with the structures and processes causing the problem in the first place.

In addition, some find the political problem to be lodged more fundamentally in the movement's guiding ecological theory. Here again, we encounter the argument that the emphasis on resilience is essentially apolitical. It says nothing, in the words of Cote and Nightingale (2012), about "transition to what, or where, or by whom and for whom," and as such ignores or hides power relations and political conflicts. Fainstein (2013; 2015) argues that resilience altogether neglects questions of social justice and thus will itself generate political tensions that block its implementation, a point she illustrates with a case study of resilience planning in New York City (Fainstein 2016). Moreover, as she writes, it does not "provide the basis for the mobilization that can ultimately change the boundaries of the politically possible." The framing of resilience does not, in the view of Alloun and Alexander (2014), "encourage the sort of regime shift, transformation and social learning required to transition to the low-carbon economy" as envisioned by the transition movement. Some even argue that it is compatible with neo-liberalism's effort to shift responsibility to the individuals in local communities and to adjust them to the fact they cannot expect help from the state (Featherstone et al. 2012: 178), a point that might be seen as a bit unfair. Without going further, it is clear enough to say that there are unresolved political issues in the Transition Town movement itself.

This has raised the question posed by critics, mainly friendly critics, as to whether one can realistically tackle the environmental crisis, a global crisis, without confronting the economic and political structures that have created and continue to create the problem (Chatterton and Cutler 2008). Whereas Transition Towns are small groups of citizens doing what they can, the ecological crisis is huge and largely out of local hands. It would hardly be new to argue that we need to change the political-economic structures—the capitalist economy and the political system that governs in its interest—if we have any real hope of reducing the massive levels of carbon output that threaten our future. As Mckibben (2014) puts it, there is a difference between an individual being a vegan and a collective movement challenging the structures of industrial farming.[7]

In short, the question posed is: How realistic is the transition movement given the size of the challenge? There is no question that it is good to engage people in environmental change. Enhancing self-reliance at the local level is both possible and important. And this the transition movement can and does do. Here we see laypeople becoming aware of the risks that confront the future, becoming concerned, and taking whatever actions that are feasible. In addition to developing do-it-yourself skills, transition initiatives can at the same time enhance community cohesion in ways that create social safety nets. People who can grow their own food and generate their own electricity surely have a better chance of surviving the worst of the crisis than those who can't.[8] Having acknowledged these ecological points, though, the movement is not about democracy. Its members are surely not against democratic governance, but the approach is formally not designed to promote or facilitate it.

Conclusion

The first part of this chapter has offered a basic overview of the growing number of climate change initiatives in both big cities and small towns with a particular focus on the role citizen participation has played in the formulation and implementation of these programs and plans. Toward this end, the emphasis was on climate change adaptation, particularly as it is now defined in terms of theories of ecological resilience. As we saw, there is a good deal of rhetoric about the role of the citizen in adaptation and resilience, but far less genuine participation in local planning practices. We also took note of the exceptions to this pattern, using Vancouver and Malmo as illustrations. Even in the best examples, however, participation appears to mainly be organized and facilitated from the top down. The second part turned to the role of the Transition Town movement, in the minds of many the most progressive local movement on the scene (Barry 2012: 20, 89–108). Moreover, by contrast, Transition Town initiatives are much more bottom-up grassroots movements devoted to the theories of permaculture and resilience.

Despite important ecological features, we also found limitations with the transition movement when it comes to democratic participation. Although democracy is not a primary emphasis in Transition Town talk about citizen engagement, either in theory or practice, we argued that there is nothing necessarily stopping participation in transition initiatives from being democratic. But democratization seldom happens automatically. Without a specific emphasis on democracy, along with a change in practices, it is unlikely to come about. As experience with a wide range of social movements shows, it usually requires encouragement and facilitation, even a form of nurturing. It would then provide a challenge for the leaders of the movement.

This is particularly the case when it comes to the theoretical emphasis on resilience. The questions of who, what, and how that are missing from the resilience discourse in environmental planning would be a suitable place to begin a more democratic engagement with transition, as we saw earlier in the case of transition management. But this is not altogether straightforward, as it brings us back to the earlier and more general question about the relationship of science to democratic politics. If resilience is an objective state of affairs, a specific target to be reached as it is usually taken to be, then we can expect the ecologists, environmental engineers, and planners to play a dominant role in discussions about ecological resilience. This would inevitably raise questions about the role of democratic deliberation in deciding questions about resilience. The scientists of resilience will surely ask why the public should be involved when it comes to questions that lend themselves to the laws of ecology. In short, what would be the boundaries of democratic deliberation related to achieving resilience? This is not a topic without a plausible answer, but we return to it in Chapter 11.

Finally, Trainer (2014), in his critique of Transition Town initiatives, envisions the political solution to the current dilemma through a grassroots self-organizing and radical participatory democracy. Holding out no hope for the existing political system, he argues that "individual citizens and communities need to build cultural, social alternatives underneath the current model." The Transition Town movement offers itself as an alternative but, in contrast to the ecovillage movement, neglects the social and cultural dimensions. We return to this point in Chapters 10 and 11.

Notes

1. See International Council for Local Environmental Initiatives (http://resilientcities2016.iclei.org/home).
2. Thanks go to Brian Walker for explaining the ecological dynamics of resilience to me in a shared taxi ride from Rotterdam to Schipol Airport in Amsterdam.
3. See www.resilience.osu.edu.
4. On Vancouver see http://vancouver.ca/green-vancouver/climate-change-adaptation-strategy.aspx.
5. On Malmo see http://www.preventionweb.net/applications/hfa/lgsat/en/image/href/2324; http://news.bbc.co.uk/2/hi/europe/5413960.stm; http://medea.mah.se/tag/malmo-living-labs//medea.mah.se/tag/malmo-living-labs/.
6. Nonetheless the challenge is substantial. British government rules appear to have sabotaged the "community energy revolution" to suit the interests of the big energy companies (Monbiot 2015).
7. I owe much to Piyapong Boossabong for showing me Transition Town projects in London.
8. Remarks made at the conference on Techno-Utopianism and the Fate of the Earth at Peter Cooper Union in New York City, October 25–6, 2014.

10

Relocalization for Sustainable Communities: Participatory Ecological Practices and Theoretical Foundations

We began this book by asking what will happen to democracy during climate crisis, especially in view of the possibility of a severe crisis. We noted along the way that the concept of democracy, where it exists in one form or another in the world, is already generally thin and often fragile. From here we took note of the fact that, confronted with the serious economic and social dislocations that climate change will bring, there will almost certainly emerge strong eco-authoritarian calls for more centralized forms of technocratic governance. Indeed, as we have seen, they have already begun to appear. To the degree that this continues, the question it raises is: how long would such an authoritarian politics last? As nobody can answer this question, which scarcely makes it less important, we are compelled to think about how we might best protect ourselves against the injustices that will surely accompany it.

This initiated the search for an alternative strategy, one that would be able to keep democratic values and practices alive during this period of hardship, however long it might endure. The best option, it has been argued, is a return to the local level. Not only does the return to localism make ecological sense on its own terms—given the future shortages of energy, food, and many other resources—it also makes sense because small face-to-face groups have always been considered the basis for authentic participatory democracy. Indeed, independent of ecological crisis, a return to the local is good for democracy generally, the sort of *authentic* democracy that writers like Gould and Jackson call for.

What is more, and equally important for the present set of concerns, there is an emerging and vibrant "relocalization" movement underway. Although it has developed more or less under the radar of the modern media and the world of academia, it is an important movement that can and should be built upon

(Norberg-Hodge 2014). This movement first and foremost involves the search for a transition to a sustainable way of life from the bottom up in contrast to a government-supported transition management strategy, as we saw in Chapter 3. In this regard, it is also an important anchor for holding on to and extending participatory democracy, or the local variant we have here called participatory governance.

Many might ask, why turn back to the local level? Is this reasonable in the age of globalization? What would be the benefits? How might we go about it? Relocalization raises a host of difficult yet potentially promising possibilities that speak to the issue at hand. In this chapter we shall outline the meaning of localization, the forms it has already taken, and some of the theories that support it in the face of environmental crisis, in particular those of Sale, Bookchin, and Bahro.

What is Relocalization?

The return to the local—or "relocalization" as it is often referred to—has already become a major theme among progressive environmental activists. Indeed, it captures an established and ongoing movement around the world.[1] According to the Post Carbon Institute, "Relocalization is a strategy to build societies based on the local production of food, energy and goods, and the local development of currency, governance and culture." The primary "goals of relocalization are to increase community energy security, to strengthen local economies, and to improve environmental conditions and social equity."[2] Relocalization, as an alternative strategy, has emerged "in response to the environmental, social, political and economic impacts of global over-reliance on cheap energy." Even though it remains on the margins of modern techno-industrial society, it has a growing number of supporters among people seeking alternatives, especially among environmentalists. Although it largely goes unnoticed in academic social science, it also relates to an important intellectual and long-standing tradition in progressive politics, and especially in the environmental social and political theory of the 1970s, which we take up in what follows.

In large part, the relocalization movement is a response to the environmental crisis; it is an effort to forge the way to a sustainable way of life. Challenging the global emphasis on international trade, finance, travel, and communication, the movement seeks to reduce the unnecessary damage to the environment caused by these long-distance movements. It is also preparation for another way of life if and when the global arrangements collapse (Hines 2004). Some would describe it as return to a simpler way of life, but one needs to be cautious here. It is simpler in the sense that the dependence on modern technologies to organize and govern our lives would be reduced. Simpler also in the sense that social circles would be smaller (but often less alienating for many). But this does not mean that a simpler life becomes easier. Indeed, it involves confronting many tasks with earlier modes of production—many of which can involve physical hardships—often including the horse and the plow. It also means that people would have to learn to deal with each other in a less competitive, more egalitarian way.

De Young and Princen (2012) define such localization as an intentional process of readjustment to our changing natural environment, or what they describe as "adapting to downshift." The basic goal is establishing a sustainable relationship between humans and the ecological world. Localization, as local sustainability, is aimed at maintaining the resources needed to support community life while promoting individual social and psychological well-being (Urry 2011). "In contrast to the placeless and faceless global," in Litfin's (2014: 157) words, "the local holds out the promise of real relationships with real people and places." Localism offers "a healthy adaptive response to a rapacious and dysfunctional globalism." Brooks (2016), in this regard, finds a growing "anti-system" interest in the pursuit of local community and community movements among young people in the U.S. as opposed to more traditional pursuits of affluent materialism.

This return to the local does not just mean moving back to the land, and in some cases it doesn't mean this at all (Willis 2016). As we shall see later, it would ideally be part of a larger effort to redesign and plan our cities and urban areas in ways that facilitate local sustainability in the cities themselves, such as neighborhoods and districts, long the position of many environmentalists and environmental planners. Genuine eco-cities would require such a redesign rather than a technological fix. But costs would be very large.

The eco-localism and relocalization projects are more often than not founded as "intentional" communities (North 2010). Intentional communities are planned residential communities designed at the outset to have a high degree of social cohesion and cooperative living. Typically sharing resources and responsibilities, the members of such communities pursue alternative lifestyles, in any case lifestyles other than those typical of the dominant techno-urban systems that define life in the West, and much of the rest of the world for that matter. The residents of such communities, whether urban or rural, often hold common views on social, cultural, and political matters, some having a spiritual orientation, Buddhism being a not uncommon commitment. Quite commonly, there is an emphasis on egalitarian values, self-sufficiency, participatory decision-making, and interpersonal development. Many such communities form to ameliorate social injustices of one sort or another. In this regard, there is no one model of an intentional community; they include co-housing communities and housing cooperatives, collective households often in urban areas, and kibbutzim, ashrams, survivalist retreats, and ecovillages more typically in rural areas.

From a similar but somewhat different angle, Schlosberg and Coles (2015: 2) explore these fast emerging environmental networks. They see them in significant part as extensions of "environmental justice movements," ranging from "community responses to a post-carbon necessity and a climate-challenged world, to an embrace of new domesticity in crafting, a range of movements offer[ing] new modes of organization, forms of resistance, and prefigurative models of democratic living, all immersed in re-formed relations with each other and the natural world." In addition to resisting unsustainable industrial practices pertaining to energy and food production, their research finds these new collective networks to be creating alternative flows of power and material resources. Locally oriented, these movements, including Transition Towns, seek to "rethink and redesign" the process and practices that create and provide for basic human needs. As community collectives, they move beyond other projects, practical and theoretical, committed to a particular concern or value—such as sustainability or environmental justice—in an effort to tie together an interrelated set of concerns, from resistance to political and economic systems captured by interest groups, the dominance of energy companies in efforts to reduce carbon dependency, and the redirection and circulation of energy and food production processes toward sustainable practices. For them, such "movements represent a new politics of sustainable materialism, an environmentalism of everyday life" (Schlosberg and Coles 2015: 2).

Localization: Renewable Energy and Sustainable Agriculture

Localization is thus already associated with very important efforts in the environmental movement, especially concerning renewable energy and sustainable agriculture, as Schlosberg and Coles explain. Increasing numbers of people are now busy "relocalizing" their communities to produce more energy, food, and goods locally. Although most of us don't know about them, there are many relocalization-oriented organizations and resources that help towns and communities to think about how to develop and implement strategies and programs to achieve local self-reliance through carbon-saving measures and food production.

Nothing is more basic to the problem of climate change than the need for energy transition. Toward this end, one of the many advantages offered by relocalization is sustainable energy security (Bomberg and McEwen 2012). Replacing vulnerable centralized energy systems with decentralized and widely distributed energy sources based on renewables—solar and wind energy in particular—provides a range of benefits from lower electrical prices to local economic development, jobs, and sustainable futures generally. Renewable energy—solar, wind, biomass, hydroelectric, and geothermal—offers an increasingly important opportunity to move away from expensive, polluting fossil fuels. In addition, research also shows that these collective renewable energy projects can help to build community cohesion and trust (Walker et al. 2010).

It is taken as an established fact that renewable energy sources are best organized and managed as decentralized systems. By placing such energy sources near their end users, there is far less need to transport energy over long distances from energy plants, avoiding relatively inefficient practices that require more resources and unnecessarily take up land. Decentralized power production "also helps to keep money from 'leaking' out of local economies" and in the process contributes to the local decentralization of political power (Norberg-Hodge 2014: 24).

In addition to renewable energy, nothing is more important to the relocalization movement than the effort to develop and expand a system of sustainable agriculture, including an emphasis on food sovereignty. The sustainable agriculture movement is perhaps the most important driver of relocalization. It is, in fact, one of the most successful grassroots movements around the world, including in particular the food sovereignty and Via Compesina movements.[3] Not only does it provide a healthier supply of food (free of harmful chemicals) in ways that dramatically reduce the ecological damage resulting from transportation, but it also helps to protect biodiversity. And most important for the discussion here, it offers people a way to become meaningfully involved in their own lives and the communities in which they live (Barry 1977). This also includes the growing urban farming movement (Mougeot 2005).

Sustainable agriculture involves a shift away from the way agribusiness grows food, transports it to marketplaces, and markets it to consumers (Schlosberg and Coles 2015: 3–5). Today, the foods we eat are transported from around the world. Indeed, this is an important component of globalization. Two thirds of farm subsidies in most countries go to big agribusinesses that rely on large-scale industrial production of food. Such subsidies are also designed to support the use of chemical pesticides, energy-intensive monoculture production, and biotechnology. According to one of the important leaders of the localization movement, Helena Norberg-Hodge (2014: 24), winner of the Alternative Nobel Prize, shifting "these expenditures towards those that encourage smaller-scale, diversified agriculture would help revitalize rural economies in both the North and the South, while promoting biodiversity, healthier soils, food security, balanced and diverse diets, and fresher food." With regard to health in particular, such local agriculture offers a solution to the growing epidemics of obesity and diabetes that are now running rampant in Western societies. "Shifting the emphasis to diversified, low-input production for local consumptions," Norberg-Hodge goes on, "would not only improve economic stability, it would also reduce the gap between rich and poor, while eliminating much of the hunger that is now so endemic in the 'developing' parts of the world."

As perhaps the most successful grassroots movement in recent times around the globe, the local food movement connects local farmers with local consumers, which benefits not only the consumer but also the local economies and ecology. Closely related to this has been the expanding demand for organic food, as organic production is best carried out by diversified small-scale farms geared to supplying local consumers. Often basic to the approach are the principles of permaculture, which prescribe how organic food production can be combined with the natural requirements of surrounding ecosystems. In Europe such local production has expanded dramatically during the past fifteen years. But the trend is found in both the global North and South, as many young people have begun to move to the land to engage in such farming. Indeed, there is now a worldwide network that promotes the effort.

The Local in the Global

The call for a return to the local is often greeted with considerable skepticism, even incredulity by some. Many see it as a largely irrelevant environmental strategy in the face of the ongoing thrust of globalization. For them the answer is global governance. We will need global governing bodies, they argue, to solve or manage today's global problems such as trade, finance, communications, and transport. How, they ask, can we deal with these problems by either ignoring them or attempting to somehow manage them from the local level?

The return to the local, however, does not—and cannot—mean that we can forget the global level (Kütting and Lipschutz 2012). Relocalization theorists acknowledge that there are powerful transnational organizations directly responsible for the coming climate crisis (Hines 2004). Indeed, they worry that these organizations can—and many surely will—continue to carry on their environmentally destructive activities until it is too late. For this reason, in their view, it is important to forge local–global networks that can continue to struggle for global change.

But what they do not accept is the idea that we can build such a strategy from the top down. Indeed, we have already witnessed the failure of such efforts, given that there are no effective global institutions strong enough to counter the subversive powers of the globalized corporate world. Without neglecting the global, the goal has to be approached from the bottom up.

Hess (2009), in this regard, shows how globalization has paradoxically given rise to such bottom-up efforts, or what he refers to as the re-emergence of the localist movement. Avoiding the romanticism often associated with—or attributed to—localist efforts to take back both economic activity and political control, he documents in considerable depth the already extensive and still growing number of projects leading to new practices such as "buy local" campaigns, community ownership of electricity, alternative transportation systems, and community media. Not least important, Hess examines what he calls "global localism," involving transnational local-to-local supply chains that can facilitate local production and local economies. Global localism, he argues, is not a magic solution to cure the ills of globalization, but it does offer an important anchor for movements struggling to construct an alternative future based on more socially just and sustainable forms of democratic governance. The Global Ecovillage Network offers an example of one such model of global localism.

These efforts, as Hess shows, provide an alternative base of power that meaningfully relates to citizens in the places where they live. One of the major problems, in this respect, is that the turn to global regulation has left a vacuum between ordinary citizens and the global political players. It is true that local citizens have nothing like the power needed to challenge the global actors, certainly as things stand, but it is also important to recognize that sustainability involves a committed and engaged citizenry. It is here that the local sustainability movement—including ecovillages, Transition Towns, eco-neighborhoods, and eco-district movements—can make important contributions. Offering citizens chances to take action where they are located can provide an important form of empowerment that feeds into the environmental effort more generally. Indeed, doing so provides a potential foundation for environmental democracy.

De Young and Princen (2012: xxi) understand that localization in itself is not necessarily a good thing. For this reason, they distinguish between what they call negative and positive localization. With negative localization they recognize that an unorganized and even desperate effort to flee unsustainable cities can lead to chaos, survivalist hoarding, militancy, warlords, and the like. This corresponds to the breakdown scenarios we examined in Chapter 1. Positive localization refers to the intentional adaptation to a future decline of cheap energy and abundant raw materials, as well as all of the social and economic problems that will accompany unsustainability. As they (2012: xxi) put it, positive localization "is a process for creating and implanting a response, a means of adapting institutions and behaviors to living within the limits of natural systems." Rather than an "end state to pursue," localization "is a way of organizing and focusing a process of transition." It is an alternative process that needs to be supported and accelerated while there are still the surpluses needed to make transitional change possible.

For them, positive localization is characterized by three features. First, it assumes that unsustainable societies will need to adapt regardless of whether their citizens are environmentally informed or not. As such, it is not based just on a moral plea to get people to localize in the name of saving the environment. Adaptation, in this view, will often happen out of necessity in advance of changing consciousness and the adoption of new value systems. Positive adaptation holds out little faith or hope in top-down centralized approaches. The centralized use of policy tools such as risk assessment, command-and-control, and incentive systems is not only seen to have failed to bring the climate crisis under control, but also offers no way to get citizens involved in helping to create sustainable communities. And third, in this latter regard, it "affirms self-organization, self-reliance, self-limitation, and self-rule" (De Young and Princen 2012: xxiii). Toward this end, it is based on the belief that people have the capability to responsibly participate in finding environmental solutions. It does this through the use of socially responsible production and trade conducted with locally appropriate technologies.

Finally, it is important to note that global environmentalism depends on local action in any case. Even if the fears of climate collapse and its dire consequences might prove to be exaggerated or false, there is in every way a need for citizen involvement in the process of adaptation that will be required. This is the case even if climate change turns out to be more a matter of discomfort than survival. And to this we can add the possibility that such local citizen involvement will strengthen democratic environmentalism.

Participatory Political Theory for Relocalization: Sale, Bookchin, and Bahro

There is nothing new about a discussion of political theory, democracy, and local governance (Denters et al. 2014). Localism has, in fact, long been considered the ideal condition for democratic government. This is built on the recognition that authentic democracy depends on face-to-face communication about things close to the lives of the affected citizens. As is well documented, large political systems make decisions removed from their citizens. At best, as a long line of theorists following Rousseau and others continue to point out, these polities rely on representative government. In these systems, citizens do not speak for themselves; they depend instead on someone else to do it for them. In most cases, the representative has never had a direct conversation with the fellow citizens he or she represents. Most typically today the citizen's opinion is to be deciphered through a representative polling sample, if at all. This is at best a "thin" form of democracy (Barber 2004).

For this reason, localization has long been a basic element of radical political movements, with a long tradition among anarchist theorists from Proudhon forward. It has also been a basic theoretical tradition and practice in the environmental movement. Not only has the environmental movement long called for local action; there has been a strong tradition of environmental political theory associated with localism. The works of Kirkpatrick Sale, Murray Bookchin, and Rudolf Bahro, among others, have been particularly important in this regard, especially in the early stages of environmentalism. But these writers largely disappeared from environmental thought. Among the several reasons for this, one is surely that they were later considered too radical for the times, especially as the focus turned to the possibility of ecologically transforming industrial production and the capitalist system. The idea that the environmental movement could press for a green capitalism pushed earlier arguments to the side. And second, the recognition that capitalist production had gone global turned attention to the need for global regulatory institutions. Few have argued against a local role per se, but the place of local governance took a back seat to the focus on global environmental politics. It was largely considered irrelevant to the big problems at hand. Even attention to the impact of climate change on cities lagged. Only more recently, as we saw in Chapter 9, has the topic begun to return to political science and urban planning.

Now, it is argued here, the time to reverse the order has come, or at least to rebalance the emphasis, given the failures to green capitalist growth and the specter of serious global turmoil ahead, whether it turns out to be a matter of serious discomforts or catastrophe. Toward this end, it is time to theoretically develop a political-ecological orientation that better addresses the portents of the coming crisis. An appropriate starting place would be a return to these earlier theorists, Sale, Bookchin, and Bahro.[4]

Revisiting Kirkpatrick Sale: Decentralization and Bioregionalism

We can best begin with Sale (1980; 2000) and his writings about decentralization and regionalism. Decentralization has long been understood as a necessary component of participatory democracy. As Sale (1980: 443) has expressed it, "the impulse to local governance... seems an eternal one and well-nigh ineradicable." The long period of nation-states has not diminished this desire. Particularly "remarkable during these long years is how this decentralist tradition remains so resilient—so resilient that every time the power of the nation-state is broken, as during wars or rebellions—immediately there spring up a variety of decentralized organizations—in neighborhoods, in the factories and offices, in the barracks and universities—that reinstitute government in local, popular, and anti-authoritarian forms." Beyond the ancient Greek small city-republics that gave rise to democracy, classical examples in modern times are found in the American Revolution and the rise of the independent localism reflected in the New England town meeting, the Paris Commune and its popular assemblies in 1871, *Räterrepublik* of Bavaria at the end of World War I, the workers' popularly elected local soviets around the time of the Russian Revolution, independent collective governments in Spain at the outbreak of the Civil War in 1936, and the independent ayatollahs in Iran in 1978 after the fall of the Shah, among others.

Sale cites Hannah Arendt (1965: 260), who pointed to the fact that local councils and societies "make their appearance in every genuine revolution throughout the nineteenth and twentieth centuries." Largely ignored by politicians, historians, political theories, and even the revolutionaries themselves, recognition of these popular councils, as "spontaneous organs of the people," have never received enough attention. More than just moments in the throes of radical change, they were, in Arendt's words, "an entirely new form of government, with a new public space for freedom which was constituted and organized during the course of the revolution itself." For Sale (1980), these popular councils reflect the "human scale of politics" associated with a long tradition of decentralized governance.

Small groups are seen to possess the necessary characteristics of authentic democratic governance. Sale points to a long tradition of political thinkers who consider the small group to be the essential requirement of democratic rule. The cornerstone for this position was laid out by an Athenian, Aristotle (1998), who wrote that for citizens to effectively choose fellow citizens for office based on merit, they would need to know the characters and competencies of the candidates and that this could only happen in relatively small settings. Centuries later Rousseau (1968), living in a Swiss canton, would write that a basic principle of the legitimately ruled political system would require that all citizens be easily brought together to judge the affairs of the day. He thus concluded that the polity should not be larger than one small town. John Stuart Mill (2004), a theorist in the nineteenth-century British Empire, similarly thought that the only arrangement that can satisfactorily deal with all the issues of state is a small one in which all can participate. Toward this end, he pointed to the importance of the small town. Even James Madison ([1787] 2003), one of the framers of the American Constitution, conceded in Federal Paper No. 10 that a "pure democracy" could only take place in a community with a small number of people. For this reason, the new government of the United States was designed as a republic, rather than a democracy per se. And in much more recent times, the prominent theorist of American democracy,

Robert Dahl (1990), stated that for a political system to be legitimate all of its basic decisions would have to be made by a gathering of citizens which would have to be small in number.

In terms of real-world experiences we can take note of the fact that in the fifth century BC, ancient Greece, the cradle of democracy, estimated about 5,000 to be the number of people who could engage in fairly regular discussions and decision-making in the *agora*. As a form of direct democracy, the citizens of Athens and other city-states participated in making legislative and executive decisions. Although participation was only open to adult male citizens, the system provided an elaborate assembly for direct participation, usually involving about 1,000 citizens at a time. All male citizens over the age of twenty could join the assembly and it was considered a moral duty to participate. To ensure a form of equality among those entitled to participate as a political official of Greek democracy, a portion were selected by a lottery (Buchstein 2009).

Another familiar example is the Swiss canton which has long engaged in forms of direct democracy. Historically, each canton was divided into communes, with the communes being composed of sovereign villages (Sale 1980). The *Landesgeminde* of these villages typically covered a population of 2–3,000 citizens who basically governed themselves. Citizens could and still can demand a popular vote to change the constitution or basic laws, and veto spending bills or laws enacted by the legislature. In modern times Switzerland has had to modify this arrangement due to population growth. Instead of eight cantons, as was the case in the fifteenth century, there are now twenty-six cantons, with some of them having populations averaging about 30,000. But others are much smaller with high percentages of participation. Nonetheless, the cantons still generally run their affairs through regular cantonal meetings open to all citizens.

One can also add that the New England town meeting originally took place in towns of around 1,000 people. As Sale explains, the New England town meeting, an influential form that helped to inspire the American Revolution, is in the minds of many one of the last remnants of direct democracy. The basic impulse was captured in these words by Henry David Thoreau in 1854: "When in some obscure country town, the farmers came together to a special town-meeting, to express their opinion on some subject which is vexing the land, that, I think, is the true Congress, and the most representative one that is ever assembled in the United States."[5] Since these earlier days the influence of these town meetings has dwindled as the states and the Federal government have steadily taken over more decisions about local functions. But the meetings still decide the policies that govern various aspects of their respective communities.

Given the prestigious lineage of this argument, and the extensive literature devoted to it, we do not need to dwell on these local democratic practices. It is enough here to just indicate that direct participatory democracy, although by far the exception to modern governance practices, is still alive in different parts of the world. For Sale, however, there is another component that needs to be added to the decentralization movement, namely bioregionalism.

Decentralization as Bioregionalism

In his book on bioregionalism Sale (2000) presents the case for bioregions as the ecologically sound, culturally diverse response to modern-day economic and ecological crisis. As a way of both protecting the land and organizing small-scale decentralized communities, he presents a framework for ecologically sound, renewable communities based on natural ecological areas, physical areas defined by such things as watersheds, habitats, terrains, specific kinds of soil, and viable population divisions. More than just a utopian theory, bioregionalism has eco-centric roots in ancient tribal understandings that guided Native Americans for centuries, but it has also been supported by an important environmental movement, Planet Drum, and a former Secretary of the U.S. Interior Department, Stuart Udall. Further, the North American Bioregional Assemblies have organized regular meetings of bioregionalists since 1984 and are given credit for having played a significant role in the emergence of green political parties in the U.S. A number of bioregions have also established their own meetings, the Ozark Plateau bioregion being an example.

Essentially, bioregionalism is an ecological and cultural system of governance based on naturally defined areas, or bioregions. As a system of self-governance, bioregionalists stress the importance of local populations, community knowledge, and practices geared to the particular area. Examples of such areas in the United States would include regions such as the Ozarks cutting across parts of Arkansas and Missouri, or a stretch of land extending from Oregon through much of Washington State.[6]

The bioregionalist movement seeks harmony between the natural environment and human cultural patterns. For this reason, some bioregionalists align themselves with the principles of Deep Ecology. Whereas traditional environmentalists focus on industry as the enemy of the natural world, bioregionalists put their emphasis on humanity as part of nature and seek to build positive relationships with their ecological surroundings. More specifically, like all localists, they support local agriculture and the consumption of locally grown foods. They also emphasize the importance of cultivating indigenous plants and animals, and the use of local building materials.

But for the present discussion, no tenet is more important than the idea that political boundaries should correspond to ecological areas. Recognizing, for example, that birds do not pay attention to the political boundaries of

New York and New Jersey, such boundaries stand in the path of meaningful environmental regulation. This integration of political and ecological boundaries, moreover, provides an ecological framework for linking together different local towns and villages, which could also provide a basis for the establishment of communal federations.

For Sale (2000: 47–8), bioregionalism is both simple and difficult. It is simple "because all of the components *are there, unhidden,* right around us, right where we live; because we know that other people, ancient and in our terms perhaps unsophisticated, understood these things and lived for uncomplicated centuries by them." Thus to collect the kind of information needed for a bioregional society is fairly easy.

It is difficult because it clashes with our standard ways of seeing and understanding the world. Bioregionalism, as Sale (2000: 48) writes, "strikes most people at first as either too limiting and provincial, or quaintly nostalgic, or wide-eyed and utopian, or simply irrelevant—or all of these things." It is clear that significant changes in attitudes and beliefs would be necessary before people would be open to the idea of bioregionalism. Although this has proven difficult for environmentalists over the past half century, changing attitudes is in any case what environmentalism is about.

Sale (2000: 49) argues that "in the bioregional paradigm we have a goal, a philosophy, and a process by which to create a world which is not only *necessary* for the continuation of our species, but is also *desirable* and *possible.*" Toward this end, he offers four criteria for bioregionalism: scale, polity, economy, and society. Proper scale for him, as we have seen, is a decentralized human scale which is defined first by the relevant bioregional eco-system. Only through limited scale can citizens see their ecological problems concretely and understand their relationships to them directly. Where they "know one another and the essentials of the environment they share, where at least the most basic information for problem-solving is known or readily available... is where governance should begin" (Sale 2000: 95).

The bioregional polity focuses on the decentralization of political power, the values of village and communal life, and the goals of egalitarianism rather than hierarchical status in family and kinship relations. Bioregional politics would center around decentralized units and the spreading of power to other units throughout the region. The only things to be done at the higher levels are necessary tasks that cannot be accomplished from below. Authority flows "upward incrementally from the smallest political unit to the largest" (Sale 2000: 94).

A bioregional economy would seek first "to maintain rather than use up the natural world and to adapt to the environment rather than to try to exploit or manipulate it, to *conserve* the sources but also the relationships and systems of the natural world; and *second* to establish a stable means of production and exchange rather than one always in flux and dependent upon growth and constant consumption" (Sale 2000: 68–9). There is no fixed size for such communities but a reasonable estimate might be about 1,000 people.

As for society, he draws on the ecological idea of symbiosis. This, he argues, "is as apt a model as any for a successful human society," which we may take "as a place where families operate in neighborhoods, neighborhoods with communities, communities with cities and cities within regions, all on the basis of collaboration and exchange, cooperation and mutual benefit" (Sale 2000: 113). In this regard, the city is not altogether rejected. Indeed, following E. F. Schumacher, he sees the countryside and the city to be in need of one another.

Sale (2000: 167) finishes by arguing that bioregionalism is achievable "because the concepts are so basically accessible" and "it can aim at people beyond the usual constituencies for social change." He notes that there have been periods when regionalism has had a strong influence on public policy and that "the bioregional idea has the potential to join what are traditionally thought of as Right and Left... because it is built upon appeals to values that, at bottom, are shared by those who identify with the two tendencies." In common, they believe "in local control, self-reliance, town meeting democracy, community power and decentralization, all basic elements in what are thought of as the traditional American—at least Jeffersonian—values" (Sale 2000: 168).

At this juncture, we turn to the work of Murray Bookchin, arguably the most important political theorist of eco-localization. Sale and Bookchin were contemporaries and drew on each other's writing. From Bookchin we get a more detailed account of what the polity of a decentralized eco-communal society could look like.

Rediscovering Murray Bookchin: Social Ecology and Eco-Communalism

There is little that is easy about either Bookchin as a man or his work on social ecology. He was a feisty fellow and his writings were both critical and provocative; there was no shortage of controversies surrounding him.[7] This concern, however, does not undercut the fact that Bookchin put forth one of the first systematic theoretical responses to the ecological crisis that is still considered by many to be one of the brilliant contributions to eco-political thought, just as it was when it first appeared, in particular his theory of social ecology (Price 2012). Given the direct relevance of his political contributions, Gundersen (1995: 1) points astonishingly to the fact that Bookchin remains more or less unknown in the academic world. Indeed, as he puts it, Bookchin is "almost wholly ignored by political scientists, including political theorists."

None of the controversy around Bookchin should overshadow the fact that much of his work still speaks directly to the social and ecological situation in which we find ourselves. Not only did he call attention to the toxic realities associated with the "synthetic society" at the same time that Rachel Carson did, he offered a trenchant analysis of social and economic forces underlying these ecological dangers. Specifically, we discover a careful analysis of the environmental degradation resulting from the capitalist mode of accumulation and exploitation, an examination of the failures of Marxist theories and practices to adequately confront this problem, the historical emergence and evolution of hierarchy and domination as more fundamental categories than social class, the irrational limits of the modern urban way of life, and a radical political program centered around a form of communalism—"eco-communes"—designed for initiating and carrying through a socio-ecological transformation (Bookchin 1990; 1992). Indeed, Price has noted that his work has strong resonance with contemporary critics of globalization such as Hardt and Negri (2000) or capitalism by Harvey (2010). As he (2012: 28) writes, Harvey "in his most recent analysis of capitalism," puts forward "a programme of change which has a 'social-ecological' dimension to it that is strikingly similar to that of Bookchin's, in both and practice."

We focus here on his political program for a localist, communal alternative. It is important to note, however, that Bookchin's (1990: 75–94) political program is based on his detailed theory of social history, which is unfortunately beyond the scope of the current discussion. It is also a theory with numerous critics. Various writers have criticized this work for different reasons. Some find a lack of empirical evidence; others find internal contradictions (White 2003). Much of it deals with his focus on the emergence of social domination and reaches back to early times before evidence and documentation were available. Thus, these writings, as Bookchin says himself, are speculative in nature. After carefully examining the criticisms against the texts, in the credible view of Price, one can conclude that while the criticisms are not always wrong, none of them fundamentally negate the basic argument, especially as it pertains to the role of hierarchy and domination.

For Bookchin (1990: 41–74), the origins of the ecological crisis can be found in the rise of hierarchy and domination, which he posits in debates with Marxists as more fundamental than social class. Hierarchy is seen to first emerge in early "organic societies" to deal with the need to distribute a surplus. Intervening in and ultimately upsetting the balance between nature and early societies, it is a process that extends throughout history to modern times in the form of capitalism. The modern state is the manifestation of hierarchy, which together with capitalism is the source of the contemporary ecological crisis. To thus think that the modern state can be transformed into a green state is considered a misunderstanding of the nature and origins of liberal capitalist society. History and the rise of modern civilization, in Bookchin's view, involve a dialectical struggle between domination and struggles for freedom.

Bookchin's (1992: 34–41) work is deeply influenced by earlier efforts to establish a genuine democratic society, in particular Athenian democracy. But rather than idealize such early efforts, he only takes insights from them in an effort to set out principles that can guide a new democratic politics of the future. The starting point for Bookchin is a distinction between politics and the state. "Authentic politics," he argues, has been lost to the business of running the state, or "statecraft" as he calls it. Whereas genuine democracy is a public activity concerned with citizen empowerment in "the domain of authentic citizenship" that emerged in early Greek cities, the original political domain has been usurped by "professional politics" focused around the state (Bookchin 1992: 285). This politics has become the domain of politicians, political parties, pollsters, policy experts, and state administrators.

Much of the problem, then, in both mainstream and radical political theory is the conflation of politics and the state. Modern political theory, in his view, has emerged to legitimate the centralized state through the concept of representative government. It gives rise to a system where citizens have to vote for people they have never met, people who have mainly taken their interests from others (Bookchin 1992: 245–6). But he is even more critical of Marxists for the belief that the working class can capture the state and use it to reshape a just society from above, the disasters of the Soviet Union being the primary case in point. At the same time, however, he criticizes anti-Marxist anarchists for condemning any form of power. For Bookchin (2007: 109) power is unavoidable, but it has to be re-anchored to local communities from where it originated.

State usurpation of power is seen to be the appropriation of local power. As he explains it, the agents and institutions of the central state have not only turned politics into a mockery, but have at the same time degraded the citizens who take an essential role in the functioning of their own community. The state, as such, is a parasite on the community that drains its resources, both material and spiritual, and thus its ability to shape its own destiny. For Bookchin, then, the struggle for power between the central state and the authentic citizen, between statecraft and genuine politics, is basic to his political program. For him, the task has to be regaining the political realm of power for the re-empowerment of democratic communities. This is for Bookchin the best strategy for countering the central state and establishing new societal systems (Price 2012: 216). It is, in his scheme, the only way to turn back the socio-ecological crises of our age.

The political program that Bookchin sets out is radical by all accounts—even utopian for many. But he offers a practical strategy to achieve it based on the transformation of existing institutions. The basic idea is to root power in decentralized community organizations through the establishment of "popular assemblies." He seeks to restructure urban governing bodies by turning their institutional arrangements into "popular democratic assembles" organized around neighborhoods, villages, and towns. Citizens in such "popular assemblies" democratically engage in decision-making through direct face-to-face discussion and deliberation.

This is for Bookchin the way to return power to the community. Through such an institutionalization of community power, the capitalist exploitation of the citizenry can be undermined. Not only would the assemblies alter the political relation of the citizens to their fellow citizens, they would also change the nature of economic life. By reconstructing municipal relations, the aim is not to nationalize the economic system, the socialist solution, but to "municipalize the economy." The municipal assemblies do this by seeking to incorporate the productive processes into the life of the local municipality. The means of production is overseen by the local assemblies.

Decisions about economic production are then made in the interests of the community, as opposed to against them. Given that the communities better know their own interests, or are in a position to find them out, this works to reintegrate the relationship between ordinary life and work. All varieties of workers—from doctors and school-teachers to carpenters and factory workers—serve in the assembly not as members of occupations and social classes, but as citizens committed to the general interest of the community. In this regard, according to Bookchin, the citizen rather than the worker becomes the agent of transformation.

The citizens, as the transformational agents of change, have to be liberated from their subjective identities as workers and consumers concerned largely with their own private, personal issues and interests. Engagement in a revitalized municipal communal life should, in the process, become a school for the formation of enlightened and empowered citizens, with the popular assemblies serving as both policy decision-making bodies and venues for educating citizens in issues related to civic and regional matters, especially those that are political and ecological.

Beyond establishing these municipal assemblies, the final organizational step in his political program involves a "confederation" linking together these democratic municipal bodies (Bookchin 2015: 67–82). For Bookchin this is founded on the recognition that these communities can never be fully self-sufficient. This is the case both economically and politically. In economic terms there would still be a need for the exchange of goods and services. But even more important for the political program is the recognition that individual and isolated municipal communes would never be able to challenge state power alone. This was, as we saw in Chapter 7, a basic motivation behind the establishment of the federated structure in Nepal.

Given its size and power, the state would never be threatened by individual groups or isolated assemblies operating in its realm; it could easily carry on with its domination. This would not be to say that the state would not at points find having these bodies and their alternative ideologies irritating. But as long as they posed no immediate threats to state power, they might be tolerated. For this reason, in the interest of being able to challenge the power of the state, as well as protecting themselves from state incursions, there would have to be a larger confederal association of communes that enables the construction of workable alternative organizational structures, or what Bookchin refers to as a "dual form of power" that can confront the power of the state. This confederation of "municipal leagues" is what would provide the basis for an oppositional stance against the central authorities (Bookchin 2015: 78). Their municipal assemblies and joint confederal councils are the organs for obtaining control over their political and economic existence. For him, this is a fundamental dimension of the ongoing opposition to centralized state power.

The essential relationship between the municipal commune and the confederation of communes rests on a practical differentiation between policymaking and policy implementation. Policy, for Bookchin, has to always be made through face-to-face discussions in the communal assembly, whereas the confederal association at the next level is designated as an administrative realm designed to carry out decisions rendered by the popular assemblies. This might rest on an overly narrow understanding of administration, as well as perhaps a weak understanding of the evolution of associations. But we return to this in Chapter 11.

Finally, to anticipate questions about connections to mainstream political theory, some will surely ask about the relationship of the eco-commune, libertarian muncipalism in particular, and the political theory of communitarianism. There are indeed some similarities here; many activities of eco-communes can be described in terms of communitarianism. But there are also differences, especially when it comes to participatory environmental governance and Bookchin's theory of libertarian municipalism. For one thing, communitarians are generally not participatory democrats, although some are surely democrats. Moreover, few democratic theorists lean toward communitarianism. If we follow Bookchin, communitarianism is mainly an alternative institutional form of ownership, a nonprofit organizational space in civil society. "By communitarianism," he writes, "I refer to movements and ideologies that seek to transform society by creating so-called alternative economic and living situations such as food cooperatives, health centers, schools, printing workshops, community centers, neighborhood farms, 'squats,' unconventional lifestyles, and the like." Communitarianism, in his view, "is often interchangeable with the word *cooperative*, a form of production and exchange that is attractive because the work is not only amiably collective but worker-controlled or worker-managed" (Bookchin 2000). Many activities that take place within eco-communes, ecovillages, and Transition Towns engage in cooperative forms of enterprise—farmers' markets and community center cooperatives, for example.

But for Bookchin the crucial difference has to do with politics and power. Communitarianism is "not a politics but a practice" whose members generally choose to work in or purchase from a cooperative enterprise. Because they try to operate and survive in a capitalist society, as he explains, many fail over time. "While working and/or living in cooperatives may be desirable in order to imbue individuals with collectivist values and concerns," as he writes, they do not "provide institutional means for acquiring collective power" needed to confront capitalism and the capitalist state. Rarely do they emerge as "centers of power," especially as they do not focus on the issues of power and partly do not have a way of mobilizing a larger group of citizens around an alternative vision of society. In contrast, his theory of libertarian municipalism, as he explains it, "is decidedly a confrontational form of face-to-face democratic, antistatist politics." Reaching outwards to both the municipality and beyond, it is fundamentally concerned with the essential question of power, who should hold it, and how it should be exercised (Bookchin 2000).

Acknowledging Rudolf Bahro: From the Logic of Emergency to Cultural Transformation?

We turn at this point to the communalism of the German political and social theorist, Rudolf Bahro, in particular his thought from the mid-1980s onwards. Bahro's career was also controversial, starting with the publication of his much-acclaimed book, *Die Alternative* (1981), which landed him in an East German prison. One can argue that he does not belong in a book in the pursuit of democratic values, certainly not as we usually think of them. But there are several reasons for including him here. One has to do with his emphasis on communal structures, including ecovillages. Another is a recognition of the need for a fundamental eco-cultural transformation as the basis of genuine self-government, and yet another for his recognition of the likelihood of eco-authoritarianism. Even when rejected as part of Bahro's prescription in the face of extreme ecological emergency, it can easily be seen as a prediction, in numerous ways aligned with the worrisome concerns put forward at the outset of this book.

To begin, there has been a rigorous and often vitriolic debate in the literature between Bahro's supporters and protagonists regarding the political nature of his contribution (Stein 1998: 201–27; Weber 2015: 229–300). For some he was an "ecofascist" or eco-authoritarian (Biehl 1995). This is especially the case for members of the German political left, who have accused Bahro of forms of authoritarianism, particularly as it relates to his earlier writings. Bahro did not deny all of the criticisms, but one can argue that his work is in many ways more complex than is recognized or accepted. Moreover, he was often misinterpreted, perhaps sometimes intentionally. The argument that he was authoritarian needs, in any case, to be understood in a particular political context relevant to major ecological destruction and the resulting crisis of civilization that he saw on the way. In this light, his alternative view offers a number of ideas about what authentic communal life requires, what could inadvertently go wrong with the communal solution, and a number of prescriptions that might well serve as predictions. In so far as this book seeks to look ahead to the politics of crisis, many of Bahro's writings—whether viewed positively or negatively—present important issues that merit reflection and discussion.

While Bahro was not a democratic theorist, he was surely political. In fact, he was one of the founders of the West German Green Party (Bahro 1984). For him, the problem with participatory democracy is that it is typically called for as a strategy within the structures of the techno-industrial "megamachine," a term he took from Lewis Mumford. In this view, it is not possible to reform our way out of the coming crisis; radical change is required instead. Thus participatory democracy, as such, does not for him rise to the occasion. In his view, we require first a deeper consciousness before a genuine transformation of the megamachine is possible, one that permits us to avoid the apocalypse which stands before us.

Bahro was deeply antagonistic to the megamachine and his works, I would argue, should be read in this light. He, like Sale and Bookchin, saw modern techno-industrial society heading for an ecological disaster of unprecedented proportions, a dysfunctional form of self-destruction with few parallels in history. But his language is even harsher. Apocryphal modernism is presented as the dead end of civilization. In this view, we confront a dire emergency that calls for extraordinary measures.

At the root of much of the dissension over Bahro's thought is his complete rejection of modernism, not only in terms of its destructive dynamics, but also in terms of its basic values and principles. Although he was not opposed to the enlightenment value of reason, he objected vehemently to its cooptation as an instrument of the megamachine. Like German theorists such as Adorno and Marcuse, he was strongly opposed to any project that tries to revive values that lend themselves to instrumental manipulation by the system of domination.

Like radical environmentalists generally, Bahro takes the modernist idea of elevating the individual over the community, as well as an understanding of the improvement of the individual's quality of life in quantitative terms, to lead to an assault on nature. The core idea running throughout his work is that the basic assumptions of modern society are not only geared to taming nature for human purposes—the assumptions of modern science and technology in particular—but that they are also in direct contradiction with the emergence of a new social formation that would allow humankind to live in harmony with the natural world. This anti-modernism, however, is not a primitive form of regression, a turn back to an earlier age. He argues instead that we have to use our reason to rediscover where we came from and to start out again on a new and more promising path, while remembering our earlier mistakes. Although this requires setting aside the modern understanding and practice of modern science and technology, the central place of individuality, and the emphasis on material progress, we can rediscover them on the road to transition, this time radically reconstructed in ways that sustain life on the planet. In a new form they can help people liberate themselves from the megamachine culture and its abstracted understandings of reason, nature, and work.

His project is thus not a revolt per se against reason and rationality as a human capacity. These intellectual faculties are needed to achieve a better awareness of the ways we are involved in both the logic of our own self-annihilation and what he called the logic of rescue (Bahro 1987). But to do this we have to overcome the one-sided instrumental thinking of economic and the technical rationality which has not only been diverted to support the reproduction of the megamachine, but also split off and become alienated from the rest of the interpersonal subjective realm.

For Bahro the commune has to be the foundation of the social alternative. Much like Bookchin, he argues that the eco-commune is the "germ-cell" of a new social form that will replace the existing bureaucratic state (Bahro 1994). Toward this end, we have to replace national and global projects with small-scale and self-reliant groups that can provide for basic needs of food, shelter, clothing, education, and health. This eco-communitarian form, and a confederal system of communes to which it gives rise, is conceived as a life-sustaining and nonhierarchical egalitarian culture in which our social, communicative, and spiritual needs can be satisfied and in which people can live in peace and harmony with both nature and themselves (Hart and Melle 1998). In short, he proposes a communitarian subsistence economy founded on deeper spiritual values.

The alternative movement, in this way, has to be more than a political project. Instead, it must involve a cultural revolutionary movement guided spiritually by a non-anthropocentric conception of the self, a new image of society, common agency, mutuality, and cooperative sharing (Frankel 1987). It is a spirituality that, like Deep Ecology, celebrates close bonds between non-human and human realms. As such, it is an orientation shared by many members of eco-communes and ecovillages, one of which Bahro started himself.

What we require, in his view, is a new cultural reconfiguration of our fundamental subjective faculties as the basis for a sustainable way of living. This new cultural level has to be grounded in a higher level of consciousness. As a precondition for an exodus from the industrial megamachine, this higher level of consciousness has to emerge among a critical mass of people. Like many transformational environmental thinkers, Bahro's goal is to promote self-exploration and self-determination (although in ways that some have likened to "New Age thinking"). The first step is thus to confront a psychological barrier. The psychological realm is conceptualized as the construction site for the alternative culture based on a new consciousness. This "ecotopian" cultural revolution presupposes an "anthropological revolution" that can lead to a radical new form of human being. This necessarily involves the socio-psychological displacement of the bourgeois form and ideals of the autonomous individual, especially the competitive ego in search of control and material comfort. The goal is to dismantle and replace the competitive modern techno-industrial mentality with a more egoless mode of existence based on personal responsibility. As Hart and Mello put it, "such a lofty project is not to be dismissed by disdainful references to the New Age theorists; it has much in common with the major Western religions and philosophical traditions when they have focused on the question of the most perfect or divine life; it also, of course, is at the center of forms of Taoism, Buddhism and Hinduism." Such thinking, in fact, has long been associated with important segments of the environmental movement and is not uncommon among members of ecovillages.

This New Age form of spirituality is one of the things that the political left has vehemently disliked about Bahro. Bahro acknowledges himself that such New Age ideas related to meditation and positive thinking will not save us. He also recognizes that this form of thinking is often co-opted by the commercialized subculture of the megamachine. But the fact that it is distorted by commercial interests does not undercut its central importance for meditation and reflection. It only needs to be reclaimed and put to the task of authentic personal and social transformation.

If German leftists have been unhappy with Bahro's spirituality, his political theory draws even more fire. Given the deep roots of the political-economic system that has to be uprooted, he rejected other radical strategies that include elements of electoral politics and the struggle for power. This he based on his experience with the Green Party. Entering electoral politics, as he foresaw, led to a weakening of the party's ecological agenda; the Green Party would gradually become more and more like other German parties in pursuit of electoral success, as it in fact did. He thus rejected the leftist pursuit of capturing power and called for a more fundamental transformation of consciousness, a major point of contention with Bookchin. This would not preclude taking over the system at some point, but it was seen rather as a precondition for the socio-ecological society that would replace the existing one. Without it, he argued, a takeover would only go astray. Ultimately, it would end up reproducing the same society in a somewhat altered form, if not create new problems altogether.

For this new cultural vision to take hold, it needs to be widely shared throughout society. Thus, Bahro was interested and willing to speak—or perhaps proselytize—to all social and political groups. This included a willingness to engage with conservative thinkers and politicians in an effort to find common ground that might facilitate changes in both ecological attitudes and cultural consciousness. Talking to the conservatives is another point that did not endear him to the German Left.

By comparison, Bahro's political program is dramatically different to those of both Bookchin and Sale. To guide his proposed system he offers two forms of governance: one he called the emergency government ("Notstandregierung") and a government of rescue ("Rettungsregierung"). This latter logic of rescue spells out the basic goal of cultural transformation as the foundation of the new ecologically sustainable, egalitarian civilization. Because this is founded on a new consciousness, he recognized that it is unlikely to arrive in time to avert the crisis. Thus, the logic of the emergency government is set forward in the name of survival, largely as an effort to buy time for the deeper cultural and personal transformation to evolve and take hold. The emergency government is, in short, designed as a defensive strategy to deal with the ecological crisis when it reaches dire proportions, such as the kind of anarchy surrounding Raskin's fortress world. With broad support in the name of human survival, this government would have negative powers—authoritative, even coercive powers—to put painful but necessary restrictions on the unsustainable activities of both businesses and citizens. That is, this government of emergency takes the reins of society to deal with the consequences of a full-fledged catastrophe.

For Bahro this emergency task ahead is far beyond the capability of parliamentary government, which he sees as an instrument of the economic megamachine responsible for the catastrophe. He is also deeply critical of Bookchin's call for electoral participation in municipal affairs. Indeed, he engaged in a vigorous debate with Bookchin about the role of electoral politics and power, both in print and as an invited guest at Bahro's seminar on social ecology at Humboldt University in Berlin, an exchange which turned and remained unfriendly.

Bahro sees the reliance on elections as a trap; in his view they only end up supporting the existing arrangements. He does not, however, call for an end of existing institutions. Parliaments along with executive bodies would continue to function, although they would now be placed under the oversight of an "Ecological Council." Made up of qualified delegates, scientists, and ecologists in particular, this council would have the responsibility of representing and protecting the eco-system, including all human, animal, and plant life, as well as minerals and other scarce resources. It would have the power to set the limits and boundaries within which the other executive and legislative institutions would function and have the right to veto policy decisions by political bodies that threatened the ecological system. The council would, moreover, make such decisions based on consensus rather than bargaining or voting. How this would work is never clarified; it would surely pose a heavy political burden on the governing process, especially given the many conflicts that will inevitably arise during a time of extreme turmoil (Stein 1998: 201–27).

Indeed, much of Bahro's political scheme is vague and has given rise to hefty criticisms from other political and social thinkers. Many have referred to this as a model for eco-dictatorship. While one can easily understand these criticisms, it is important to keep in mind that Bahro introduces this as an emergency government to deal with extreme tragedy. It is advanced in the name of nothing less than survival on the planet. In this regard, it is not entirely unrelated to what other leading survivalist thinkers such as Lovelock seem to call for, not to mention Shearson and Smith who would substitute a technocratic elite for democracy. For Bahro the effectiveness of this interim form will depend on how deeply anchored and widespread a new ecologically oriented spiritual vision is in the hearts and minds of the people, how deeply committed they are to ecological rescue as a step toward a new consciousness. But a key question here is whether such an eco-authoritarian system will facilitate the next step. Experiences with dictatorships like the ones that followed the Russian or Chinese Revolutions lead one to a fair amount of skepticism.

Nothing in Bahro's work has contributed to more heated controversy than this dictatorial ecological council (which he also described as the "salvation council"). Indeed, in Germany it remains today a topic of dissertations and books. For him the only way to penetrate and destroy the deeply rooted institutional structures of the unsustainable political economy is through strong leadership. In fact, Bahro argues that this kind of transformation will require a form of charismatic leadership that can lead people from the emergency to the cultural transformation of rescue. According to his understanding of historical change, times of great transformation have always depended on the presence of charismatic transformational leaders who personify the new cultural vision. Given Germany's Nazi history, such an argument is not surprisingly denounced by the German Left. But to make his case, Bahro points beyond Hitler and Mao Tse-tung to Gandhi and Mandela. Insofar as the members of the alternative society have a new cultural consciousness, based on mature and liberated selves with high moral character living in egalitarian communes, such as an ecovillage, he believes the worry about dictatorship can be set aside. Given this optimistic level of high character development, it is not difficult to understand the deep concerns that this theory has raised.

At this point, we can conclude the discussion of Bahro by acknowledging that he speaks to the crisis ahead of us, even if in very controversial ways. Toward this end, he serves not only as an intellectual foil, but also as a source for critical reflection. We thus turn in the next and final chapter of the book to a more specific assessment of the political theory of such eco-localist approaches.

Conclusion

Following the previous discussion of Transition Towns and ecovillages, we began this chapter by presenting the relocalization movement as a vibrant practical reality that should be supported and extended, underscoring in particular its emphasis on renewable energy and sustainable agriculture. The discussion also took up the relation of relocalization to globalization; we saw that it is not a choice of either/or, as it is too often posed. The call is rather for a return to the local on its own terms without ignoring the fact that the fight against global environmental degradation remains as important as ever. Indeed, as Hopkins and many others have pointed out, the struggle for sustainability depends as much on local action as it does on other higher levels of government.

The chapter then turned to three theorists whose work in various ways presaged the contemporary relocalization project. The writings of Sale, Bookchin, and Bahro were seen to offer the contemporary movement strong theoretical undergirding, both socio-ecological and political. The earlier environmental contributions of these three environmentalists speak directly to the political-ecology of relocalization and the climate crisis more generally. Although lost during decades focused on environmental globalism, the sort of decentralized governance advocated by environmental localists has a long history of its own and is well worth recovering. The importance of ecological consciousness, ideas about decentralization and bioregionalism, and the practices of eco-communes, we have argued, can—even will—become much more salient during a period of severe ecological stress in the not-all-that-distant future. In addition to offering both a deep critique of the current environmental crisis and an alternative approach that should be part of the solution, this return to the local is good for democratic life more generally. The local face-to-face level is the fount of authentic participatory democracy even in the best of times.

In Chapter 11 we seek to connect these themes to academic political theory. We have argued that environmental political theory should take a renewed interest in these ideas and activities, given their relevance to both political thought generally and environmental political thought specifically. While these theorists—let alone relocalization activists—have not been formal political theorists, the ideas they put forward are not as strange to mainstream democratic political theory as they might at first seem. Even if they are not the dominant themes, they have significant reference points in the field. The task then in the next and final chapter is to draw out these connections—bridges between the two literatures—while offering a discussion of the argument as a whole.

Notes

1. See www.relocalize.net.
2. See http://www.postcarbon.org/relocalize.
3. La Via Campesina is an "international movement which brings together millions of peasants, small and medium-size farmers, landless people, women farmers, indigenous people, migrants and agricultural workers from around the world." It depends on decentralized "small-scale sustainable agriculture as a way to promote social justice and dignity." As a grassroots movement, "it strongly opposes corporate driven agriculture and transnational companies that are destroying people and nature." Representing roughly 200 million farmers, it is made up of more than 160 organizations in more than seventy countries from the Americas, Africa, Europe, and Asia. Established in 1993, "it is an autonomous, pluralist and multicultural movement, independent from any political, economic or other type of affiliation" that opposes agribusiness and the globalization of agriculture policy. Widely recognized by other local and global social movements, La Via Campesina is a "main actor in the food and agricultural debates" by international organizations such the Food and Agriculture Organization (FAO) and the UN Human Rights Council. http://viacampesina.org/en/index.php/organisation-mainmenu-44/what-is-la-via-campesina-mainmenu-45.
4. One could differ on the order of the presentation of these theorists. We begin with Sale, as he establishes the background for much of what follows in the discussion of Bookchin. Some might argue that the section could begin with Bahro, moving from the most abstract and theoretical to the more concrete. But Bahro is not an altogether easy fit here. When it comes to participatory democracy, his arguments can be out of place, given his call for a transitory period based on what many have

described as a form of eco-authoritarianism. But his emphasis on challenging the megamachine, the fundamental importance of political and social consciousness for a genuine participatory collective, the eco-commune as the germ of a new society, and a need for a transitory ecological council raise important considerations—directly and indirectly—that underscore the difficulty of the challenge and are thus worth taking up.

5. See http://www.transcendentalists.com/slavery_in_ma.htm accessed 4.13.215.
6. See http://www.encyclopediaofarkansas.net/encyclopedia/entry-detail.aspx?entryID=7680.
7. Bookchin was himself responsible for some of this controversy, as he did manage to make intellectual enemies. Much of the interest or value of Bookchin's work was lost as a result of two sets of polemical exchanges, first with the Deep Ecologists and second with the anarchists of the 1960s and 1970s, or those he referred to as "lifestyle anarchists." In both cases the exchanges were at times rather ugly, turning more toward a crude caricature of Bookchin's character and his motives, portraying him as a dogmatic authoritarian who sought to dominate radical politics for personal motivations. Bookchin's often aggressive polemical style, it can be said, was in part responsible for these conflicts. Price (2012) has illustrated how these developments, evolving over two decades, led unfortunately to an overshadowing and distortion of his theoretical work. As the unflattering images took hold, other scholars turned away from him, the result of which was a fading of his ideas into relative obscurity. He was often written off as a cantankerous man with a self-serving agenda, whose work was said to be filled with contradictions and errors.

11

Sustaining Democracy in Hard Times: Participatory Theory for Local Environmental Governance

In this final chapter we explore ideas taken up in the previous chapters and relate them to political theory. Up to this point, these ideas have been presented as theoretical contributions to both participatory governance and the relocalization movement. As we have already noted in Chapters 6 to 8, participatory governance bears an inherent relationship to mainstream political theory, in so far as it is a form of deliberative participatory politics with connections to the theory of deliberative democracy. We argued, however, that participatory governance, geared to citizen empowerment, is more politically grounded in real-world political struggles. As a mode of participatory democracy in civil society, often related to oppositional social movements, participatory governance includes elements from deliberative democratic theory, but is also a practical innovation in deliberative governance. It is distinguished from other perspectives on deliberative democracy by the ways it has moved beyond advisory consultation to shaping or determining policy outcomes. We also examined how the theory and practices of participatory governance have been applied to environmental policymaking, offering a number of specific case illustrations. Briefly we presented the examples of participatory budgeting in Brazil and people's planning in Kerala, India before presenting two larger case studies, one concerned with the community forest movement in Nepal and the other with the ecovillage movement.

In this chapter, we seek to extend the theoretical perspective more specifically to a number of important but relatively neglected traditions in democratic political theory, especially as they relate to ideas taken from Bookchin and Sale. In particular, these involve the theories of associative democracy, insurgent democratic politics, and participatory or democratic expertise. These orientations are provided as steps in search of a broader environmental political theory that addresses the democratic struggles that can be anticipated during the socio-ecological crisis ahead.

With regard to Sale, Bookchin, and Bahro, none of them were formal academic political theorists, though all three were engaged in advancing political thought. One could also say the same thing for the ecovillagers who theorize about participatory decision-making based on practical experiences. But the former are not part of standard university courses on political theory and the practices of the latter are not topics on the conventional scholarly research agenda. There are, however, bridges that can connect such eco-local ideas to formal academic political theory, including environmental political theory. The task in this concluding chapter is to draw out and underscore these connections. Associative democracy, as we shall see, is related to political organization and the other two, insurgent politics and participatory expertise, pertain to political struggle. While this is not the place to engage in elaborate discussions of these linkages, it is important to locate the radical localist theories of Sale, Bookchin, and Bahro in the literatures of political theory and participatory governance, with the hope that others will pick up these threads and develop them more fully.

In what follows we first point to similarities between the theory of associative democracy and a confederation of eco-communes. We then introduce the concept of insurgent politics or insurgent democracy as an alternative way to think about democratic politics, an orientation closely related to the politics of libertarian municipalism and associative democracy. Finally, we present the theory and practice of participatory expertise as both a democratic and "liberatory" challenge to technocratic politics and as an approach to the development of appropriate environmental technologies. While these themes intersect with one another, they are at the same time distinct. For this reason, we divide this chapter more formally into three separate sections before returning to a general discussion of common themes.

Eco-Communes and Associative Environmental Democracy

In the search for an alternative model in political theory that resonates with this emphasis on eco-localism, we can turn to the theory of associative democracy. Although Bookchin's writings do not generally fit into anyone else's categories, his theory of libertarian municipalism can without too much difficulty be seen as a form of—even in a sense a forerunner of—modern theories of associative democracy. Bookchin's libertarian municipalism can be understood as a radical form of associative democracy applied to the environmental challenge—that is, a form of associative environmental democracy. Bookchin did not rely on academic political theory as such, but one can plausibly argue that when it comes to rescuing or preserving democracy during an ecological catastrophe ahead, political theory would do well to take on board his unique theory of participatory democracy.

Associative democracy has never been a mainstream orientation in political theory, but it has been advanced by important political theorists. Moreover, Achterberg and Morrison have suggested associative democracy as an approach to environmental democracy, although neither have worked out a conception of associative environmental democracy. In an early volume on democracy and the environment Achterberg (1996) pointed to the potential relevance of associative democracy for sustainability. Short of that, though, he did not carry through by empirically or analytically connecting these two concerns. While Bookchin is not mentioned in Achterberg's discussion of associative democracy, his work points to the usefulness of the potential connection. Morrison (1995: 140) similarly makes this connection, but he also briefly mentions Bookchin and Bahro. As for Bookchin himself, while he does not employ the terminology of associative democracy per se, the central concepts and theoretical delineations in his theory of libertarian municipalism come close to capturing its concerns. This theory constitutes a radical understanding of associative democracy, which comes as no surprise given the early influence of radical anarchist theorists on Bookchin's thought, Proudhon in particular.

While this is not the place to engage in a full-scale delineation of associative democracy, it is possible to point to the conceptual understandings that underpin this useful linkage. The first has to do with the fundamental role of associations as the basic units of political activity, taken to refer to groups of people with similar views, goals, and concerns. In general, as de Tocqueville famously argued, associations are a basic element of democratic activity, liberal as well as participatory. Modern democratic theory takes voluntary associations to be the political foundation of its emphasis on pluralism. Not only do such associations represent and express the diverse interests and concerns in the public realm, but the multiplication of such groups serves as a guard against the emergence of tyrannical government.

For associative democracy the emphasis is on a detachment from both the state and the market. As basic units of civil society—whether neighborhood groups, social clubs, arts and crafts associations, political interest groups, professional societies, or transition movements—associations can be seen as devoted to practices that have an internal good (O'Neill 1993; Wright 1995). Through the activities of such associations citizens learn to see their well-being in ways that are independent of money, power, or status. Moreover, engagement in such activities is passed along from one generation to the next. Engaged in a multigenerational experience, citizens can come to recognize the meaning of their own lives in terms of an ongoing social tradition. In the process, as O'Neill explains, they can learn to appreciate the difference between acquiring individual possessions and contributing to the public good. Such learning is a precondition basic to Bookchin's radical transformation of modern capitalist society.

In the theory of associative democracy this role of associations is expanded to play an even more central role. Although there are differences among theorists, all of the theories of associative democracy are presented as alternatives to hierarchies and markets in that they promote cooperation among citizens. Structurally, the emphasis is on political decentralization, both territorially and functionally. Spread across these organizational divisions are a large number of mutually associated, overlapping groups that reflect a wide range of shared interests. These groups can freely organize as long as their activities are compatible with the rights of other associations. This is seen to be the basis for the advancement and support of shared interests, cultural as well as economic, and as a way of determining and implementing both basic goals and practical policies. Following the principle of subsidiarity, public tasks are delegated to the lowest administrative levels compatible with the effective conduct of those tasks. In the communal conception of associational life put forward by Bookchin, these local bodies are the essential agents for societal transformation.

Associative democracy is scarcely a new idea. Despite the fact that the theory has never managed to achieve prominence outside of a relatively small group of radical political thinkers, it has a rather long pedigree. Its neglect surely has to with the radical, transformative implications of the theory for the modern state which it exists to challenge. In fact, it builds on a combination of anarchist ideas in nineteenth-century political thought and Guild Socialism in the UK in the 1920s as developed by C. D. H. Cole (Cole 1920; Hirst 1989; Hirst 1994). In more contemporary discussions some approaches lean toward bolstering representative democratic government, helping to rebalance the increasing centralization of the state, reducing the need for complicated top-down forms of social and economic regulation, and countering the dominance of corporate business (Westall 2011). Working within the normative framework of existing political systems, such associations cooperatively compete with one another in ways that inform and facilitate the pursuit of their objectives and goals. By doing this, they supply—directly and indirectly—the social system with additional methods of political representation (individuals being represented by associations), as well as with an alternative means of policy regulation and implementation. In Streeck's words, it is about "socially responsible self-governance of functional groups" (1995: 188).

For some writers, such as Cohen and Rogers (1992), associations are mainly viewed as a useful support of representative government. Although created in civil society rather than by governments, they see such associations as secondary groups helping to facilitate and support many of the tasks of liberal representative democracy. Hirst (1994) and Martell (1992), on the other hand, advance associative democracy as a way to revitalize more social democratic traditions. That is, whereas Cohen and Rogers see associations as secondary organizations, Hirst turns the order around and views them as the primary associations. Hirst preserves strong economic aspects of the guild ideology, but he broadens the focus toward applications within modern political systems. In his theory associations are spontaneously arising voluntary groups that democratically govern themselves (Hirst 1994: 19). Associations are built up, or spontaneously emerge, "from below," with the central state having minimal power over their existence and structure.

Hirst believes that the functioning of such groups would be a solution for the shortcomings of both the liberal capitalist economy and majoritarian parliamentary representative democracy; he considers associative democracy to be a "vital supplement"—rather than substitute—for the latter (Hirst 1994; 1997). In fact, Hirst states that associative democracy can strengthen representative institutions and thus representative democracy because freeing the central institutions from an overload of activities allows them to concentrate better on their main functions, to supply society with basic laws to govern social action, to monitor the provision and administration of public services, to hold political officials accountable for their actions, and to protect the interests and rights of the citizenry. In fact, for Hirst the relations between authorities and such associations should take place on the local or regional level, with (where possible) a role for the national state level (Hirst 1994: 39).

Yet others such as Martell (1992) and Mathews (1989) seek to extend the guild socialism of C. D. H. Cole (1920) to promote a more radical conception of cooperative decentralization that stresses mutuality and self-organization. In this latter category we would find Bookchin, influenced by his early interest in anarchist theories of mutuality and autonomous self-regulation. For Bookchin, as well as Sale and Bahro, the basic association takes the form of the commune.

Associative democracy, according to Westall (2011), "envisions decision-making and collaboration between different groupings, whether within the economy or society (for example, formal networks of multi-stakeholders, cross-sectoral partnerships, etc.)." As a challenge to today's structurally biased pluralism, it combines "a need for collective agreements and solidarity with the recognition of and appropriate responses to diversity and difference" (Westall 2011). Spreading decision-making widely within and across economic and social organizations, associative democracy is seen to lead to better informed and more effective decisions based on higher degrees of legitimacy.

For Hirst, this is the result of greater information flows, greater understanding of differing viewpoints, higher levels of accountability, trust, increased social inclusion, individual freedom, and empowerment.

Policy goals and other social and political activities emerge from negotiations among the relevant associations. In the more state-oriented theories, coordination takes place between decentralized structures, markets, and central political coordination through the usual channels of representative democracy. Much of the literature is not very specific on how these negotiations would take place. But, as we have seen, this is less the case with Bookchin. He put forward his confederation of associations as the collective political organ responsible for coordinating the goals of the communes. It is a structure rather similar to the associated network of local community users in Nepal discussed in Chapter 7. These structures need to be understood as a full-scale challenge to, if not rejection of, representative government.

Compared to most of the literature on associative democracy, Bookchin's (1986; 1987; 1992) theory much more radically rejects the state in all of its forms. His writings specify a radical form of participatory associative democracy, described as a "revolutionary project" that rejects representative democracy as a contradiction in terms. Bookchin's approach to decentralization, as we saw, focuses not on balancing central and local power, but rather on altogether taking power back from the central government. Based on his socio-historical analysis, the state is considered a parasite on community life, emptying the community of its resources, both material and spiritual, and systematically undercutting its powers and legitimacy to determine its own destiny (Bookchin 1987). In short, Bookchin rejects the state structures that Cohen, Rogers, and Hirst are trying to make more democratic. For him this is not an option for constructing a genuine democratic society. In effect, whereas they seek to make associative democracy relevant for contemporary representative government, Bookchin refuses to accept representative government. For many, this has rendered Bookchin beyond the realm of the realistic. That, however, is not necessarily a problem for the present discussion, as we have speculated on a period in which the state has either turned away from democracy or has perhaps collapsed. The argument here is that Bookchin's libertarian municipalism, as a form of associative democracy, can be understood as a radical transformative theory for democratic political struggles in the face of increased authoritarian politics from the center.

Bookchin's concept of an associational confederation, however, like the other coordinating mechanisms advanced by associative democratic theorists, tends to suffers from a potentially serious weakness. As we saw in Chapter 10, conceptualization of a central committee that would only be an administrative body concerned with implementation seems both worrisome and unlikely.

First, it fails to recognize that central administration is already a form of political activity, one that often hides behind the pretense of neutrality. Moreover, it is unclear that this body, as an administrative unit, would be capable of guiding action to protect the members of the confederation from incursions from the outside (or perhaps dissension or revolt from within). Bookchin, like associative democratic theorists in general, tends to somewhat neglect this problem as he is focused more on fears of losing local democracy to a system of central representation. Indeed, this is a classical dilemma in the Rousseauian tradition and Bookchin has reason to worry about it. It is always a risk, but the risk would be lower in egalitarian systems than in systems based on a spectrum of inequalities, a problem that plagues less radical theories of associational democracy. Some theorists have sought to confront this with a call for a guaranteed minimum income (Hirst 1994: 180). For Bookchin, the problem can be compensated for by the fact that these communities would be small enough to permit local citizens to maintain a fairly close connection to their representatives and vice versa. Is this the solution? This is hard to answer. We can only conclude by saying that Bookchin runs up against a classic problem in political theory. But he is not alone; it is a question that still needs further attention by political thinkers.

Finally, with regard to the confederal structure basic to libertarian municipalism, we can see how Sale's theory of decentralized bioregionalism supplies a territorial foundation for Bookchin's confederated eco-communes. Within bioregions there could be eco-communes of various sorts linked together in one fashion or another. All would have their distinctive features and would be contributing to the overall experiment in sustainability. Following Bookchin, these eco-communes might well form confederated structures to both communicate and coordinate activities with one another and to supply political support, even protection against an unfriendly state. The case of the association of forest users in Nepal provides an example of what such a relationship might look like. Protecting themselves from the encroachments of the state and changing state forestry policies were the very reasons the association of small communities across the country was established.

Some mainstream scholars, it is worth noting, are beginning to recognize that regionalism offers a better political-ecological strategy than both nation-states and globalization (Majone 2014). There is, in fact, nothing to stop inventive communities from introducing and experimenting with such regions and some, as we saw in Chapter 10, already have. Indeed, we can expect people under climate stress to become increasingly inventive. And it is just here that regional associations of eco-communes, federated or confederal, could become a practical option. One could also easily image eco-neighborhoods or districts in cities becoming part of such an association, which would build connections between urban and rural areas.

Insurgent Politics and Democratic Struggles in Civil Society

Associative democracy, then, is a civil society effort at self-governance. As a rejection of the state—the dominant role of the state in any case—it relates to the emphasis in green thought on the essential importance of civil society as the fount of creativity and innovation. An argument presented in Chapter 5, this theory is based on strong reservations if not rejection of the central state, including the possibility of the green state. It is true, as Eckersley and Barry have argued, that the state is where the power is. At the same time, however, the pace of greening is not very convincing, especially when measured against the time frame of the evolving climate crisis. Maybe in the best of worlds, there might be a green shift in state politics. But we no longer have much time to wait. We could say that we are thus left with civil society. In acknowledging this, though, we should not take it to be an unfortunate lesser strategy. Civil society, as we have seen, was the source of the environmental movement and it remains a primary source of creativity, as participatory environmental governance, the Transition Town movement, and the ecovillage network have shown. In this regard, the theories of libertarian municipalism and associative democracy recognize the limited possibilities—both ecological and democratic—posed by a reliance on the central state and have thus sought to build parallel structures for political-ecological renewal.

It is quite likely, as we have already argued, that the state will not appreciate the emergence of such alternative structures, at least over time, especially those that would take the state to be the problem, if not the enemy. For this reason, it is also important to consider the opposition to and incursions against such alternative structures on the part of the state, especially authoritarian states. Under some circumstances—particularly those posed by centralized authoritarian systems of governance—politicians at the center may not relish having alternative participatory systems of governance within their boundaries, even if on the margins. Various experiences in Latin America and Asia bear this out. In the case of China, for example, political officials closed down a thriving ecovillage with no explanation. Having received a very large numbers of visitors from other parts of the world interested in seeing what was considered an impressive ecovillage project, also supported by the global ecovillage network, it was viewed as a source of irritation, even possibly as a subversive activity.[1]

In the future, states with strong militaries can quickly see such parallel structures as threats to their legitimacy. Their response could take various forms, from pre-emptory political strategies to military action. There is no way to know about this in advance, but we can assume that eco-communes may need to take this into consideration. It was, in fact, a situation that the community forest movement in Nepal confronted. As small communities were unable to fend off the incursions of the central state, they formed a federated network to give them power in numbers. Bookchin and Bahro both deal with this point. For them the unsustainable state is the enemy, from which such projects need to protect themselves. Toward this end, the concepts of insurgent politics and insurgent democracy speak directly to the challenge.

Plausibly, the theory and practice of insurgent democracy offer a deeper source of insight and strategy regarding many of the central political challenges ahead. A concept that already speaks to many struggles around the world today, we already saw versions of it in Nepal and Kerala, India. It has also served as a counter-strategy in places around Latin America and in Egypt and it describes well the innovative politics of such movements as Occupy Wall Street in New York and the Arab Spring. As Miguel Abensour (2011) argues, insurgent politics captures the activities of many of the most important waves of democratic engagement around the world struggling against a repressive state. In fact, the conceptualization will most likely be a better source of guidance for democratic struggles during climate crisis than the focus on theoretical texts that offer general discursive perspectives on broad concepts such as political freedom, social justice, and deliberative democracy. It is not that these theoretical perspectives are irrelevant, but they seldom speak to the particular circumstances of the struggles on the ground. In the real world of politics these concepts are interpreted in terms of ongoing political struggles.

Associative and Insurgent Democracy

Although none of the contemporary writers call attention to it, associative democracy bears a close relationship to the politics of insurgent democracy. Insofar as insurgent democracy is a political concept that describes democratic struggles against an oppressive state, it speaks directly to the efforts of associative democracy to challenge the centralized state's undemocratic structures. As Dryzek points out, democracy never comes from the top down, but rather is rooted in participatory struggles in civil society to democratize the state. Insofar as the central state will as likely as not find the theory and practices of associative democracy a threat to its own power, it would surely counter such efforts. Such resistance must, in short, come from below.

Theorists of associative democracy seldom devote much attention as to how to bring it about. Given that it is a theory that challenges the dominant power relations, it will most likely be viewed as a threat by many among the dominant elites who benefit from the state. In this regard, the fact that associative democracy would involve a power struggle tends to be neglected. Here Bookchin is a major exception. In his work such associations emerge as an alternative form of politics in direct confrontation with the existing power structures. Again, as described earlier, the community forest movement moves the struggle from theory to practice.

Even though there is opposition to the central state, there are today few, if any, credible political challenges on the horizon that can shake its powerful hold. Thus, associative democracy only remains an interesting theory, despite its potential relevance. This could perhaps be a reason for neglecting the question of how to bring it about. But reluctance could change with crisis conditions of the type that climate change threatens to impose. Under circumstances that would suggest the need for a transformative project, the struggles might well involve a politics of insurgency.

Insurgent democracy, as a concept, is relatively unfamiliar to most people in the West. In significant part, it has been largely employed to describe political struggles in developing countries, especially in the southern hemisphere. It is most often identified as the form of politics in the major revolutionary oppositional movements of the previous century, whether in India, Cuba, Vietnam, the Congo, or the Philippines, and has been examined in detail in the histories of insurgent struggles in parts of Latin America and Africa. In these places, the realization that distorted electoral processes could not bring about basic social change has led to political revolt. Indeed, since the end of World War II, there have been regular insurgent rebellions and uprisings throughout the world (Sanger 1967). Focusing on South America and Africa, Wood (2003) applies the term to democratic struggles in social systems structured by oligarchic alliances of privileged business and state elites that practice and maintain exclusionary regimes based on unequal distributions of wealth and resources. Or, in other words, the sort or political situation that will most likely dominate many parts of the world during an eventual climate catastrophe.

Abensour (2011) points to the worldwide wave of insurgent democratic activity directed against the state in more recent years. In Cairo, New York, or Washington, people protest against a state that is unresponsive to their needs and demands. The Arab Spring in Egypt, for example, involved massive protests against a powerfully monolithic state that forcefully repressed its own citizens and their demands for recognition of their own sovereignty. Responding to manipulated or fraudulent elections and various strategies to shield or protect the state against opposition from below, such movements have waged a battle against powerful economic and military forces. Demanding that the government ease its repressive grip on society, such insurgencies have called for more egalitarian political and social systems through a reduction of the power and influence of state bureaucracies, banks, and corporations. These arguments were also easy to recognize in the "political revolution" called for in the United States during the presidential candidacy of Senator Bernie Sanders and his followers.

The project of insurgent democracy, like associative democracy, is thus oriented to diminishing the size and sway of the central state. Recognizing that political activity extends beyond state structures, insurgent citizens' movements realize their social nature by joining with others in what can at times be personally dangerous struggles. Citizens, in effect, step out of their everyday roles and, in the act of asserting themselves as members of civil society, protest and reject the instrumental behaviors that advertently or inadvertently defend or preserve the state and its support for particular class interests. Insurgent democracy, like libertarian municipalism and associative democracy, thus involves an expanded understanding of what it means to be both human and political, with people acting in ways that transcend established societal roles. For writers like Bookchin it is also what it means to be "a citizen."

A turn to insurgent democracy in civil society will strike some as a limited form of democratic politics. They will argue, as we saw in Chapter 5, that the political focus has to be on democratizing state institutions and greening state policy. Beyond the fact that such a strategy does not appear to be a realistic option, it can also depend on how one understands both democracy and the state. If one follows Wolin (1994), who was surely one the most important democratic theorists of our time, insurgent politics can be interpreted as the essence of democratic politics throughout history. For Wolin, political theory has long made the mistake of focusing discussions about democracy on political institutions, such as legislatures, parties, and chief executives. He argues, much like Rancière (2015) and Abensour (2011), that democracy has never existed as a fixed political form or location. Rather than a state practice, democracy should be understood as political action that stands up against unjust governmental practices. In this conceptualization, democratic engagement throughout history is properly understood in terms of political experiences involving ordinary citizens in their roles as political actors mainly outside of formal government.

Although Rancière comes from quite a different theoretical perspective, he too offers a similar challenge to the way we generally think about democracy. Focusing on political movements from the ground upward, he finds democratic struggles to be rooted in questions related to the distribution of equality. Equality is theorized as the presupposition and point of departure of those engaged in progressive struggles, rather than the outcome of the political process. In his view, democratic political impulses, always occurring within particular circumstances and conditions, inevitably come up horizontally from below rather than vertically from above. As he (1995: 49) puts it, "democracy is the community of sharing," both as "membership in a single world which can only be expressed in adversarial terms, and a coming together which can only occur in conflict." Democratic politics does not happen very often because, in his sense of the term, it is generally discouraged. But insurgent outbreaks of "dissensus" and "disagreement" can happen anywhere (Rancière 2015; 2001). Never fully realized as such, democracy is not an end state which can be institutionalized, at least on a large scale. Democracy is about struggles that are always in the process of becoming. May (2010) illustrates this broader understanding of democracy with a number of case studies, including the Zapatista movement in Mexico and Palestinian struggles against Israel.

This is a view shared by Bookchin, who insisted on a distinction between politics on the one hand and the institutions of statecraft on the other. For Bookchin politics and Hellenic democracy were expropriated by the state as a means of controlling the behavior of the citizenry (Bookchin 2015). Genuine politics thus has to be reappropriated in the form of libertarian municipalism and returned to correspond to its original Greek understanding; that is, as the organization and guidance of the community or polity through direct face-to-face assemblies of citizens (Bookchin 1991). This original conception of politics was lost in modern times, a historical reality lamented by both Bookchin and Wolin. Whereas it refers today to the activities of legislatures, party leaders, bureaucracies, and the military, the "Hellenic notion of politics as public activity" is what provides the possibility for political beings to engage in self-discovery of shared concerns and the ways in which they can be achieved through public action.

Understood as such, the history of democracy has at best been episodic, unstable, and temporary. In fact, in the views of Wolin, Rancière, and Bookchin, such a process of institutionalization signals its attenuation or demise. In Wolin's (1994: 19) words, which could have been written by Bookchin, "leaders begin to appear; hierarchies develop; experts of one kind or another cluster around the centers of decision; order, procedure, and precedent displace a more spontaneous politics." Gone is the radical, revolutionary politics that gives rise to the great democratic moments that mark political history.

Insurgent Democracy as Fugitive Democracy

To capture this understanding, Wolin (1994) advanced the concept of "fugitive" democracy. The idea of fugitive democracy points to the ways that the modern-day exercise of power has turned democratic politics into momentary and evanescent encounters. Far from an institutionalized form of deliberative democracy, fugitive democracy speaks to the ways that modern societies experience "breakouts" of democratic politics, often fleeting but at times profoundly defining events. Calling for political redress, such outbreaks are episodic, such as street demonstrations, mass meeting protests, petitions and newspaper articles, or writing political graffiti on official buildings. Informal and often spontaneous, Wolin argues, they are initiated for collective action to alleviate inequalities in power and wealth that permit those with money and education to dominate the governmental processes. Often tumultuous and disruptive, insurgent fugitive democracy expresses moments of frustration, even at times rage and violence. They are episodic acts born out of necessity and generally more "improvisational" than they are "institutional" (Wolin 1994).

Fugitive democracy does not, however, preclude deliberation. During the Occupy Wall Street movement, for example, participants discussed and debated issues daily. As one observer put it, "when I went to Occupy in Zuccotti Park, what I saw was not people protesting per se, but people engaging in what they called 'learning circles'" (Rushkoff, cited by Been 2016). They were learning things about the ways capitalism works, the use of alternative currencies, how lenders work, and more. Basically, as Rushkoff explains, they set up something of a "graduate school" with "seventy concurrent seminars." It is what Choudry (2015) describes as "learning activism" in social movements.

In Wolin's view, many of the important events in political history are fugitive in character. The American Revolution, for example, is generally told as a story of political leaders, statesmen, and generals. But the political struggle leading up to the War of Independence was just as much dependent on episodic protests of small farmers, artisans, workers, shopkeepers, women, and seamen (Wolin 2008: 254). Further, Wolin (2008: 288–9) offers the human and environmental disaster caused by Hurricane Katrina in New Orleans as a contemporary illustration. Examining how citizens rose materially and empathetically to deal with the disaster, he wrote:

> That response was a political act on behalf of commonality. While the administration's vaunted "Homeland Security" agencies and a highly disciplined White House floundered, there was a spontaneous outpouring of aid, financial and material, from ordinary citizens, civic and religious groups, and local governments from all parts of the nation. It was as though the United States could express democracy only by bypassing a national government preoccupied with distant fantasies of being democracy's agent to the world.

The fact that New Orleans had a limited supply of basic human necessities was something that ordinary citizens everywhere could understand and empathize with. There was no one location or deliberative forum—municipal or state government—that called for a democratic response. The story was rather more about a civil society outpouring of sentiment that for Wolin is the most essential political dynamic of democracy. In his view, it is a defining feature.

We find here justification for Dryzek's emphasis on civil society as the source of political creativity, discussed in Chapter 5. Fugitive democracy is a theoretical perspective that casts further doubt on the possibility of turning a green state into a budding environmental democracy. Indeed, we can recognize the politics of social movements, environmental movements in particular, to be a form of insurgent breakout democracy. Radical environmental movements episodically organize and pursue a wide range of disruptive activities, such as Greenpeace boarding whaling vessels, people chaining themselves to trees, mass demonstrations and speakouts calling for the mitigation of global warming, local citizens protesting the contamination of the local water supply, among other actions. As political actions originating in the fragmented arenas of civil society, such events are the incendiary initiatives that keep the democratic flame alive in the face of the centralizing powers that invariably coalesce around the special political interests of the privileged minorities. True today, the emergence of such actions will be even more evident under eco-authoritarian regimes.

Understood this way, insurgent democracy is interpreted as something that "is always potentially present," given the ongoing and always indeterminate struggle for freedom and equality. As Abensour explains, one can never know in advance when or where such insurgencies will spring up, create public arenas, and talk back to the state. Insurgent breakout democracy thus provides an important conceptualization of a contemporary political phenomenon. In Abensour's view, it usefully permits us to understand situations such as the Occupy Wall Street movement or the Arab Spring within a broader historical perspective, such as the uprising in Budapest in 1956 or even the struggles of the Paris Commune in 1871. Such movements, we argue here, should compel social and political theorists to revisit their democratic theories, focusing on agents of action, issues, struggles, and methods.

To the question as to how insurgent democracy could be anything other than oppositional, periodic, and exceptional in nature, the answer might plausibly be found in associative democracy, an open and flexible practice grounded in opposition to the state. It would potentially be a way in which popular struggles could give rise to more institutionalized alternative forms of democratic participatory governance. For some this might involve adapting anarchist principles of organization. Loosely organized, they would have a better chance of maintaining the sort of fugitive spontaneity that can challenge the existing state and its legal and administrative power structures. For members of ecovillages it is more than a possibility. Outside of the mainstream, such oppositional action is generally a part of their way of life.

Viewed this way, insurgent politics is not something to come from political theorists. Rather, theorists themselves need to take their lead from the actors. Such struggles pose a fundamental challenge relevant to the politics of climate change, especially as social injustices start to become more and more recognized. In this regard, fugitive democracy at the barricades, especially as a social learning experience, will be as much or more important than what can be learned about democracy in the classroom (Brecher 2016).

This is a point that comes through again and again in Bookchin's writings. From the perspective of fugitive democracy, Bookchin is a model theorist coming out of the radical tradition of protest. He, along with Bahro and Sale, clearly expresses the transformative character of such a political project aimed at undermining an unsustainable central state. In his variant of what could otherwise be called associative democracy, Bookchin recognizes that the state will often not look kindly on having such participatory localist efforts on its periphery and will at times take measures to squash them. His theory is grounded in the transformative idea that power should be taken back from the state, while fully realizing that the state will not look upon this in a friendly way.

One can ask whether insurgent democracy is anything more than protest. Occupy Wall Street had an important—even powerful—impact on public discourse, but it was unable to effectively organize itself as a strategic opposition capable of presenting a coherent program of ideas that could be advanced as a political agenda. While this is true, it is not the whole story. An assessment of the impact of the movement depends on one's understanding of the relation of such protest movements to the state. For one thing, it is surely the case that Occupy Wall Street is part of a movement that gave force to Senator Bernie Sanders' surprisingly successful anti-Wall Street presidential campaign. Even more importantly, if we recognize the relationship between insurgency politics and the state to be dialectical, we can develop a better understanding of the dynamics of movement politics that will speak to the situation ahead. In this view, the state is largely a regulatory and administrative control structure, what Foucault (1991) captured in his theory of "governmentality." State institutions can only be opened up and changed by protest—serious protest—under certain circumstances (Piven and Cloward 1971). If the state responds, adopting and coalescing around certain changes, it invariably closes again, slowly but steadily. It is a process of state change administered and monitored by hierarchies of power, career politicians, and professional experts. At a later point, the cycle of protest and change starts again in response to new circumstances and the problems they bring forth. This corresponds to Wolin's and Rancière's argument that the state is never democratic as such and that democratic initiatives to alter the state come from civil society movements. Indeed, insurgent politics rather than the state per se is the primary force that drives democratic political change and social justice, a dynamic that can be documented by an examination of movements for change throughout history.

This understanding of democratic politics would surely speak to the politics of climate crisis, especially if eco-authoritarians capture the state. Even if one rejects fugitive democracy's understanding of the state, insurgent politics would under such circumstances become as salient and important as it will be contentious. That is, even if the theory doesn't capture all of democratic politics, especially as it pertains to the state, it will be an important part of an understanding of democratic theory and practices during a climate crisis.

Confronting Technocratic Eco-Authoritarianism

Appropriate Technologies and Participatory Expertise

As we have contended from the outset, environmental expertise and technological innovation will need to play an essential role in the struggle to mitigate and adapt to climate crisis (Stirling 2011). As discussed throughout, there is no way to avoid ecological disaster without the assistance of new technologies, especially renewable energy technologies. But will it be possible to avoid a system dominated by an anti-democratic form of technocratic politics? This is an issue of special importance for the topic of democracy for two interrelated reasons. The first has to do with the fact that technology is generally believed to be developed without regard to political processes. It is understood to be the stuff of engineers who apply scientifically tested principles. So understood, there is no place for public opinion or citizen participation. Rather, in short, the process of technological innovation is seen to be governed by the best technical decisions.

The second reason relates to a technical bias built into this process. While it is largely the case that the specifically technical aspects—nuts and bolts—of new innovations do not necessarily require public or political involvement, there are nonetheless many questions related to the development of such technologies that involve social and political issues. Most obvious is the question of goals. What is the purpose of these technologies? Who judges what those goals are? How are they built into research designs? Who determines whether they are attained, and who are the winners and losers? And so on. In the decision processes as they generally operate, however, these questions and issues are typically either not taken up or blurred by overriding technical considerations. In part, this is deliberate, as technological innovation has been managed by industry, which has an interest in depoliticizing technological decisions. This can be achieved in part because most members of the public do not understand the details of technologies and are blinded by the complexity of the issues that are related to them. In fact, however, the techno-scientific process is laden with various subjective judgments, including norms and values, that are seldom transparent and visible (Fischer 2000).

Some have thus posed the awkward question: if citizens cannot understand the issues, how can we speak of democracy (Fischer 2009)? This is a question that dates back to Plato but even more importantly in modern times to John Dewey. In the 1920s, Dewey (1927) recognized that the emergence of a technological society required special attention to the relationship of citizens to experts. And, in recent times, there have been a number of innovations in deliberative processes designed to assist citizens to take up policy issues. Citizen juries and consensus conferences, as we have seen, bring citizens and experts together to explore and deliberate such issues. This has also raised important questions about the relation of technical to social issues and about the nature of expertise more generally, all of which have been important in social scientific research. But as deliberative procedures, these issues are typically removed from political decision-making processes and remain advisory at best. The question then is: Who will control technological development and for what purposes?

This is scarcely a new question; a range of theorists have dealt with it for the past fifty years, if not more (Fischer 1990). Indeed, Bookchin (1986) and Bahro (1994), drawing on Mumford, maintained that much of the problem rests on the fact that technological development serves the techno-industrial megamachine at the expense of other interests and values. Toward this end, Bookchin posited the need to transform "a technology for survival into a technology for life."[2] Such a "technology for life," as he saw it, will not only assume the vital role of integrating a network of eco-communities, but also as a "liberatory technology" would keep open the doors for radical social change (Bookchin 1986: 107–61).

In the future, the issue of technology will surely become even more pressing. Many have spoken of the need for something like a Manhattan Project to deal with the ecological crisis, which surely brings the questions of survival to the fore (Gore 1992). Indeed, technology innovation was perhaps the biggest theme at the Paris CO21 conference. As we saw in Chapters 3 and 4, however, movement in this direction remains disappointing. It is not that there have been no relevant technological innovations; indeed, there are regular reports in the media on the search for new eco-technologies. But the movement is too slow. And at the same time, the focus is on finding a technical fix that sidesteps the underlying social-ecological issues—in particular the issues related to transformative change—that drive the crisis, geoengineering being the extreme example.

Eco-Technological Politics and Participatory Technology Assessment

Over the past thirty years there have been numerous efforts to find ways to make these technological decisions both wiser and more socially just. One of the most important efforts has been the participatory technology assessment movement. Participatory technology assessment emerged first in the United

States in the 1990s as an extension of the call for more public participation; it sought to reorient traditional technology assessment as part of a broader participatory movement. Specifically, the goal has been to make technology policies more democratic by extending the assessment process beyond experts and political elites to include the voices of the public. Such a more inclusive process would make decisions more socially and environmentally compatible by involving both the users and those affected (Joss and Bellucci 2002).

This new space for a broader range of participants in the technology decision-making processes, including a role for ordinary citizens, has been considered a "promising way to promote direct interaction among members of the general public, interest groups, professional experts and policy makers in multi-actor spaces with the general aim of democratising S&T governance" (Gabels 2007: 301). Over a period of a decade or so there were in fact many efforts to employ one form or another of the methodology, especially in Europe. The development and introduction of this assessment procedure were based on two interrelated assumptions. It was considered to be an approach both to supporting and extending participation in government, if not participatory governance per se, and for the democratization of expertise. Taken together, it is seen to enhance knowledge about technologies and their relationship to relevant values, forging new opportunities for dealing with technologically based conflicts, increasing the motivation of the participants involved, creating a learning process, and offering technology developers a better awareness of both stakeholder and consumer ideas and interests. And, last but not least, such assessment procedures increase the chances of supplying greater legitimacy, transparency, and accountability in socio-technical policy decisions.

After ten or fifteen years of experimentation with participatory technology assessment in the U.S. and Europe, the initial enthusiasm waned. This was due, first and foremost, to its inability to bring about the promised changes. Critics argued that it was based on a "romantic" notion of democratic politics and questioned the idea that citizens have better knowledge than the appropriate experts (Weingart 2001). There was also too little attention paid to who participates and why. For the most part, these experiments in participation lacked direct linkages to the decision-making processes of the established political arrangements and, moreover, the concepts about the democratization of science were often vague, when not ideological.

All of these criticisms have merit. But one can argue that many of the failures have resulted from the effort to use participatory technology assessment to transform systems of relatively undemocratic institutions that characterize modern governmental systems. As such, these methods were employed in settings that were not very favorable to the kinds of decisions that they could potentially generate. Indeed, conservative politicians have been fairly hostile to technology assessment in general, let alone participatory technology assessment.[3] In fact, in many cases they have succeeded in eliminating the practice altogether. Some of the disappointment with the limited outcomes can be considered to be based on naïve expectations. At the same time, however, some of it can also be viewed as strategic. Short of changing the system, participatory technology assessment has helped to bring a wider range of perspectives into public discussion, as well as to better educate the citizenry. These do not in and of themselves alter the system but they can supply pressures that help to lead to larger changes over time (Sclove 2010).

Localization and Appropriate Technologies: The Case for Participatory Expertise

The results or the practices, however, could be more promising in a different setting. And here, as an extension of our discussion of localization and the call for a technology for life, we find the alternative technology movement set into motion by the German-born British economist E. F. Schumacher (1973). When it comes specifically to questions about technology, the localization movement has an especially important tradition to draw on, namely Schumacher's famous "Small is Beautiful" philosophy and its focus on "appropriate technologies." Based on his extensive travels in the developing world, Schumacher formulated a theory of local economic production that can be considered one of the forerunners of the localization movement, indeed one of its founding ideas (Schumacher 1973). In the 1970s Bookchin's own Institute of Social Ecology at Ramapo College in New Jersey became an important center for the investigation of and experimentation with appropriate technologies, including the call for liberatory technologies.

For Schumacher the central idea is that local production for local needs is the most rational and sustainable mode of economic and social life. In support of this fundamental premise, he pioneered an alternative approach to technology based on the idea that technologies should be technically and ecologically appropriate to the scale of the communities that would use them. The goal of these "intermediate technologies" is to reduce the complexity of the technical instruments, thus making them more accessible and easier to use.[4]

Appropriate technology is an orientation that fits with another line of work on expertise, the democratization of expertise or the variant we will refer to here as "participatory expertise." Some of the work on participatory expertise and various forms of collaborative expertise has generally emerged to support and facilitate citizen conferences and participatory technology assessment. But, as we saw in the earlier discussion of the community forest movement in Nepal, there is another more radical variant—participatory research—that speaks to the goals of relocalization and the local politics of alternative social movements (Adridge 2015). Drawing on the work of such theorists and practitioners as Freire (1970; 1973), Reason (1994), and Tandon (1988), the method of participatory research has sought in the name of participatory democracy and social justice to contribute to social movements' struggles for empowerment and self-help (Coles 2016; Fischer 1990; 2000).

Experiments in participatory research are, in significant part, the products of political activists and progressive intellectuals identified with Third World communities and social movements in more advanced industrial countries that have sought to develop a new kind of expert grounded in efforts to empower ordinary citizens to make their own action-oriented decisions (Fischer 2000). Toward this end, participatory researchers have experimented with new ways to democratize the expert's relationship to the citizen or client (Chambers 1997; Fischer 2009), including the development and practice of "popular epidemiology" (Novotny 1994; Brown and Mickkelsen 1990). They have sought to do this by transforming the top-down methods of positivist expertise into a bottom-up participatory methodological orientation (Reason 1994). Addressing the argument that professional experts have more often than not aligned themselves with elite interests, the approach is designed to counter the techno-bureaucratic and elitist politics that dominate the modern unsustainable society. Participatory inquiry, as such, constitutes a method for bringing citizens and their local knowledges directly into the decision-making processes (Ottinger and Cohen 2011). By facilitating cooperative work among experts and community members in helping to understand and solve local problems, the approach is designed to democratize the research process by empowering the participants. The method seeks first to understand the problems of communities and their members within their own social-cultural context and the particular "logic of the situation" to which it gives rise (Heron 1981: 158). Then, beyond analyzing the socio-cultural logic of action, it seeks to link the contextual situation to the larger social structure (Fischer 1990). It represents, as such, an effort to interpret the context in terms of the more fundamental structures of social domination that shape it.

The participatory researchers at ForestAction in Nepal, as we saw, have employed the method to first help local forest communities to understand their social plight and, second, to motivate them to participate in the development of a federated network that accumulated their power in such a way that they could, as a movement, have a political impact on the power structure in Kathmandu. Not only did the effort have an important impact on forestry policy in the country, it helped to reshape the deliberative practices of forestry in other forest-covered countries.

This focus on the assessment of technology and the democratization of expertise offers a mutually beneficial relationship for both political theory and science and technology studies. These are two fields of inquiry that have mainly failed to interact with one another, at the expense of both (Strassheim 2015). For political theory the ability of the citizen to participate meaningfully in discussions and decision-making about complex technological issues of the times is crucial to democratic theory. At the same time, too many scholars of science and technology studies fail to grasp the dynamics of the political process. The difficult questions concerning the possibility of democratizing expertise represent a very appropriate topic that can—and should—bring them together.

Can Ecological Science and Technology be Democratic?

Participatory Environmental Expertise

Participatory expertise also speaks to the critical question we raised in Chapter 2: Does ecological science and eco-technological development exclude or impede democratic politics? There, Goodin's argument that ecological substance and democratic agency do not mix was examined. While the tension between them appears to be unbridgeable in the first instance, the question has yet to be decided once and for all. In significant part, the answer depends on how one understands the democratization of expertise.

The question of democratization and the role of the citizen in matters related to technical expertise is one of considerable significance for democratic politics, as eco-technological innovation is a necessary component of a sustainable transition. In Chapter 3, we took note of the fact that technology development and democratic decision-making involve different logics—the technical logic of efficiency versus the political logic of compromise—suggesting that these two forces are incompatible at a basic epistemological level (Fischer 2009).

From the perspective of participatory expertise, however, the key question can be reformulated. It may be the case that technological development as engineering has little or no room for democratic opinion, but this argument misses the fact that the legitimacy and implementation of technologies do indeed hinge on social and political questions. Technology, as such, has different impacts on different social groups and thus is not neutral in its societal implications.

In bringing technological experts and citizens together in policy-oriented discussions about eco-technologies, the discussion primarily focuses on these public inputs, both social and political (Fischer 2000). While such deliberation is not about technical considerations per se, the outcomes can be referred back to the processes of technology design. In the interest of effective policy decision-making, the goal of participatory policy expertise is to bring these two different tasks together in a mutually beneficial deliberative interaction.

From this orientation, then, the question can be recast to recognize that in the world of political action policy decisions about how we want to relate to the outcomes of ecological research and eco-technological processes raise questions in need of answers. Most fundamentally, we need to decide how we want to relate or orient ourselves to the outcomes and implications of these scientific and technical processes. When it comes to these questions, it is not just a matter of technical science; we need to ask how we prefer to live together in view of what otherwise appear to be pressing, unavoidable—even at times inevitable—imperatives. It is here that we can in fact speak meaningfully of democratic participatory expertise. In this practice, the social and political considerations that come into play are recognized not to be an essential part of the domain of the ecological expert. While the technical development of the ecological innovations is largely a matter for engineers, we recognize here that they have no privileged position in matters related to societal goals and values. In this regard, engineers are themselves just citizens among citizens. To proceed without the social and political judgments of the citizenry as a whole is to court failure, whether it comes earlier or later.

Finally, a few more general words about the role of such political and moral judgments. There is an ongoing debate about the role of the scientist—social or natural—with regard to value judgments related to climate change. Some argue that scientists should simply stick to the facts and refrain from expressing normative or moral judgments. This point of view is based on a narrow division of science into specializations that methodologically exclude more comprehensive judgments on the phenomena under investigation. For others, such as Hansen (Milman 2015), this neglect of the larger picture in the face of crisis is itself immoral. From this perspective value neutrality—at least in the strict sense of the term—can best be understood as a myth. All empirical statements are, in this view, based on particular normative assumptions and for the most part they depend on theoretical orientations that wittingly or unwittingly support particular views. Oreskes and Conway, as we saw in the Introduction, argue that part of the failure to adequately raise concerns about climate change has been the result of this scientific commitment to positivism and its belief in the strict separation of empirical and normative judgments.

Underlying this concern is a worry that science will be politicized, which has indeed already happened. There is here, however, an alternative approach to facts and values (Fischer 1995). Rather than advancing particular social or political judgments, experts can follow a proposal initially put forth by Dewey (1927) and lay out the implications of competing positions, empirical and normative. The task of the participatory policy expert, in this regard, should be to bring the facts and values together in ways that illustrate what different configurations of the two would mean for particular policy decisions pertaining to specific situations (Fischer 2006: 223–36). Indeed, this is an orientation embodied in the methods of "deliberative policy analysis" developed by critical policy scholars (Hajer and Wagenaar 2003; Fischer and Boossabong 2016). It has, as such, resonance with Burawoy's (2005) concept of a "public" social science, in which the empirical findings and theoretical perspectives of social science could be brought to bear on important public issues, climate change being one of them. Instead of just engaging other social scientists in academic discussions, social scientists could rigorously inform public debates about the facts related to a problem but also explain what they mean in terms of current political judgments, alternative policy possibilities, and normative judgments generally. It is a task that should not be confused with political advocacy.

In just this regard, one could also suggest that participatory expertise might well offer a way to revise Bahro's concept of an ecological council. Bahro proposed his council not as a permanent arrangement, as many have taken it to be, but rather as an interim strategy in the context of survival. Some, as we have seen, argue that a turn to governmental oversight by ecological experts, while less than optimal, offers a more attractive alternative than leaving power in the hands of powerful transnational corporate elites and the political elites that do their bidding. In this case, the idea of an ecological council could become much more politically attractive if it were organized around the ideas and participatory practices of democratic expertise. For Bahro the ecological council would depend as much or more on an ecological commitment—a new ecological consciousness—as it does scientific ecological expertise. While a change in consciousness surely has to be a part of the larger environmental project, it does not eliminate the need for discussion and deliberation. In terms of democratic governance, a discursive turn to participatory expertise could give a boost to democratization on its own terms, as such a council would be part of the larger political system.[5] The democratization of such an ecological council would seem unlikely given the lack of genuine democracy at this level. But there could be an alternative. One might well conceive of a democratic ecological council as a project for the confederation of eco-communes.

Conclusion

In this final chapter we presented four basic theoretical issues related to the relocation movement and the contributions of Sale, Bookchin, and Bahro that relate directly and indirectly to theoretical concerns in democratic political theory. In addition to participatory environmental governance, the issues of associative democracy, democratic insurgency, and the democratization of expertise were seen to better prepare environmental political theory for climate crisis and to suggest how they might be integrated into more mainstream theoretical discourses. We have not developed these themes in great detail, as there are already substantial literatures that do that. In this regard, there is in general nothing new about either the issues raised by relocation or the theoretical connections to them. The purpose here is to call for a return to issues raised by earlier environmental thinkers that have either been neglected or forgotten. As the political and social situation changes under the conditions of climate crisis, it has been argued here, these theories and practices speak directly to times ahead.

There is no expectation here that there will be a wide-scale political transformation to associative environmental democracy or that eco-communities will save the day. Rather, once again, the argument here is that a civil society-based parallel association of eco-communities should be part of the diverse mix that will be needed to deal with the trouble ahead. No one doubts the need for active local citizens. But even more importantly for the discussion at hand, these ideas speak to concerns about democratic prospects during hard times, including a rise of eco-authoritarianism. It is in this context that the practices of local democratic governance have a chance of surviving, and even potentially thriving. Indeed, climate crisis aside, small-scale communities have always been considered to be the fount of authentic democratic life. For this reason, the idea stands on its own even in the best of times. It holds out the possibility of rejuvenating the democratic ideal in a post-democratic world of democratic deficits.

Notes

1. See https://secure.avaaz.org/en/petition/Chinese_President_Xi_Jinping_Save_Chinas_largest_ecovillage/?pv=10.
2. Illich (1973) spoke similarly of such alternative technologies as the "tools of conviviality."
3. The most prominent example was the elimination of the Office of Technology Assessment by a Republican-controlled Congress in the United States.
4. In South Devon in the UK there is a Schumacher College that offers lectures, workshops, and short courses. The college confers master's degrees in Holistic Science, Economics for Societal Transition, and Sustainable Horticulture and Food Production, with lectures offered by some of the most prominent environmental thinkers of our time, such as Fritjof Capra, James Lovelock, Arne Ness, Wolfgang Sachs, and Vandana Shiva. It also has strong practical research commitments to

low-carbon, sustainable farming, agroforestry, and conservation. https://www.schumachercollege.org.uk.

5. Bahro, in fact, did not see his ecological council as constituted only by scientists. Indeed, he included a range of non-scientists. But he says little about how these people would work together.

Conclusion

It is always difficult to conclude a long and detailed theoretical discussion. Thus it is important here at the end to be clear about where we have been and how we want to be understood. Throughout the eleven chapters we have engaged in an exploratory search for an answer to the question: what will the democratic prospects be during a full-scale climate crisis? It is an exploration that took us from the global to the local level. Having taught global environmental politics for many years, a topic that places most of its emphasis on global agreements, it was not always easy to put aside the hope for strong global agreements. Indeed, like many observers, I was reluctant to shift my primary focus to eco-relocalization and the ecovillage movement. But after a long and sober assessment of the current state of climate change politics, it seemed like the time to hedge bets had come and to start rethinking an all-too-easy set of assumptions about the circumstances that future generations will face, and the fate of democratic governance even more specifically. It was only after field research and a careful examination of various existing local initiatives that I came to embrace the conclusions to which my search was leading. What we will do about the climate challenge remains unclear, but the relocalization movement shows that we are not without alternatives. This is especially the case when it comes to preserving the democratic practice of participatory environmental governance.

The Main Arguments in Four Parts

We began in Part I, "Climate Change, Crisis, and the Future of Democracy," with a delineation of climate change and various crisis scenarios to which it has given rise, including "fortress world." To put it succinctly, it has been difficult to find a "silver lining." Indeed, we are now regularly bombarded by reports indicating that the situation is getting worse, faster and faster. Since 2000 we have had fifteen of the sixteen warmest years since scientists began keeping records in the latter part of the nineteenth century. The current assessment portends, as Christoff (2016: 781) has put it, "a social catastrophe of unimaginable proportions." Further, financial estimates of the costs to respond and adapt to these new conditions are catastrophic as well.

There are those who still say that we do not yet have proof about the causes of the warming and should hold off on speaking about action to deal with climate change. But waiting is very risky. As Bromwich (2016) has put it, "climate change is never going to announce itself by name." What can be said with a high degree of certainty, however, is that the current disasters are what climate scientists expect to happen. In this view, a failure to make serious preparations to deal with them, especially while we still have the economic resources to do so, is to court a risky folly of the first order. But largely speaking, this is just what we appear to be doing.

We completed this first part with two chapters that examined the political responses to the crisis—democratic and authoritarian—and explored the question of whether the logics of ecology and democracy exist in a fundamental tension with one another. That is, can they be happily wedded? Or is environmental democracy an oxymoron? The answer here was yes and no, depending on how one understands the linkage between them. If it is about the ways we want to orient ourselves and respond, the answer is yes.

We then presented the dominant eco-modern search for a technological fix, largely designed—wittingly or unwittingly—to sidestep questions of environmental justice and democracy. While technological innovation remains an important goal, both to mitigate and adapt to the long-term consequences of climate change, the narrow focus on such a fix was seen problematically to neglect the socio-ecological dimensions of the problem. We also discussed the theory and practice of a prominent Dutch approach to transition management. Although it constitutes an impressive effort to bring about a sustainable energy transition, it evolved largely as a top-down process, participatory rhetoric aside. It would not be unfair to say that these approaches lean more toward a central political hand rather than democratic environmental governance.

Part II, "Democratic Prospects in the Face of Climate Crisis," took up questions concerning environmental democracy, ecological citizenship, and the green state. We examined the main themes in the literature of environmental political theory as well as practical experiments in deliberative democracy. While these were deemed to be commendable contributions, they are largely theoretical efforts far removed from the real world of politics. As political realities stand, none of them would have a chance of fending off forms of eco-authoritarianism. As for the idea of democratizing a green state, the concept was seen to lack political traction, especially when considered against current political realities moving in the opposite direction, mainly democratic deficits and post-democracy, including simulative democracy. The discussion left us wondering what a more relevant environmental political theory for the future during climate turmoil would look like, one that could be useful to following generations who will confront the brunt of the crisis.

These considerations motivated the search for a more realistic alternative. Part III, "Environmental Democracy as Participatory Governance," investigated the possibilities presented by the theory and practice of participatory governance, an approach seen to be better grounded in real-world democratic struggles. We illustrated the argument with the case of participatory governance practiced by the community forest movement in Nepal and the politics and governance of the global ecovillage movement. These were seen to be models of an ecopraxis that offer better chances of either standing up to and/or withstanding the political challenges of eco-authoritarianism. While they are not put forth as solutions to climate crisis, these eco-communal arrangements were seen to make important—even essential—contributions to both the realization of an ecological transition and the preservation of democratic environmental practices. In particular, the practices of participatory environmental governance, anchored to civil society, would seem to hold out a reasonable chance of preserving and sustaining democratic values and practices during hard times. It is an argument that can apply to both the global north and south, as the case of Senegal suggested.

Finally, in Part IV, "Making Theory Matter," we more directly took up the question underlying much of what has gone before: namely, how can we engage and revitalize democratic environmental theory in ways that speak to needs of a different time that is surely coming? We began first with a critique of the apolitical politics underlying the scientific orientation—ecological resilience—that is shaping most of the contemporary thinking about urban environmental adaptation, eco-cities, and the Transition Town movement. While there is a growing recognition in cities and urban areas generally of the need to plan for measures to deal with the eventual consequences of climate change, most of what has been introduced is technically oriented to finding and employing technological solutions. Guided by eco-modernization and the ecological concept of resilience, these approaches have largely ignored or neglected the social and political implications of the coming turbulence, many of which are themselves inherent components contributing to the crisis.

This led us to consider the alternative contemporary relocalization movement, emphasizing in particular sustainable agriculture, renewable energy, and eco-communal living. Having made this turn, it was not hard to rediscover earlier environmental thinkers who had begun working out theoretical arguments to undergird an eco-local participatory conception of society that resonates with these current initiatives on the ground. Here we examined

Sale's focus on human scale and decentralized bio-regions, Bookchin's theory of libertarian municipalism, and Bahro's writings on eco-communes and their contributions to ecological consciousness, a precondition for genuine socio-ecological transformation.

In the final chapter we sought to engage contemporary environmental political theory by identifying theoretical bridges that can bring these ideas into discursive interaction with the dominant theoretical discourses. In addition to the theories of participatory governance, we pointed to associative democracy, insurgent democratic politics, and participatory expertise as concepts with established connections to political theory. While these are scarcely dominant themes in the discipline, they have been recognized and advanced by leading political theorists. In some cases they can be relatively directly connected to the dominant theory of deliberative democracy. Indeed, participatory governance can be seen as a specific contribution to deliberative democratic practices. As such, the introduction of concepts and practices drawn from these alternative theories can serve to help forge a more political conceptualization of deliberative politics.

Climate Change as Crisis

Throughout we have seen the way climate change is bringing about a set of circumstances that will push us to the limits of what we know about our ecological systems and how we might best respond, socially and politically. In particular, what might be the role of democratic governance? It is this lack of understanding and knowledge about how to respond that led us to refer to climate *crisis*.

Some of the skeptics and deniers will say that referring to climate "crisis" is to exaggerate, especially given the uncertainty of the evidence. One could better refer instead, some say, to "climate warming," "climate disruption," or "climate destabilization," among other terms. Indeed, it is sometimes said that words like crisis will only frighten people into inaction. But one can also argue that we are now beyond that point, not to mention that most people are already doing little or nothing to help alleviate the problem. What is more, the goal is not to frighten people but rather to awaken them. Down the road, it will no longer be what people can bring themselves to accept, but instead what they and following generations will be forced to deal with. We have chosen here to speak of climate crisis because this conceptualization incorporates the underlying social and political dimensions of the challenge. To speak of climate disruption mainly points only to ecological concerns; it does not in and of itself capture the fact that our human and natural relationships are out of balance—created and accelerated by Western consumer lifestyles and an industrialist capitalist system of production. As we explained in Chapter 1, crisis means that the established practices of our broader societal systems—including global systems—are working in ways that themselves create threats to the goals and priorities of ecological stability. Crisis, as such, portends a turning point in the face of a foreboding time. If the crisis cannot be managed or resolved, the result will be socio-ecological disruption and possibly catastrophe, as is the case already with nations in the South Pacific. The climate crisis thus results from internal social and political factors related to the way we live on the planet as well as external ecological causes.

Climate crisis is then also a political crisis. Given that the governing institutions that could guide society toward a more sustainable future have failed to rise to the occasion, questions have been raised about the ability of our governments to confront the task at hand, especially democratic governance. It is not that nothing is being done about climate change, but rather it is a question of whether steps can be taken fast enough to avert dire consequences. As we have argued, we are running out of time. Up to this point, as one observer has put it, political actors have avoided tackling the core of the climate crisis. Mainly they have been nibbling around the margins of the problem, mostly hoping for technical solutions. One could point here again to the Paris COP 21 conference; it is hard to identify any convincing initiatives in the first year after the meeting attributable to the agreement, despite the fact that the formal agreement has now been signed by the requisite number of countries.

Not only have few governments taken the kinds of action necessary to ameliorate the climate challenge, some continue to impede measures required to either slow or adapt to the accelerating build-up of CO_2 that is warming the atmosphere, such as continuing to build coal-fired power plants. Included here are the major nations generally taken to be democracies. Not surprisingly, this has led to worries about their inability to take on the task. In the environmental literature, many say that more democracy has to be part of the solution to climate change, which would surely be the case in a better world. But, as we saw at the outset, a growing number of voices have questioned the ability of existing democracies to deal with this problem before it is too late. The failures of democracies to respond have given rise to technocratically oriented calls for more eco-authoritarian political arrangements in the name of saving a way of life, if not survival itself. Indeed, this is all the more worrisome given the fact that authoritarianism is on the rise around the world.

In Search of a Strategy: Eco-Localism

The best solution to avoid climate crisis would be to mount a full-scale socio-ecological transformation in pursuit of a sustainable way of existence on the planet, globally and locally. In view of the urgency of such a transformation, McKibben (2016) has called for a "War on Climate Change." It is a struggle that we indeed need to engage in. And it should begin sooner rather than later. But it is not yet happening anywhere—and certainly unlikely to happen in the time frame imposed by the looming climate crisis ahead.

What is more, such a full-scale strategy—a war against climate change—will almost certainly result in forms of strong authority rather than democratic governance. As Zahariadis (2015) has pointed out, a lack of time in the face of the urgency will undercut democratic decision-making. Most likely what we will get, at some point down the road, is a top-down effort by political, military, and economic elites to coordinate and plan the use of increasingly scarce resources, probably much like the eco-modern technocratic strategy that Giddens has suggested, as we discussed in Chapter 3. As a form of environmental guardianship, in the name of emergency, it will mainly involve a technocratic emphasis on central planning and technological fixes, with questions of environmental justice and democracy indefinitely suspended. In short, it portends a stronger degree of authority, if not authoritarianism, many signs of which are already visible.

Thus, in pursuit of the democratic prospect given of these probabilities, we have explored other directions in search of an alternative strategy, a contribution to Plan B to use Lester Brown's term. Moving away from the theoretical focus on democratizing an environmental state, this search led us to eco-localism and the relocation movement generally. This strategy, as we saw, speaks to both the ecological and democratic issues at hand. Skeptics will continue to say that such an approach is unrealistic—even idealistic—especially in a globalized world. But it is important to keep in mind that the charge of being unrealistic starts to lose both its intellectual and political traction once we recognize that we are anticipating an unimaginable ecological catastrophe.

This turn to the local, as we have argued, is thus anything but unrealistic. For one thing, local activism has always been seen as a necessary part of a full-scale global strategy to deal with climate change; it is written into basic documents on sustainable development. The problem is rather that it has too often been neglected or subordinated to other concerns. Further, when the local level is considered, it is mainly understood narrowly as a component of a multi-level global system, an understanding that fails to capture not only the significance of localism on its own terms but also its potential for energetic activities, social creativity, and political imagination. And finally, but not least, we have seen that such eco-localism is in fact thriving around the globe, even if operating under the conventional radar. That is, it does not have to be invented; it is up and running. It only needs to be (re)discovered, supported, and sustained.

Localism's ecological contributions to climate crisis can be substantial. Having looked for alternatives that might be available during times of dire ecological stress, even possible survival for many, the relocalization movement can take on new meaning for desperate people. This may be difficult to see now, but the relearning of skills from earlier ways of life based on less materialism may well be one of the few viable options for many. As our understanding of modern life changes, many will likely flee uninhabitable cities in search of alternative ways of getting by. Rather than a choice, this may emerge then as a necessity. And it is just here that the relocalization movement, and Global Ecovillage Network in particular, will suddenly become more visible and potentially attractive as options that have many useful things to teach and pass along.

But even more importantly for present purposes, eco-localism also offers a venue for preserving a semblance of democracy during dark times. If we are serious about the democratic way of life during climate crisis—or under normal circumstances, for that matter—we need to anchor democratic practices to venues outside of an unsustainable elitist state and its bureaucratic institutions. How, we have asked, might we be able to preserve and sustain democratic values and practices under dire ecological stress? Toward this end, we identified ecovillages, eco-neighborhoods, and Transition Towns as sites for refocusing the case for participatory environmental democracy. Despite standard arguments about the need for representative government, genuine democracy—strong democracy—has always required small numbers of people who can more or less address one another face to face. The relocation movement and ecovillages, in this regard, are seen as places where the practice of democratic governance has a reasonable chance in the face of an eco-authoritarian challenge. Linked up with the kind of civil society politics that has long been an inherent part of environmentalism, as we saw in Chapter 5, the practices of participatory democratic governance also provide a bridge to environmental political theory, including deliberative democracy.

It is not that we will all move to ecovillages, or even that all ecovillages are democratic. But it is easier to democratize eco-communes than it is to democratize the liberal-capitalist state. The point is that these movements are on the whole producing valuable participatory insights and practices that will most likely be useful in the future. Whether or not we will follow such a path, which surely remains politically uncertain, eco-localism nonetheless shows that we do have a concrete and ready alternative when the question turns to eco-democratic governance. On the one hand, it is a fallback position available during hard times, but it can at the same time be a way to revitalize an authentic democratic way of life. Moreover, a particularly attractive option is the urban variant of the ecovillage or eco-neighborhood, given that the majority will be unable to escape the polluted cities. For this reason, as we have argued, we should prioritize the passage of public policy initiatives that can both encourage their establishment and facilitate the burden of their activities.

To be clear, as indicated earlier, democratic localism and participatory governance are not *the* solutions to climate change. Moreover, it is surely the case that localism and participation will not be attractive to everyone, just as is the case now. But the issue presented here is different. Future generations will be confronting a situation that is unattractive from the outset. When the predicted hardships of crisis arrive, questions related to coping economically and politically with a different reality will for many overtake questions of everyday comfort. In many places, lifestyle choices will be significantly reduced, even dramatically so in others. This will particularly be the case in large parts of the developing world, though the developed world will be affected as well.

Dealing with the situation ahead will require many different types of strategies coming from different directions at all levels of government. One of them will have to remain focused on the global level. As indicated from the outset, the emphasis on democratic relocalization is not an argument for forgetting about the problems posed by the global system. In so far as the transnational movement of capital, goods, and people continues to grow, this struggle has to be carried on. Indeed, people in the relocalization movement are among the most active environmentalists protesting global climate change. Relocalization is, as Hess (2009) explains, a call for global eco-localism. And, at the same time, eco-communes are schools for ecological citizenship. As Bahro argued, a new kind of ecologically conscious citizen is the precondition for a democratic sustainable transition.

The purpose of this volume, then, has been first and foremost to broaden the discourse in environmental political theory. A relevant democratic theory, we have maintained, will need to find ideas and issues that capture the imaginations of democratic activists willing to take on the pressing political issues brought forth by climate crisis. The contemporary theoretical emphases on greening the democratic state and environmental deliberative democracy are not irrelevant, but they draw up short. While theoretically commendable, they are more ideal than practical in the face of current challenges posed by political-ecological transition, let alone those that will be confronted under serious ecological duress. We thus turned to participatory environmental governance anchored to civil society, a more politically engaged and practical model which incorporates the efforts of eco-localism, including ecovillages as we saw in Chapter 8, but also connects to the political ideas of earlier radical environmentalists such as Sale, Bookchin, and Bahro. The argument has been to bring these progressive thinkers back in. In particular, ideas about associative environmental democracy, insurgent environmental politics, and participatory environmental expertise offer important insights for theoretical rejuvenation. These ideas, of course, do not provide all the answers or solutions, but they provide good groundwork for starting a much-needed discussion.

Without reservations, we have supported the argument that democracy is ultimately important for a sustainable society. But we have done it with an important caveat. The case for democracy cannot be spelled out in relatively ideal terms of conventional democratic political theory. The democratic prospect in times ahead will depend on the ability to make the case in terms of immanent political and social realities associated with credible agents of democratic change. Short of that, political theory leaves open an intellectual vacuum between theory and practice that eco-authoritarians will easily step into to take us in undemocratic directions.

Eco-authoritarianism will only perpetuate the ecological problem. While their environmental guardians can for a while protect the lifestyles of elite citizens, perhaps in the form of a fortress world, it is highly unlikely that authoritarianism will be able to come to grips with the underlying normative and empirical assumptions of the corporate-dominated consumer societies that have created and continue to drive us toward climate crisis. For this reason, it is important to stand with those who call for more democracy, particular democracy in the name of environmental justice. We have argued in this work, however, that the best chances for saving democratic governance along the way will come from local initiatives beyond the sphere of large bureaucratic governments. It is important to concede that not all such local activities will be democratic; some may even follow theories of political anarchism, with others falling prey to demagoguery. In this regard, there are no guarantees. But, as is the case today, some will in fact preserve the democratic practices of participatory environmental governance during life under climate crisis.

In this view here, then, the initiatives on the part of concerned activists in civil society will have the best chances of securing this essential democratic component of a strategy for sustainability (Brecher 2016). Perhaps they will be able along the way to influence stymied institutions of the state; that would be a welcome development. But waiting for bureaucratic state institutions to act, as Hopkins (2013) has made clear, is no longer something we can afford to do. Local efforts where we live will not stop global warming, but they can make the future more manageable and meaningful for many. Such efforts, as parallel structures, already exist as part of the relocalization movement, including

Transition Towns and ecovillages. But they need more attention and support, intellectually, politically, and materially. It is our hope that the discussion in these pages can contribute to this goal. Hopefully, even if only in a small way, it will be a contribution to the democratic struggle to create a more sustainable future, if not survival, over the long haul.

References

Abensour, M. (2011). *Democracy against the State*. London: Polity Press.

Achterberg, W. (1996). "Sustainability and Associative Democracy," in W. M. Lafferty and J. Meadowcroft (eds.), *Democracy and the Environment: Problems and Issues*. Cheltenham: Edward Elgar, 157–74.

Achterberg, W. (1996). "Sustainability, Community and Democracy," in B. Doherty and M. de Geus (eds.), *Democracy and Green Political Thought: Sustainability, Rights and Citizenship*. London: Routledge, 170–87.

Adger, W. N. and Jordan, A. (2009). *Governing Sustainability*. Cambridge: Cambridge University Press.

Adhikari, B. (2009). "Reducing Emissions from Deforestation and Degradation: Some Issues and Considerations," *Journal of Forest and Livelihood*, 8 (1): 14–24.

Agrawal, A. (2005). *Environmentality: Technologies of Government and the Making of Subjects*. Durham: Duke University Press.

Aldrich, D. P. (2012). *Building Resilience: Social Capital in Post-Disaster Recovery*. Chicago: University of Chicago Press.

Aldridge, J. (2015). *Participatory Research*. Bristol: Policy Press.

Alloun, E. and Alexander, S. (2014). *The Transition Movement: Questions of Diversity, Power, and Affluence*. Simplicity Institute Report, 14. http://simplicityinstitute.org/wp-content/uploads/2011/04/TransitionMovement.pdf

Andersen, I. E. and Jaeger, B. (1999). "Scenario Workshops and Consensus Conferences: Towards More Democratic Decision-Making," *Science and Public Policy*, 26: 331–40.

Archer, D. and Rahmstorf, S. (2010). *The Climate Crisis: An Introductory Guide to Climate Change*. Cambridge: Cambridge University Press.

Arendt, H. (1965). *On Revolution*. New York: Viking Press, 254–85.

Arias-Maldonado, M. (2013). "Rethinking Sustainability in the Anthropocene Environmental Politics," *Environmental Politics*, 22 (3): 428–46. http://dx.doi.org/10.1080/09644016.2013.765161

Aristotle (1998). *Politics* (translated by C. D. C. Reeves). Indianapolis: Hackett Publishing, Book VII, Chapter 4.

Association of Climate Change Officers (2012). *Defense, National Security, and Climate Change: Building Resilience and Identifying Opportunities related to Water, Energy and Extreme Events*, Washington, DC: Association of Climate Change Officers.

Austin, J. E. and Bruch, C. E. (eds) (2000). *The Environmental Consequences of War: Legal, Economic, and Scientific Perspectives*. Cambridge: Cambridge University Press.

Avritzer, L. (2012). "The Different Designs of Public Participation in Brazil: Deliberation, Power Sharing and Public Ratification," *Critical Policy Studies*, 6 (2): 113–27.

Baber, W. F. and Bartlett, R. V. (2005). *Deliberative Environmental Politics: Democracy and Ecological Rationality*. Cambridge: MIT Press.

Bache, I. and Chapman, R. (2008). "Democracy through Multilevel Governance? The Implementation of the Structural Funds in South Yorkshire," *Governance*, 21 (4): 397–418.

Bäckstrand, K. J., Kahn, A., Kronsell, E. and Loevbrand, E. (eds) (2010). *Environmental Politics and Deliberative Democracy: Examining the Promise of New Modes of Governance*. Cheltenham: Edward Elgar.

Bader, V. (2001). "Problems and Prospects of Associative Democracy: Cohen and Rogers Revisited," *Critical Review of International Social and Political Philosophy*, 4 (1): 31–70.

Bahro, R. (1981). *The Alternative in Eastern Europe*. London: Verso.

Bahro, R. (1984). *From Red to Green*. London: Verso.

Bahro, R. (1987). *Logik der Rettung: Wer Kann die Apokalypse Aufhalten?* Stuttgart: Weitbrecht.

Bahro, R. (1994). *Avoiding Social and Ecological Disaster: The Politics of World Transformation*. Bath: Gateway Books.

Baiocchi, G. (2003). "Participation, Activism, and Politics: The Porto Alegre Experiment," in A. Fung and E. O. Wright (eds.), *Deepening Democracy*. New York: Verso, 77–102.

Bang, J. M. (2005). *Ecovillages: A Practical Guide to Sustainable Communities*. Gabriola Island, Canada: New Society Publishers.

Barber, B. (2004). *Strong Democracy: Participatory Politics for a New Age*. Berkeley: University of California Press.

Barber, B. R. (2013). *If Mayors Ruled the World: Dysfunctional Nations, Rising Cities*. New Haven: Yale University Press.

Barnett, J. and Adger, W. N. (2007). "Climate Change, Human Security, and Violent Conflict," *Political Geography*, 26 (6): 639–55.

Barns, I. (1996). "Environment, Democracy and Community," in F. Mathews (ed.), *Ecology and Democracy*. London: Frank Cass, 101–33.

Barringer, F. (2014). "Climate Change Threatens to Disrupt the Ranges of Birds," *New York Times*, September 8. http://www.nytimes.com/2010/03/13/science/earth/13birds.html?_r=0

Barry, J. (1999). *Rethinking Green Politics*. London: Sage.

Barry, J. (2001). "Greening Liberal Democracy: Theory, Practice and Political Economy," in J. Barry and M. Wissenburg (eds.), *Sustaining Liberal Democracy: Ecological Challenges and Opportunities*. London: Palgrave, 59–81.

Barry, J. (2008). "Toward a Green Republicanism: Constitutionalism, Political Economy and the Green State," *The Good Society*, 17 (2): 3–11.

Barry, J. (2012). *The Politics of Actually Existing Unsustainability*. Oxford: Oxford University Press.

Barry, J. and Eckersley, R. (eds). (2005). *The State and the Global Ecological Crisis*. Cambridge: MIT Press.

Barry, W. (1977). *The Unsettling of America: Culture and Agriculture*. Oakland: Sierra Club.

Been, E. A. (2016). "Q&A: Douglas Rushkoff, The Growth Trap," interviewed by E. A. Been, *Village Voice*, March 16–22, LXI, (11): 13.

Beeson, M. (2016). "Environmental Authoritarianism and China," in T. Gabrielson, C. Hall, J. M. Meyer, and D. Schlosberg (eds.), *The Handbook of Environmental Political Theory*. Oxford: Oxford University Press, 520–32.

Beck, U. and Beck-Gernsheim, E. (2002). *Individualization: Institutionalized Individualism and its Social and Political Consequences*. London: Sage Publications.

Beierle, T. C. and Cayford, J. (2002). *Democracy in Practice: Public Participation in Environmental Decisions*. Washington, DC: Resources for the Future.

Berger, B. (2013). "The Accidental Theorist: Diana Mutz & Deliberative Democracy," *Critical Review*, 25 (2): 181–98.

Berman, M. (2000). *The Twilight of American Culture*. London: Gerald Ducksworth.

Bernstein, S. (2002). *The Compromise of Liberal Environmentalism*. New York: Columbia University Press.

Bhatta, C. D. (2007). "Civil Society in Nepal: In Search of Reality," *Contributions to Nepalese Studies*, 34 (1): 45–57.

Bhattarai, A. M. and Khanal, D. R. (2005). *Communities, Forests and Law of Nepal: Present State and Challenges*. Kathmandu: Center of International Law.

Biehl, J. (1995). *Ecofascism: Lessons from the German Experience*. Edinburgh and San Francisco: AK Press.

Biehl, J. (2011). *Mumford Gutkind Bookchin: The Emergence of Eco-Decentralism*. Porsgrunn, Norway: New Compass.

Biel, R. (2012). *The Entropy of Capitalism*. London: Haymarket Books.

Blaug, R. (1999). *Democracy, Real and Ideal*. Albany: State University of New York Press.

Blühdorn, I. (2013). "The Governance of Unsustainability: Ecology and Democracy after the Post-Democratic Turn," *Environmental Politics*, 22 (1): 16–36.

Blühdorn, I. (2016). "Sustainability—Post-Sustainability—Unsustainability," in C. Gabrielson, C. Hall, J. M. Meyer, and D. Schlosberg (eds.), *The Handbook of Environmental Political Theory*. Oxford: Oxford University Press, 259–73.

Bohman, J. (1996). *Public Deliberation: Pluralism, Complexity, and Democracy*. Cambridge: Cambridge University Press.

Bomberg, E. and McEwen, N. (2012). "Mobilizing Community Energy," *Energy Policy*, 51: 435–44.

Bookchin, M. (1982). *The Ecology of Freedom*. Palo Alto: Cheshire Books.

Bookchin, M. (1986). *Post-Scarcity Anarchism*. Montreal: Black Rose Books.

Bookchin, M. (1987). *The Rise of Urbanization and the Decline of Citizenship*. San Francisco: Sierra Club Books.

Bookchin, M. (1990). *Remaking Society: Pathways to a Green Future*. Boston: South End Press.

Bookchin, M. (1991). "Libertarian Municipalism: An Overview," *Green Perspectives*, (24). http://dwardmac.pitzer.edu/Anarchist_Archives/bookchin/gp/perspectives24.html

Bookchin, M. (1992). *Urbanization without Cities: The Rise and Decline of Citizenship*. Montreal: Black Rose Books.

Bookchin, M. (1996). *Toward an Ecological Society*. Montreal: Black Rose Books.

Bookchin, M. (2000). "Thoughts on Libertarian Municipalism," *Left Green Perspectives*, 41. http://social-ecology.org/wp/1999/08/thoughts-on-libertarian-municipalism/

Bookchin, M. (2007). *Social Ecology and Communalism* (ed. by E. Eiglad). Edinburgh: AK Press.

Bookchin, M. (2015). *The Next Revolution: Popular Assemblies and the Promise of Direct Democracy*. New York: Verso.

Bookchin, M. with Biehl, J. (1997). *The Politics of Social Ecology: Libertarian Municipalism*. Montreal: Black Rose Books.

Borgerson, S. G. (2008). "Arctic Meltdown: The Economic and Security Implications of Global Warming," *Foreign Affairs*: March/April, 63–7.

Bourdieu, P. (1977). *Outline of a Theory of Practice*. Cambridge: Cambridge University Press.

Brand, R. and Fischer, J. (2012). "Overcoming the Technophilia/Technophobia Split in Environmental Discourse," *Environmental Politics*, 22 (2): 235–54.

Brand, U. (2010). "Sustainable Development and Ecological Modernization—The Limits to a Hegemonic Policy Knowledge," *Innovation—The European Journal of Social Science Research*, (2): 135–52.

Braun, B. (2000). "Producing Vertical Territory: Geology and Governmentality in Late Victorian Canada," *Cultural Geographies*, 7 (1): 17–46.

Braun, B. (2003). *The Intemperate Rainforest*. Minneapolis: University of Michigan Press.

Brecher, J. (2016). *Climate Insurgency: A Strategy for Survival*. London: Routledge.

Bressen, T. (2012). "Consensus Decisions-Making, What, Why, How." http://treegroup.info/topics/consensus-in-sharing-law.pdf

Brint, S. (1994). *In an Age of Experts*. Princeton: Princeton University Press.

Broder, J. M. (2012). "Climate Change Threatens Migratory Birds, Report Says," *New York Times*, March 10.

Broers, L. (2005). *The Triumph of Technology: The BBC Reith Lectures 2005*. Cambridge: Cambridge University Press.

Bromley, D. W. (2005). "The Poverty of Sustainability: Rescuing Economics from Platitudes," *Agricultural Economics*, 32 (1): 201–10.

Bromwich, J. (2016). "Flooding in the South Looks a Lot like Climate Change," *New York Times*. http://www.nytimes.com/2016/08/17/us/climate-change-louisiana.html?_r=0

Brooks, D. (2016). "The Great Affluence Fallacy," *New York Times*. http://www.nytimes.com/2016/08/09/opinion/the-great-affluence-fallacy.html?_r=0

Brown, L. R. (2003). *Plan B: Rescuing a Planet Under Stress and a Civilization in Trouble*. New York: W.W. Norton.

Brown, P. and Mickkelsen, E. J. (1990). *No Safe Pace: Toxic Waste, Leukemia and Community Action*. Berkeley: University of California Press.

Bryant, R. (2001). "Political Ecology: A Critical Agenda for Change?" in N. Castree and B. Brawn (eds.), *Social Nature: Theory Practice and Politics*. Malden, MA: Blackwell Publishers, 151–69.

Buchstein, H. (2009). *Democratie und Lotterie: Das Los als politisches Entscheidungsinstrument von der Antike bis zur EU*. Frankfurt: Campus Verlag.

Buck, J. and Villines, S. (2007). *We the People: Consenting to a Deeper Democracy*. Washington: Sociocracy Info.

Bulkeley, H. and Betsill, M. (2003). *Cities and Climate Change: Urban Sustainability and Global Environmental Governance*. London: Routledge.

Burawoy, M. (2005). "For Public Sociology," *America Sociological Review*, 70: 4–28.

Burnell, P. J. (2012). "Democracy, Democratization and Climate Change: Complex Relationships," *Democratization*, 19 (5): 813–42.

Busby, J. W. (2007). "Climate Change and National Security: An Agenda for Action," *Foreign Affairs*, 32. New York: Council on Foreign Relations.

Carius, A., Dabelko, D. G. and Wolf, A. T. (2004). "Water, Conflict, and Cooperation," *Policy Brief: The United Nations and Environmental Security*, 10: 61–6.

Carpini, M. X. D., Cook, F. L. and Jacobs, L. R. (2004). "Public Deliberation, Discursive Participation, and Citizen Engagement: A Review of Empirical Literature," *Annual Review of Political Science*, 7: 315–44.

Carson, R. (1962). *Silent Spring*. New York: Houghton Mifflin.

Carter, N. (2010). *The Politics of the Environment: Ideas, Activism, Policy*. Cambridge: Cambridge University Press.

Castree, N. and Braun, B. (eds.). (2011). *Social Nature: Theory, Practice and Politics*. Malden, MA: Blackwell Publishers.

Cato, M. S. and Hillier, J. (2010). "How Could We Study Climate-Related Social Innovation? Applying Deleuzean Philosophy to Transition Towns," *Environmental Politics*, 19: 869–87.

Catton, W. R. (1982). *Overshoot: The Ecological Basis for Revolutionary Change*. Urbana: University of Illinois Press.

Chambers, R. (1997). *Whose Reality Counts? Putting the First Last*. London: Intermediate Technology Publications.

Chambers, S. (2003). "Deliberative Democratic Theory," *Annual Review of Political Science*, 6: 307–26.

Chaudhary, P. and Arysal, K. P. (2009). "Global Warming in Nepal: Challenges and Policy Imperatives," *Journal of Forest and Livelihood*, 8 (1): 4–13.

Chhotray, V. and Stoker, G. (2010). *Governance Theory and Practice: A Cross-Disciplinary Approach*. London: Palgrave Macmillan.

Chatterton, P. and Cutler, A. (2008). "The Rocky Road to A Real Transition: The Transition Towns Movement and What It Means for Social Change." Leeds, UK: Trapese Collective. https://www.transitionculture.org/2008/05/15/the-rocky-road-to-a-real-transition-by-paul-chatterton-and-alice-cutler-a-review/

Choudry, A. (2015). *Learning Activism: The Intellectual Life of Contemporary Social Movements*. Toronto: University of Toronto Press.

Christian, D. L. (2003). *Creating a Life Together: Practical Tools to Grow Ecovillages and Intentional Communities*. Gabriola Island, BC, Canada: New Society Publishers.

Christoff, P. (1996). "Ecological Citizens and Ecologically Guided Democracy," in B. Barry and M. de Gues (eds.), *Democracy and Green Political Thought: Sustainability, Rights, and Citizenship*. London: Routledge, 151–69.

Christoff, P. (2016). "The Promissary Note: COP 21 and the Paris Climate Agreement," *Environmental Politics*, 15 (5): 765–87.

CNA Corporation (2007). *Climate Change and National Security: An Agenda for Action*. CSR (32), November.

Cohen, D. K. M. (2010). "Reaching Out for Resilience: Exploring the Approaches to Inclusion and Diversity in the Transition Movement." https://transitionnetwork.org/resources/reaching-out-resilience-exploring-approaches-inclusion-and-diversity-transition-movement

Cohen, J. and Rogers, J. (1992). "Secondary Associations and Democratic Governance," *Politics and Society*, 20 (4): 393–472.

Cohen, R. (2016). "The Death of Liberalism," *New York Times*, Op-Ed. April 14.

Cole, C. D. G. (1920). *Guild Socialism Re-stated*. London: Leonard Parsons.

Coles, R. (2016). "Environmental Political Theory and Environmental Action Research Teams," in T. Gabrielson, C. Hall, and D. Schlosberg (eds.), *The Oxford Handbook of Environmental Political Theory*. Oxford: Oxford University Press, 131–43.

Connelly, J. and Smith, G. (2003). *Politics and the Environment: From Theory to Practice*. London: Routledge.

Cooke, B. and Kothari, U. (2001). *Participation: The New Tyranny?* London: Zed Books.

Cooke, P. and Morgan, K. (1998). *The Associational Economy: Firms, Regions, and Innovation*. Oxford: Oxford University Press.

Corkery, M. (2016). "Banks Pull Back on Funding Goal: Once a Driver of Profits, Now Too Risky a Bet," *New York Times*, March 1.

Cote, M. and Nightingale, A. J. (2012). "Resilience Thinking Meets Social Theory: Situating Social Change in Socio-ecological Systems (SES) Research," *Progress in Human Geography*, 36 (4): 475–89.

Crosby, N. and Nethercut, D. (2005). "Citizen Juries: Creating a Trustworthy Voice of the People," in F. Gastil and P. Levine (eds.), *The Deliberative Democracy Handbook*. San Francisco: Jossey-Bass, 111–19.

Crouch, C. (2004). *Post-Democracy*. Cambridge: Polity.

Cullen, H. (2016). "What Weather Is the Fault of Climate Change?" *New York Times*, March 11. http://www.nytimes.com/2016/03/11/opinion/what-weather-is-the-fault-of-climate-change.html?_r=0

Curato, N. (2014). "Participation without Deliberation: The Crisis of Venezuelan Democracy," *Democratic Theory*, 1 (2): 113–21.

Dahal, D. R. (2001). *Civil Society in Nepal: Opening the Ground for Questions*. Kathmandu: Centre for Development and Governance.

Dahl, R. (1989). *Democracy and Its Critics*. New Haven: Yale University Press.

Dahl, R. (1990). *After the Revolution?* New Haven: Yale University Press.

Dahl, R. (1994). "A Democratic Dilemma: Systems Effectiveness and Citizen Participation," *Political Science Quarterly*, 109 (1): 23–34.

Dalby, S. (2002). *Environmental Security*. Minneapolis: University of Minnesota Press.

Daly, H. (1977). *Steady-State Economics*. Washington: Island Press.

Daly, H. (2001). *Ecological Economics: Principles and Applications*. Washington, DC: Island Press.

Darier, E. (1999). "Foucault and the Environment: An Introduction," in E. Darier (ed.), *Discourses of the Environment*. Malden, MA: Blackwell Publishers, 1–34.

Davenport, C. (2014). "Pentagon Signals Security Risk of Climate Change," *New York Times*, October 13. http://www.nytimes.com/2014/10/14/us/pentagon-says-global-warming-presents immediate-security-threat.html

Davenport, C. (2016a). "Paris Climate Deal Passes Milestone as 20 More Nations Sign On 'Without the Faintest Idea How They're Going to Achieve the Goals.'" *Before It's News*, September 22. http://beforeitsnews.com/environment/2016/09/paris-climate-deal-passes-milestone-as-20-more-nations-sign-on-without-the-faintest-idea-how-theyre-going-to-achieve-the-goals-2562663.html

Davenport, C. (2016b). "Nations Agree to Cut Use of Harmful Coolant," *New York Times*, October 15: 1.

Davies, J. (2011). *Challenging Governance Theory: From Network to Hegemony*. Bristol: Policy Press.

Davis, J. H., Landler, M. and Davenport, C. (2016). "Obama on Climate Change: The Trends are 'Terrifying'," *New York Times*. http://www.nytimes.com/2016/09/08/us/politics/obama-climate-change.html?_r=0

Dawson, J. (2006). *Ecovillages*. Cambridge: Green Books.

De Geus, M. (1996). "The Ecological Restructuring of the State," in B. Doherty and M. de Geus (eds.), *Democracy and Green Political Thought: Sustainability, Rights, and Citizenship*. London: Routledge, 188–211.

De-Shalit, A. (2000). *The Environment: Between Theory and Practice*. Oxford: Oxford University Press.

De Tocqueville, A. (2003). *Democracy in America*. 13th edition. New York: Penguin.

De Young, R. and Princen, T. (eds.) (2012). *The Localization Reader: Adapting to the Coming Downshift*. Cambridge: MIT Press.

Deakin, N. and Taylor, M. (2002). "Citizenship, Civil Society and Governance." Reut Institute. http://reut-institute.org/data/uploads/Articles%20and%20Reports%20from%20other%20organizations/deakin-taylor-civil%20society%20UK.pdf.

Death, C. (2011). "Summit Theatre: Exemplary Governmentality and Environmental Diplomacy in Johannesburg and Copenhagen," *Environmental Politics*, 20 (1): 1–19.

Deen, T. (2008). "Climate Change Deepening World Water Crisis: Global Policy," *Forum*, March 19. https://www.globalpolicy.org/component/content/article/212environment/45362-climate-change-deepening-world-water-crisis.html

Demirovic, A. (1994). "Ecological Crisis and the Future of Democracy," in M. O'Connor (ed.), *Is Capitalism Sustainable: Political Economy and the Politics of Ecology*. New York: Guilford Press, 253–74.

Denters, B., Goldsmith, M., Ladner, A., Mouritzen, P. E. and Rose, L. E. (2014). *Size and Local Democracy*. Cheltenham: Edward Elgar.

Devall, B. and Sessions, G. (1985). *Deep Ecology*. Salt Lake City: Gibbs Smith.

Dewey, J. (1927). *The Public and Its Problems*. New York: Swallow.

Dhital, N., Paudel, K. P. and Ojha, H. (2003). "Inventory of Community Forests in Nepal: Problems and Opportunities," *Journal of Forest and Livelihood*, 3 (1): 62–6.

Di Liberto, D. (2015). "Indian Heat Wave Kills Thousands." Climate.gov, June. https://www.climate.gov/news-features/event-tracker/india-heat-wave-kills-thousands

Diamond, J. (2011). *Collapse: How Societies Choose to Fail or Succeed*. Revised Edition. New York: Penguin Books.

Diesing, P. (1962). *Reason in Society: Five Types of Decisions in their Social Contexts*. Urbana: University of Illinois Press.

Dobson, A. (1990). *Green Political Thought*. London: Unwin Hyman.

Dobson, A. (2003). *Citizenship and the Environment*. Oxford: Oxford University Press.

Dobson, A. and Sáiz, A. V. (2005). "Introduction," *Environmental Politics*, 14 (2): 157–62.

Dryzek, J. S. (1987). *Rational Ecology: Ecology and Political Economy*. London: Blackwell.

Dryzek, J. S. (1990). *Discursive Democracy: Politics, Policy and Political Science*. Cambridge: Cambridge University Press.

Dryzek, J. S. (1995). "Political and Ecological Communication," *Environmental Politics*, 4 (4): 13–30.

Dryzek, J. S. (2000). "Insurgent Democracy: Civil Society and State," *Deliberative Democracy and Beyond: Liberals, Critics, Contestations*. Oxford: Oxford University Press, 81–114.

Dryzek, J. S. (2005). *The Politics of the Earth: Environmental Discourse*. Oxford: Oxford University Press.

Dryzek, J. S. (2006). *Deliberative Global Politics*. Cambridge: Polity Press.

Dryzek, J. S. (2009). "The Australian Citizens' Parliament: A World First," *Journal of Public Deliberation*, 5 (1): 1–9.

Dryzek, J. S. (2016). "Institutions for the Anthropocene: Governance in a Changing Earth System." *British Journal of Political Science*, 46 (4): 937–56.

Dryzek, J. S., Downes, D., Hunold, C. and Schlosberg, D. with H. Hernes (2003). *Green States and Social Movements: Environmentalism in the U.S., U.K., Germany, and Norway*. Oxford: Oxford University Press.

Dryzek, J. S., Norgaard, R. B. and Schlosberg, D. (eds.) (2011). *The Oxford Handbook of Climate Change and Society*. Oxford: Oxford University Press.

Duit, A., Feindt, P. H. and Meadowcroft, J. (eds.) (2016). "Special Issue: Greening Leviathan? The Emergence of the Environmental State," *Environmental Politics*, 25 (1): 1–23.

Durnova, A. (2015). "Lost in Translation: Expressing Emotions in Policy Deliberation," in F. Fischer, D. Torgerson, A. Durnova, and M. Orsini (eds.), *Handbook of Critical Policy Studies*. Cheltenham: Edward Elgar, 222–40.

Dutschke, R. and Gollwitzer, H. (1980). *Mein Langer Marsch: Reden, Schriften und Tagebucher aus Zwanzig Jahren*. Berlin: Rororo Aktuell.

Dyer, G. (2010). *Climate Wars: The Fight for Survival as the World Overheats*. Oxford: Oneworld Publications.

Eckersley, R. (1987). "Green Politics: A Practice in Search of a Theory." Paper delivered at the Ecopolitics II Conference, University of Tasmania, May 22–5, 1987.

Eckersley, R. (1992). *Environmentalism and Democratic Theory: Toward an Ecocentric Approach*. Albany: State University of New York.

Eckersley, R. (2004). *The Green State: Rethinking Democracy and Sovereignty*. Cambridge: MIT Press.

Economist Magazine (2010). "Geoengineering: Lift Off," November 6: 99–100.

Economist Magazine (2011). "Welcome to the Anthropocene," May 26. http://www.economist.com/node/18744401

Economist Magazine (2015). "If All Else Fails," November 28. http://www.economist.com/news/special-report/21678963-man-made-global-cooling-scary-may-become-necessary-if-all-else-fails

Edelman, M. J. (1985). *The Symbolic Uses of Politics*. Urbana: University of Illinois Press.
Elliott, D. (2016). *Balancing Green Power: How to Deal with Variable Energy Sources*. Bristol: IOP Publishing.
Elstub, S. and McLaverty, P. (eds.) (2014). *Deliberative Democracy: Issues and Cases*. Edinburgh: University of Edinburgh Press.
Esty, D. (2014). "Bottom-Up Climate Fix," *New York Times*, September 21.
Eurobarometer Survey Series (2012). http://www.icpsr.umich.edu/icpsrweb/ICPSR/series/26
Faber, D. (1998). *The Struggle for Ecological Democracy: The Environmental Justice Movement in the United States*. New York: Guilford.
Fainstein, S. (2013). "Resilience and Justice." Melbourne: University of Melbourne, Melbourne Sustainability Society Institute, Research Paper Series, No. 2.
Fainstein, S. (2015). "Resilience and Justice," *International Journal of Urban and Regional Planning*, 39: 157–67.
Fainstein, S. (2016). "Resilience and Justice in New York City." Unpublished paper.
Farrell, J. (2010). "Dirty Coal, Clear Future." *Atlantic Monthly*. http://www.theatlantic.com/magazine/archive/2010/12/dirty-coal-clean-future/308307/
Featherstone, D., Ince, A., Mackinnon, D., Strauss, K. and Cumbers, A. (2012). "Progressive Localism and the Construction of Political Alternatives," *Transactions of the Institute of British Geographers*, 37: 177–82.
Feldman, M. S. and Khademian, A. M. (2007). "The Role of the Public Manager in Inclusion: Creating Communities of Participation," *Governance*, 20 (2): 305–24.
Ferguson, K. (1985). *The Feminist Case Against Bureaucracy*. Philadelphia: Temple University Press.
Filkins, D. (2016). "The End of Ice," *New Yorker*, April 4.
Fischer, F. (1980). *Politics, Values and Public Policy: The Problem of Methodology*. Boulder, CO: Westview Press.
Fischer, F. (1990). *Technocracy and the Politics of Expertise*. Newbury Park, CA: Sage.
Fischer, F. (1995). *Evaluating Public Policy*. Belmont, CA: Wentworth.
Fischer, F. (2000). *Citizens, Experts, and the Environment: The Politics of Local Knowledge*. Durham: Duke University Press.
Fischer, F. (2003). *Reframing Public Policy: Discursive Politics and Deliberative Practices*. Oxford: Oxford University Press.
Fischer, F. (2006). "Participatory Governance as Deliberative Empowerment: The Cultural Politics of Discursive Space," *The American Review of Public Administration*, 36 (1): 19–40.
Fischer, F. (2009). *Democracy and Expertise: Reorienting Policy Inquiry*. Oxford: Oxford University Press.
Fischer, F. (2015). "Environmental Democracy," in K. Bäckstrand and E. Lövbrand (eds.), *Research Handbook on Climate Governance*. Cheltenham: Edward Elgar, 435–45.
Fischer, F. and Black, M. (1995). *Greening Environmental Policy: Toward a Sustainable Future*. New York: St. Martins Press.
Fischer, F. and Boossabong, P. (2016). "Deliberative Policy Analysis," in J. Dryzek et al. (eds.), *Handbook of Deliberative Democracy*. Oxford: Oxford University Press, forthcoming (early 2018).

Fischer, F. and Gottweis, H. (eds.) (2012). *The Argumentative Turn Revisited: Public Policy as Communicative Practice*. Durham: Duke University Press, 343–70.

Fischer, F. and Mandell, A. (2012). "Transformative Learning in Planning and Policy Deliberations: Probing Social Meaning and Tacit Assumptions," in F. Fischer and H. Gottweis (eds.), *The Argumentative Turn Revisited*. Durham: Duke University Press, 343–70.

Fischer, F. (2012). "Participatory Governance," in D. Levi-Faur (ed.), *The Oxford Handbook of Governance*. Oxford: Oxford University Press: 457–71.

Fishkin, J. S. (1996). *The Voice of the People: Public Opinion and Democracy*. New Haven, CT: Yale University Press.

Fishkin, J. S. (2009). *When the People Speak: Deliberative Democracy and Public Consultation*. Oxford: Oxford University Press.

Fleming, T., Ghilardi, L. and Napier, N. (2006). "Rethinking Small Places—Urban and Cultural Creativity: Examples from Sweden, the USA and Bosnia-Herzegovina," in D. Bell and M. Jayne (eds.), *Small Cities: Urban Experience Beyond the Metropolis*. London: Routledge, 145–65.

Flint, J. and Raco, M. (2011). *The Future of Sustainable Cities: Critical Reflections*. Bristol: Policy Press.

ForestAction, Nepal (2010). *Annual Report: ForestAction Nepal*. Patan: ForestAction, Nepal.

Forsyth, T. (2015). "Vulnerability," in K. Bäckstrand and E. Lövbrand (eds.), *Research Handbook on Climate Governance*. Cheltenham: Edward Elgar, 164–74.

Foster, J. B. (2015). "Marxism and Ecology: Common Fronts of a Great Transition," *Great Transition Initiative*. http://tratarde.org/wp-content/uploads/2015/11/Foster-Marxism-and-Ecology-common-Fonts-2015.pdf

Foucault, M. (1972). *The Archeology of Knowledge*. New York: Pantheon.

Foucault, M. (1973). *The Order of Things*. New York: Vintage.

Foucault, M. (1980). *Knowledge/Power: Selected Interviews and Other Writings*. New York: Pantheon.

Foucault, M. (1991). "Governmentality," in G. Burchell, C. Gordon, and P. Miller (eds.), *The Foucault Effect*. London: Harvester Wheatsheaf, 87–104.

Frankel, B. (1987). *Post-Industrial Utopians*. Madison: University of Wisconsin Press.

Franzke, J. and Kleger, H. (2012). *Burgerhaushalte: Chanzen und Grenzen*. Berlin: Edition Sigma.

Freire, P. (1970). *Pedagogy of the Oppressed*. New York: Seabury Press.

Freire, P. (1973). *Education for Critical Consciousness*. New York: Seabury Press.

Fung, A. (2006). "Varieties of Participation in Complex Governance," *Public Administration Review*, 66 (1): 66–75.

Fung, A. and Wright, E. O. (eds.) (2003). *Deepening Democracy: Institutional Innovations in Empowered Participatory Governance*. New York: Verso.

Gabels, A. (2007). "Citizen Participation and System Effectiveness? Assessing Participatory Technology Assessment in Representative Democracy," *Interdisciplinary Information Sciences*, 13 (1): 103–16.

Gabrielson, T., Hall, C., Meyer, J. M. and Schlosberg, D. (eds.) (2016). *The Oxford Handbook of Environmental Political Theory*. Oxford: Oxford University Press.

Gail, W. B. (2016). "A New Dark Age Looms." *New York Times*, April 19. http://www.nytimes.com/2016/04/19/opinion/a-new-dark-age-looms.html

Galbraith, J. K. (1967). *The New Industrial State*. Boston: Houghton Mifflin.

Garside, N. (2013). *Democratic Ideals and the Politicization of Nature: The Roving Life of a Feral Citizen*. New York: Palgrave Macmillan.

Gastil, J. (2008). *Political Communication and Deliberation*. Los Angeles: Sage Publications.

Gastil, J. and Levine, P. (eds.) (2005). *The Deliberative Democracy Handbook: Strategies for Effective Civic Engagement in the 21st Century*. San Francisco: Jossey-Bass.

Gaventa, J. (2006). "Towards Participatory Governance," *Currents*, 29: 29–35.

Geels, F. W. and Schot, J. (2007). "Typologies of Socio-technical Transition Pathways," *Research Policy*, 36: 399–417.

Gellner, D. (ed.) (2007). *Resistance and the State: Nepalese Experiences*. Oxford: Berghahn.

Giddens, A. (1990). *The Consequences of Modernity*. Palo Alto: Stanford University Press.

Giddens, A. (2009). *The Politics of Climate Change*. London: Polity.

Gillis, J. (2014). "US Climate Has Already Changed, Study Finds, Citing Heat and Floods," *New York Times*, May 6. http://www.nytimes.com/2014/05/07/science/earth/climate-change-report.html?_r=0

Gillis, J. (2016). "Flooding of Coast, Caused by Warming, Has Already Begun," *New York Times*. http://www.nytimes.com/2016/09/04/science/flooding-of-coast-caused-by-global-warming-has-already-begun.html?_r=0

Gomes, R. C. and Secchi, L. (2015). "Public Administration in Brazil: Structure, Reforms, and Participation," in A. Massey and K. Johnston (eds.), *The International Handbook on Public Administration and Governance*. Cheltenham: Edgar Elgar, 225–41.

Goodin, R. (1992). *Green Political Theory*. Cambridge: Polity Press.

Gore, A. (1992). *Earth in the Balance: Ecology and the Human Spirit*. New York: Houghton Mifflin.

Gould, C. (2013). "Beyond the Dual Crisis: From Climate Change to Democratic Change." Humans and Nature. http://www.humansandnature.org/democracy-carol-gould

Green, J. (2015). "Nuclear Lobbyists' Epic COP21 Fail. Our Next Job? Keep Their Hands off Climate Funds." http://www.theecologist.org/News/news_analysis/2986693/nuclear_lobbyists_epic_cop21_fail_our_next_job_keep_their_hands_off_climate_funds.html

Greer, J. M. (2014). *Decline and Fall: The End of Empire and the Future of Democracy in 21st Century America*. Gabriola, Island, CA: New Society Publishers.

Griggs, S., Norval, A. and Wagenaar, H. (eds.) (2010). *Practices of Freedom: Decentred Governance, Conflict and Democratic Participation*. Cambridge: Cambridge University Press.

Grin, J., Rotmans, J. and Schot, J. (2010). *Transitions to Sustainable Development*. New York: Routledge.

Groenlund, K., Baechtiger, A. and Setälä, M. (2014). *Deliberative Mini-Publics: Involving Citizens in the Democratic Process*. Colchester, UK: ECPR Press.

Gundersen, A. G. (1995). "Bookchin: Our Marx?" Paper prepared for delivery at the Annual Meeting of the American Political Science Association, Chicago, IL, August 31–September 3.

Gutmann, A. and Thompson, D. (2004). *Why Deliberative Democracy?* Princeton: Princeton University Press.

Haas, P. (1992). "Introduction: Epistemic Communities and International Policy Coordination," *International Organization*, 46: 1–35.

Habermas, J. (1970). *Toward a Rational Society*. Boston: Beacon Press.

Hajer, M. and Dassen, T. (2014). *Smart About Cities: Visualizing the Challenge for 21st Century Urbanism*. The Hague: PBL Publishers.

Hajer, M. A. (1995). *The Politics of Environmental Discourse*. Oxford: Oxford University Press.

Hajer, M. A. and Wagenaar, H. (eds.) (2001). *Deliberative Policy Analysis: Understanding Governance in the Network Society*. Cambridge: Cambridge University Press.

Hansen, J. (2009). *Storms of My Grandchildren: The Truth About the Coming Climate Catastrophe and Our Last Chance to Save Humanity*. London, UK: Bloomsbury.

Hardin, G. (1968). "The Tragedy of the Commons," *Science*, 162 (3859): 1243–8.

Hardt, M. and Negri, A. (2000). *Empire*. Cambridge: Harvard University Press.

Hart, J. and Melle, U. (1998). "On Rudolf Bahro," *Democracy and Nature: The International Journal of Inclusive Democracy*, 4 (2/3). http://www.democracynature.org/vol4/hart_melle_bahro.htm

Harvey, D. (2010). *The Enigma of Capital and the Crisis of Capitalism*. London: Profile Books.

Harvey, D. (2012). *Rebel Cities: From the Right to the City to the Urban Revolution*. London: Verso.

Hayes, C. (2014). "The New Abolitionism: Averting Planetary Disaster Will Mean Forcing Fossil Fuel Companies to Give Up at Least $10 Trillion in Wealth," *The Nation*, 298 (19), May 22: 12–18.

Hayward, T. (2000). "Constitutional Environmental Rights: A Case for Political Analysis," *Political Studies*, 48 (3): 558–72.

Hayward, T. (2013). "Why Taking the Climate Challenge Seriously Means Taking Democracy More Seriously." Humans and Nature. http://www.humansandnature.org/democracy-tim-hayward

Hazen, S. (1997). "Environmental Democracy." http://www.ourplanet.com/imgversn/86/hazen.html

Heilbroner, R. (1974). *An Inquiry into the Human Prospect*. New York: Harper and Row.

Heinberg, R. (2007). *Peak Everything: Waking up to the Century of Declines*. Gabriola Island: New Society Publishers.

Heinelt, H. (2010). *Governing Modern Societies: Towards Participatory Governance*. London: Routledge.

Hendriks, C. (2005). "Consensus Conferences and Planning Cells: Lay Citizen Deliberations," in F. Gastil and P. Levine (eds.), *The Deliberative Democracy Handbook*. San Francisco: Jossey-Bass, 80–110.

Hendriks, C. M. (2016). "Coupling Citizens and Elites in Deliberative Systems: The Role of Institutional Design," *European Journal of Political Research*, 55: 43–60.

Heron, J. (1981). "Philosophical Basis for a New Paradigm," in P. Reason and J. Rowan (eds.), *Human Inquiry: A Sourcebook of New Paradigm Research*. Chichester, UK: John Wiley.

Hess, D. J. (2009). *Localist Movements in a Global Economy: Sustainability, Justice and Urban Development in the United States*. Cambridge, MA: MIT Press.

Hester, R. T. (2006). *Design for Ecological Democracy*. Cambridge: MIT Press.

Hickman, L. (2010). "James Lovelock: Humans are too Stupid to Prevent Climate Change," *The Guardian*. http://www.theguardian.com/science/2010/mar/29/james-lovelock-climate-change

Hines, C. (2000). *Localisation: A Global Manifesto*. London: Earthscan.

Hines, C. (2004). *A Global Look at the Local: Replacing Globalisation with Democratic Localisation*. London: International Institute for Environment and Development.

Hirst, P. (1989). *The Pluralist Theory of the State: Selected Writings of C. D. H. Cole, J. N. Figgis, and H. J. Laski*. London: Routledge.

Hirst, P. (1994). *Associative Democracy*. Cambridge: Polity Press.

Hirst, P. (1997). *From Statism to Pluralism*. London: University College Press.

Hiskes, A, N. and Hiskes, R. (1986). *Science, Technology and Policy Decisions*. Boulder: Westview Press.

Hobolt, S. (2012). "Citizen Satisfaction with Democracy in the European Union," *Journal of Common Market Studies*, 50: 88–105. http://ssrn.com/abstract=2002521

Holling, C. S. (1973). "Resilience and Stability of Ecological Systems," *Annual Review of Ecological Systems*, 4: 1–23.

Holling, C. S. (2004). "From Complex Regions to Complex Worlds," *Ecology and Society*, 9 (1). http://www.ecologyandsociety.org/vol9/iss1/art11/

Holmgren, D. (2009). *Future Scenarios: How Communities Can Adapt to Peak Oil and Climate Change*. White River Junction, VT: Chelsea Green Publishing.

Homer-Dixon, T. (1991). "On the Threshold: Environmental Changes as Causes of Acute Conflict," *International Security*, 16 (2): 76–116.

Homer-Dixon, T. (2006). *The Upside of Down: Catastrophe, Creativity and the Renewal of Civilization*. Washington, DC: Island Press.

Hopkins, R. (2008). *The Transition Handbook: From Oil Dependency to Local Resilience*. Cambridge: Green Publishing Co.

Hopkins, R. (2013). *The Power of Just Doing Stuff: How Local Action Can Change the World*. Cambridge: Green Books.

Horkheimer, M. and Adorno, T. W. (2002). *The Dialectic of Enlightenment*. Stanford: Stanford University Press.

Huesemann, M. and Huesemann, J. (2011). *Techno-Fix: Why Technology Won't Save Us or the Environment*. Gabriola Island: New Society Publishers.

Hull, J., Ojha, H. and Paudel, K. P. (2010). "Forest Inventory in Nepal: Technical Power or Social Empowerment?" in A. Lawrence (ed.), *Taking Stock of Nature: Participatory Biodiversity Assessment for Policy, Planning and Practice*. Cambridge: Cambridge University Press.

Hulme, M. (2009). *Why We Disagree About Climate Change: Understanding, Controversy, Inaction and Opportunity*. Cambridge: Cambridge University Press.

Hulme, M. (2014). *Can Science Fix Climate Change? A Case Against Climate Engineering*. Oxford: Polity Press.

Humphrey, M. (2007). *Ecological Politics and Democratic Theory: The Challenge of the Deliberative Ideal*. London: Routledge.

Hutt, M. (ed.) (2004). *Himalayan 'People's' War: Nepal Maoist Rebellion*. London: Hurst.

Illich, I. (1973). *Tools for Conviviality*. New York: Harper and Row.

Ingelhart, R. (1971). *The Silent Revolution: Changing Values and Political Styles among Western Publics*. Princeton, NJ: Princeton University Press.

Innis, M. (2016). "Scientists Protest Cuts and Commercialization at Australian Climate Center," *New York Times*, February 27. http://www.nytimes.com/2016/02/28/world/australia/cape-grim-climate-change-research.html?_r=0

Isaac, J. (1995). "The Strange Silence of Political Science," *Political Theory*, 23 (4): 636–52.

Isaac, T. H. with Franke, R. (2000). *Local Democracy and Development: People's Campaign for Decentralized Planning in India*. New Delhi: Left World Press.

Isaac, T. H. and Heller, P. (2003). "Democracy and Development: Decentralized Planning in Kerala," in A. Fung and E. O. Wright (eds.), *Deepening Democracy*. New York: Verso, 77–102.

James, C. (2009). "The Politics of Transition," *Arena Magazine*, 101: 18–19.

Jamieson, D. (2014). *Reason in a Dark Time: Why the Struggle Against Climate Change Failed—and What It Means for Our Future*. New York: Oxford University Press.

Jänicke, M. (1996). "Democracy as a Conditions for Environmental Policy Success: The Importance of Non-Institutional Factors," in W. M. Lafferty and J. Meadowcroft (eds.), *Democracy and the Environment: Problems and Issues*. Cheltenham: Edward Elgar, 71–85.

Jasanoff, S. (1990). *The Fifth Branch: Science Advisors as Policymakers*. Harvard: Harvard University Press.

Jessop, B. (2003). "Governance and Metagovernance: On Reflexivity, Requisite Variety and Requisite Irony," in H. P. Bang (ed.), *Governance as Social and Political Communication*. Manchester: Manchester University Press, 101–2.

Jessop, B. (2015). "The Symptomatology of Crises, Reading Crises and Learning from Them: Some Critical Realist Reflections," *Journal of Critical Realism*, 14 (3): 238–71.

Jha, P. (2015). *Battles of the New Republic: A Contemporary History of Nepal*. London: C. Hurst and Co.

Johnson, G. F. (2015). *Democratic Illusion: Deliberative Democracy in Canadian Public Policy*. Toronto: University of Toronto Press.

Joss, S. (1995). "Evaluating Consensus Conferences: Necessity or Luxury," in S. Joss and J. Durant (eds.), *Public Participation in Science: The Role of Consensus Conferences in Europe*. London: Science Museum, 89–108.

Joss, S. (2011). "Eco-Cities: The Mainstreaming of Urban Sustainability; Key Characteristics and Driving Factors," *International Journal of Sustainable Development Planning*, 6 (3): 268–85.

Joss, S. and Bellucci, S. (eds.) (2002). *Participatory Technology Assessment*. London: University of Westminster Centre for the Study of Democracy.

Joss, S. and Durant, J. (eds.) (1995). *Public Participation in Science: The Role of Consensus Conferences in Europe*. London: Science Museum.

Jowit, J. (2008). "Is Water the New Oil?" *Guardian*. http://www.theguardian.com/environment/2008/nov/02/water

Kaldor, M. (2005). "The Idea of Global Civil Society," in G. Baker and D. Chandler (eds.), *Global Civil Society: Contested Futures*. New York: Routledge, 103–12.

Kallis, G. (2015). "The Degrowth Alternative." http://www.greattransition.org/publication/the-degrowth-alternative#sthash.egCsrdZW.dpuf

Kates, R. W., Parris, T. M. and Leiserowitz, A. A. (2005). "What is Sustainable Development? Goals, Indicators, Values, and Practice," *Science and Policy for Sustainable Development*, 47 (3): 8–21.

Keane, J. (2009). *The Life and Death of Democracy*. London: Pocket Books.

Kenis, A. (2016). "Ecological Citizenship and Democracy: Communitarian Versus Agonistic Perspectives," *Environmental Politics*, 25 (6): 949–70.

Keulartz, J. (1999). "Engineering the Environment: The Politics of Nature Development," in F. Fischer and M. A. Hajer (eds.), *Living with Nature: Environmental Politics as Cultural Discourse*. Oxford: Oxford University Press, 83–102.

Khwaja, A. I. (2004). "Is Increasing Community Participation Always a Good Thing?" *Journal of European Economic Association*, 2: 2–3.

Kimani, N. (2010). "Participatory Assumptions of Environmental Governance in East Africa," *Law, Environment and Development Journal*, 6 (2): 202–15.

Klare, M. (2002). *Resource Wars: The New Landscape of Global Conflict*. New York: Holt.

Klein, N. (2014). *This Changes Everything: Capitalism vs. the Climate*. New York: Simon and Schuster.

Klijn, E. H. (2008). "Governance and Governance Networks in Europe. An Assessment of Ten Years of Research on the Theme," *Public Management Review*, 10: 505–25. http://dx.doi.org/10.1080/14719030802263954

Kluver, L. (1995). "Consensus Conferences at the Danish Board of Technology," in S. Joss and J. Durant (eds.), *Public Participation in Science*. London: Science Museum, 41–9.

Koehane, R. O. (2015). "The Global Politics of Climate Change: Challenge for Political Science," *PS*, 1: 19–26.

Kolbert, E. (2014). *The Sixth Extinction: An Unnatural History*. New York: Henry Holt and Co.

Korinek, R. and Straßheim, H. (2013). "Mehrebenenkoordination in der Krise: Der Fall EHEC" (Multi-level Coordination in Times of Crisis: The Case of EHEC), Paper for the annual meeting of the section on "Policy Analysis and Administrative Science" of the German Association for Political Science, University of Bamberg, March 1.

Koontz, T. (1999). "Citizen Participation: Conflicting Interests in State and National Agency Policy Making," *The Social Science Journal*, 36 (3): 441–58.

Koontz, T. M. and Thomas, C. W. (2009). "What Do We Know and Need to Know about the Outcomes of Collaborative Management?" *Public Administration Review*, 66: 111–21.

Krause, R. M. (2014). "Climate Policy Innovation in American Cities," in Y. Wolinsky-Nahmias (ed.), *Changing Climate Politics*. Los Angeles: Sage, 82–107.

Krauss, C., Myers, S. L., Revins, A. C. and Romero, S. (2005). "As Polar Ice Turns to Water, Dreams of Treasure Abound," *New York Times*, October 10. http://www.nytimes.com/2005/10/10/science/10arctic.html?pagewanted=all&_r=0

Kreuter, J. (2015). "Climate Engineering—A Technofix to Solve the Problem of Climate Change?" *KIB*. http://www.kib.be/articles/1103/climate-engineering-a-technofix-to-solve-the-problem-of-climate-change

Krugman, P. (2016). "What About the Planet?" *New York Times*, October 7. http://www.nytimes.com/2016/10/07/opinion/what-about-the-planet.html?_r=0

Krupp, F. (2014). "Don't Just Drill, Baby—Drill Carefully: How to Make Fracking Safer for the Environment," *Foreign Affairs*, 93 (3): 15–20.

Kunstler, J. (2006). *The Long Emergency: Surviving the Converging Catastrophes of the 21st Century*. London: Atlantic Books.

Kunze, I. (2003). "'Bildet Gemeinschaften—oder geht unter!' Eine Untersuchung selbstverwalteter, subsistenter Geminschaftsprojekte und Oekerdorfer in Deutschland—Modelle fuer eine zukunftsfaehige Lebensweise." Institut fuer Geographie, Universitaet Muenster, Diplomarbeit.

Kunze, I. (2009). *Soziale Innovationen fuer eine zukunfitige Lebensweise: Gemeindschaften und Oekerdorfer als experimentierend Lernfelder fuer sozial-oekologische Nachhaltigkeit.* Muenster: Eco-Transfer-Verlag.

Kunze, I. (2012). "Social Innovations for Communal and Ecological Living: Lessons from Sustainability Research and Observations in Intentional Communities," *Journal of the Communal Studies Association*, 32 (1): 50–67.

Kunze, I. (2015). "Ecovillages: Isolated Islands or Multipliers of Social Innovations?" Blog on website of TRANSIT EU-FP7 research project: http://www.transitsocialinnovation.eu/blog/ecovillages-isolated-islands-or-multipliers-of-social-innovations, October 9.

Kunze, I. and Avelino, F. (2015). Transformative social innovation narrative of the Global Ecovillage Network. https://www.academia.edu/15251014/Kunze_I._and_Avelino_F_2015_Transformative_social_innovation_narrative_of_the_Global_Ecovillage_Network._TRANSIT_EU_SSH.2013.3.2-1_Grant_agreement_no_613169

Kurzweil, R. (2005). *The Singularity is Near*. New York: Penguin Group.

Kütting, G. and Lipschutz, R. (eds.) (2009). *Environmental Governance: Power and Knowledge in a Local-Global World*. London: Routledge.

Laclau, E. and Mouffe, C. (2001). *Hegemony and Socialist Strategy: Towards a Radical Democratic Politics*. London: Verso.

Lafferty, W. M. and Meadowcroft, J. (eds.) (1996). *Democracy and the Environment: Problems and Prospects*. Cheltenham: Edward Elgar.

Laird, F. N. (1990). "Technocracy Revisited: Knowledge, Power, and the Crisis of Energy Decision-Making," *Industrial Crisis Quarterly*, 4 (1): 49–61.

Lalitpur, S. (2012). *Community Forest Management in Nepal*. Kathmandu: Forest Action.

Lama, M. S. (2011). *Forest Action Review: An Independent Review of a Decade of Forest Action in Nepal (2001–2010)*. Kathmandu: ForestAction.

Lasswell, H. D. (1941). "The Garrison State," *The American Journal of Sociology*, 46 (4): 455–68. http://www.jstor.org/stable/2769918

Lederer, M. and Kreuter, J. (2015). "'Crisis Talk' and the Wizards of Climate Engineering." Unpublished Paper. University of Münster.

Lemke, Thomas. (2004). "Foucault, Governmentality, and Critique," *Rethinking Marxism*, 14 (3): 49–64.

Lempinen, E. W. (2013). "Climate Change, Arctic Resource Boom Create New Challenges for US Policy and Diplomacy," *American Association for the Advancement of*

Science, January 14. http://www.aaas.org/news/climate-change-arctic-resource-boom-create-new-challenges-us-policy-and-diplomacy

Levi-Faur, D. (ed.) (2012). *The Oxford Handbook of Governance*. Oxford: Oxford University Press.

Lewis, M. W. (1992). *Green Delusions: An Environmentalist Critique of Radical Environmentalism*. Durham: Duke University Press.

Litfin, K. (2014). "Localism," in C. Death (ed.), *Critical Environmental Politics*. London: Routledge, 156–64.

Litfin, K. T. (2009). "Reinventing the Future: The Global Ecovillage Movement as Holistic Knowledge Community," in G. Kütting and R. Lipschutz (eds.), *Environmental Governance: Knowledge and Power in a Local-Global World*. London: Routledge: 124–42.

Litfin, K. T. (2013). *Ecovillages: Lessons for Sustainable Community*. Cambridge: Polity.

Loorbach, D. (2007). *Transition Management: New Mode of Governance for Sustainable Development*. Utrecht: International Books.

Lovelock, J. E. (1979). *Gaia: A New Look at Life on Earth*. Princeton: Princeton University Press.

Luke, T. W. (1997). *Ecocritique: Contesting the Politics of Nature, Economy, and Culture*. Minneapolis: University of Minnesota Press.

Luke, T. W. (1999a). "Eco-Managerialism: Environmental Studies as a Power/Knowledge Formation," in F. Fischer and M. Hajer (eds.), *Living with Nature: Environmental Politics as Cultural Discourse*. Oxford: Oxford University Press, 103–20.

Luke, T. W. (1999b). "Environmentality as Green Governmentality," in E. Darier (ed.), *Discourses of the Environment*. Malden, MA: Blackwell Publishers, 121–51.

Luke, T. W. (2009). "An Apparatus of Answers: Ecologism as Ideology in the 21st Century," *New Political Science*, 31 (4): 487–98.

Luke, T. W. (2011). "Environmentality," in J. S. Dryzek, R. B. Norgaard, and D. Schlosberg (eds.), *The Oxford Handbook of Climate Change and Society*. New York: Oxford University Press, 96–112.

Lundmark, C. (1998). "Eco-Democracy: A Green Challenge to Democratic Theory and Practice," *Research Report* 2, Umea, Sweden: Umea University.

Machin, A. (2012). "Decisions, Disagreements and Responsibility: Towards an Agonistic Green Citizenship," *Environmental Politics*, 21 (6): 847–63.

Madison, J. (1787 and 2003). "Federalist 10," in A. Hamilton, J. Madiso, and J. Jay (eds.), *The Federalist Papers*. New York: Signet.

Mahajan, Y. R. (2010). "Nanotechnologies to Mitigate Global Warming," *NanoWerk Spotlight*. http://www.nanowerk.com/spotlight/spotid=16126.php

Mair, P. (2013). *Ruling the Void: The Hollowing of Western Democracy*. London: Verso.

Majone, G. (2014). "Policy Harmonization: Limits and Alternatives," *Journal of Comparative Policy Analysis*, 16 (1): 4–21.

Malena, C. (ed.) (2009). *From Political Won't to Political Will: Building Support for Participatory Governance*. Sterling, VA: Kumarian Press.

Malette, S. (2009). "Foucault for the Next Century: Eco-governmentality," in S. Binkley, S. Capetillo, and J. Capetillo (eds.), *A Foucault for the 21st Century: Governmentality, Biopolitics and Discipline in the New Millennium*. Cambridge: Cambridge Scholars Publishing.

Mansbridge, J. (1999). "Everyday Talk in Deliberative Systems," in S. Macedo (ed.), *Deliberative Politics: Essays on Democracy and Disagreement*. New York: Oxford University Press, 211–39.

Mansuri, G. and Vijayendra, R. (2012). *Localizing Development: Does Participation Work?* Washington, DC: World Bank Research Report.

Marshall, T. H. (1950). *Citizenship and Social Class and Other Essays*. Cambridge: Cambridge University Press.

Martell, L. (1992). "New Ideas of Socialism," *Economy and Society*, 21 (2): 152–72.

Mason, M. (1999). *Environmental Democracy*. London: Earthscan.

Mathews, F. (ed.) (1996). *Ecology and Democracy*. London: Frank Cass.

Mathews, J. (1989). *Age of Democracy: The Politics of Post-Fordism*. Melbourne: Oxford University Press.

Mathiesen, K. (2015). "Extreme Weather Already on the Increase Due to Global Warming, Study Finds," *The Guardian*, April 27. http://www.theguardian.com/environment/2015/apr/27/extreme-weather-already-on-increase-due-to-climate-change-study-finds

May, T. (2010). *Contemporary Political Movements and the Thought of Jacques Rancière: Equality in Action*. Edinburgh: Edinburgh University Press.

McHugh, K. (2013). "Jimmy Carter: 'America No Longer has a Functioning Democracy'," *The Daily Caller*. http://dailycaller.com/2013/07/17/jimmy-carter-america-no-longer-has-a-functioning-democracy

McKibben, B. (2012). "Global Warming's Terrifying New Math," *Rolling Stone*. http://www.rollingstone.com/politics/news/global-warmings-terrifying-new-math-20120719

McKibben, B. (2016). "A World at War: We are Under Attack from Climate Change—and Our Only Hope is to Mobilize Like We Did in WW II," *New Republic*, August 15. https://newrepublic.com/article/135684/declare-war-climate-change-mobilize-wwii

Mead, D. (2012). "Is Bioengineering More Efficient Humans the Solution to Climate Change?" *Motherboard*. http://motherboard.vice.com/blog/is-bioengineering-more-efficient-humans-the-solution-to-climate-change

Meadowcroft, J. (2012). "Greening the State," in P. Steinberg and S. Van Deveer (eds.), *Comparative Environmental Politics*. Cambridge, MA: MIT Press, 56–88.

Meaney, T. and Mounk, Y. (2014). "What Was Democracy?" *The Nation*. http://www.thenation.com/article/what-was-democracy/

Mendonça, R. F. and Ercan, S. A. (2015). "Deliberation and Protest: Revealing the Deliberative Potential of Protest Movements in Brazil and Turkey," in F. Fischer, D. Torgerson, A. Durnova, and M. Orsini (eds.), *Handbook of Critical Policy Studies*. Cheltenham: Elgar Publishing, 205–21.

Met Office (2014). *A Global Perspective on the Recent Storms and Floods*. http://www.metoffice.gov.uk/research/news/2014/uk-storms-and-floods.

Middlemiss, L. (2014). "Individualised or Participatory? Exploring Late-modern Identity and Sustainable Development," *Environmental Politics*, 23 (69): 929–46.

Mill, J. S. (1861 and 2004). *Considerations on Representative Government*. Whitefish, MT: Kiesinger Publishing, LLC.

Milman, O. (2015). "James Hansen, Father of Climate Change Awareness, Calls Paris Talks 'a Fraud'," *The Guardian*. http://www.theguardian.com/environment/2015/dec/12/james-hansen-climate-change-paris-talks-fraud

Misar, D. (2005). *Participatory Governance Through NGOs*. Jaipur: Aalekh Publishers.

Mitchell, R. E. (2006). "Green Politics or Environmental Blues? Analyzing Ecological Democracy," *Public Understanding of Science*, 15: 459–80.

Mitchell, T. (2013). *Carbon Democracy: Political Power in the Age of Oil*. London: Verso.

Monbiot, G. (2013). "The End of Nuclear Power? Careful What You Wish For," *The Guardian*. https://www.theguardian.com/commentisfree/2013/feb/04/end-of-nuclear-careful-what-you-wish-for

Monbiot, G. (2015). "With This Attack on Community Energy the Big Six Win Out Over 'Big Society'," *The Guardian*, January 26. http://gu.com/p/456ve/sbl

Morrison, R. (1995). *Ecological Democracy*. Boston: South End Press.

Morse, E. L. (2014). "Welcome to the Revolution: Why Shale Is the Next Shale," *Foreign Affairs*, May/June. http://www.cfr.org/energy-and-environment/welcome-revolution/p32907

Moss, T. and Newig, J. (2010). "Multilevel Water Governance and Problems of Scale: Setting the Stage for a Broader Debate," *Environmental Management*, 46 (1): 1–6.

Mouffe, C. (2005). *On the Political*. London: Routledge.

Mougeot, L. (ed.) (2005). *Agropolis: The Social, Political and Environmental Dimensions of Urban Agriculture*. London: Earthscan.

Müller, J. (2016). "The EU's Democratic Deficit and the Public Sphere," *Current History*, 115 (79): 83–8.

Mutz, D. C. (2006). *Hearing the Other Side: Deliberative versus Participatory Democracy*. New York: Cambridge University Press.

Myers, N. (1989). "Environment and Security," *Foreign Policy*, 74: 23–41.

Myers, N. (1989). *Environmental Security: What's New and Different?* http://www.envirosecurity.org/conference/working/newanddifferent.pdf

Nadesan, M. (2016). *Crisis Communication, Liberal Democracy and Ecological Sustainability*. Lanham, MD: Rowman and Littlefield.

Nagel, S. (1979). "Policy Analysis Explosion," *Society*, 16 (5): 9–10.

Newig, J. (2007). "Does Public Participation in Environmental Decisions Lead to Improved Environmental Quality? Towards an Analytical Framework," *International Journal of Sustainability Communication*, 1 (1): 51–71.

Newig, J. and Fritsch, O. (2009). "Environmental Governance: Participatory, Multi-level and Effective?" *Environmental Policy and Governance*, 19: 197–214.

Newig, J. and Fritsch, O. (2009). "More Input—Better Output: Does Citizen Involvement Improve Environmental Governance?" in I. Blühdorn (ed.), *In Search of Legitimacy: Policy Making in Europe and the Challenge of Societal Complexity*. Farmington Hills: Opladen, 205–24.

Nightingale, A. and Ojha, H. (2013). "Rethinking the Power of Authority: Symbolic Violence and Subjectivity in Nepal's Terai Forest," *Development and Change*, 44 (1): 29–51.

Nightingale, A. (2005). "The Experts Taught Us All We Know: Professionalization and Knowledge in Nepalese Community Forestry," *Antipode*, 37 (3): 581–604.

Noble, C. (1987). "Economic Theory in Practice: White House Oversight of OSHA Health Standards," in F. Fischer and F. J. Forester (eds.), *Confronting Values in Policy Analysis: The Politics of Criteria*. Newbury Park: Sage, 266–84.

Norberg-Hodge, H. (2014). *Localization: Essential Steps to an Economics of Happiness*. Berkeley, CA: Local Futures.

Nordhaus, T. and Shellenberger, M. (2013). "Going Green? Then Go Nuclear," *Wall Street Journal*. http://www.wsj.com/articles/SB10001424127887323716304578482663491426312

North, P. (2010). "Eco-Localisation as a Progressive Response to Peak Oil and Climate Change—a Sympathetic Critique," *Geoform*, 41: 585–94.

Norton, A. and Stephens, T. (1995). *Participation in Poverty Assessment*. Environmental Department Papers Participation Series, Social Policy and Resettlement Division, Washington, DC: World Bank, 2: 11–176.

Novotny, P. (1994). "Popular Epidemiology and the Struggle for Community," *Capitalism, Nature and Society*, 5 (2): 29–42.

O'Brien, G. and O'Keefe, P. (2014). *Adaptation to Climate Risk: Beyond Fragmented Responses*. London: Routledge.

O'Neill, J. (1993). *Ecology, Policy and Politics: Human Well-Being and the Natural World*. New York: Routledge.

Offe, C. (2009). "Governance: An 'Empty Signifier'," *Constellations*, 16 (4): 550–62.

Ojha, H. (2002). *A Critical Assessment of Scientific and Political Aspects of the Issue of Community Forest Inventory in Nepal*. Kathmandu, Nepal: ForestAction.

Ojha, H. (2006). "Techno-bureaucratic Doxa and Challenges for Deliberative Governance: The Case of Community Forestry Policy and Practice in Nepal," *Policy and Society*, 5 (2): 11–176, 1–6.

Ojha, H. R. (2008). *Reframing Governance: Understanding Deliberative Politics in Nepal's Terai Forestry*. New Delhi: Adroit.

Ojha, H. R. (2009). "Civic Engagement and Deliberative Governance: The Case of Community Forest Users' Federation, Nepal," *Studies in Nepali History and Society*, 14 (2): 303–34.

Ojha, H. (2011). "The Evolution of Institutions for Multi-Level Governance of Forest Commons: The Case of Community Forest User Groups Federation in Nepal." Paper presented at the Workshop on Political Theory, Indiana University, Bloomington, Indiana.

Ojha, H. (2013). "Counteracting Hegemonic Powers in the Policy Process: Critical Action Research on Nepal's Forest Governance," *Critical Policy Studies*, 7: 242–62.

Ojha, H. R., Paudel, N. S., Banjade, M. R., McDougall, C. and Cameron, J. (2010). "The Deliberative Scientist: Integrating Science and Politics in Forest Resource Governance in Nepal," in L. German, L. Ramisch, and R. Verma, (eds.), *Beyond the Biophysical: Knowledge, Culture, and Politics in Agriculture and Natural Resource Management*. Heidelberg: Springer, 167–91.

Ojha, H. R., Paudel, N., Khatri, D. and Bk, D. (2012). "Can Policy Learning be Catalysed? Ban Chautari Experiment in Nepal's Forest Sector," *Journal of Forest and Livelihood*, 10 (1): 1–27.

Ojha, H. R., Timsina, N. P., Chhetri, R. B. and Paudel, K. P. (2008). *Knowledge Systems and Natural Resources: Management, Policy and Institutions in Nepal.* New Delhi: Cambridge University Press India.

Ojha, H. R. Timsina, N. P. and Khanal, D. R. (2007). "How are Forest Policy Decisions Made in Nepal?" *Journal of Forest and Livelihood,* 6 (1): 1–16.

Ophuls, W. (1977). *Ecology and the Politics of Scarcity.* San Francisco: Freeman.

Ophuls, W. and Boyan, A. S. (1992). *Ecology and the Politics of Scarcity Revisited: The Unraveling of the American Dream.* San Francisco: Freeman.

Oreskes, N. and Conway, E. M. (2010). *The Merchants of Doubt: How a Handful of Scientists Obscured the Truth on Issues from Tobacco Smoke to Climate Warming.* New York: Bloomsbury Press.

Oreskes, N. and Conway, E. M. (2014). *The Collapse of Western Civilization: A View from the Future.* New York: Columbia University Press.

Orr, D. (2009). *Down to the Wire: Confronting Climate Collapse.* Oxford: Oxford University Press.

Ortmann, S. and Thompson, M. R. (2016). "China and the Singapore Model," *The Journal of Democracy,* 1: 39–48.

Osmani, S. R. (2007). "Participatory Governance: An Overview of the Issues and Evidence," in S. R. Osmani (ed.), *United Nations, Participatory Governance and the Millennium Development Goals.* New York: United Nations: 1–48.

Ottinger, G. and Cohen, B. (eds.) (2011). *Technoscience and Environmental Justice: Expert Cultures in Grassroots Movements.* Cambridge: MIT Press.

Owen, D. and Smith, G. (2015). "Survey Article: Deliberation, Democracy and the Systemic Turn," *Journal of Political Philosophy,* 23 (2): 213–34.

Paehlke, R. (1995). "Environmental Values for a Sustainable Society: The Democratic Challenge," in F. Fischer and M. Black (eds.), *Greening Environmental Policy: The Politics of a Sustainable Future.* New York: St. Martins, 129–41.

Pahl, G. (2007). *The Citizens-Powered Energy Handbook: Community Solutions to a Global Crisis.* White Junction River, VT: Chelsea Green Publishing Company.

Pandey, N. P. (2010). *New Nepal: The Fault Lines.* London: Sage Publications.

Papadopoulus, Y. and Warin, P. (2007). "Are Innovative, Participatory and Deliberative Procedures in Policy Making Democratic and Effective?" *European Journal of Political Research,* 46: 445–72.

Paris Tech Review Editors (2013). "Ecodistricts: A Sustainable Utopia?" *Paris Tech Review.* www.paristechreview.com

Parkinson, J. and Mansbridge, J. (2012). *Deliberative Systems—Deliberative Democracy at the Large Scale.* Cambridge: Cambridge University Press.

Parr, A. (2009). *Hijacking Sustainability.* Cambridge: MIT Press.

Parr, A. (2012). *The Wrath of Capital: Neoliberalism and Climate Change Politics.* New York: Columbia University Press.

Pateman, C. (2012). "Participatory Democracy Revisited," *Perspectives on Politics,* 10 (1): 7–19.

Paterson, M. (2016). "Political Economy of the Greening of the State," in G. T. Hall, C. Meyer, and D. Schlosberg (eds.), *Oxford Handbook of Environmental Political Theory.* Oxford: Oxford University Press, 475–90.

Paudel, K., Green, K., Ojha, H. and Barnes, R. (2007). "Challenges to Participation: Lessons from Participatory Action Research with Community Forest User Groups in Nepal," *Journal of Forest and Livelihood*, 6 (1): 70–8.

Paudel, K. P. and Ojha, H. R. (2007). "From Imposed Indicators to Co-Creating Meanings in Nepal," in I. Guijt (ed.), *Negotiated Learning: Collaborative Learning in Resource Management*. Washington, DC: Resources for the Future, 54–78.

Paudel, N. S., Banjade, M. R. and Dahal, G. R. (2008). "Handover of Community Forestry: A Political Decision or a Technical Process?" *Journal of Forest and Livelihood*, 7 (2): 27–35.

Pettibone, L. (2015). *Governing Urban Sustainability: Comparing Cities in the USA and Germany*. Farnham: Ashgate.

Pilkington, E. (2007). "I Am Creating Artificial Life, Declares U.S. Gene Pioneer," *Guardian*. https://www.theguardian.com/science/2007/oct/06/genetics.climatechange

Piven, F. F. and Cloward, R. (1971). *Regulating the Poor: The Functions of Public Welfare*. New York: Pantheon.

Platteau, J. (2007). "Pitfalls of Participatory Development," *Participatory Governance and the Millennium Development Goals*. New York: United Nations, 127–62.

Pogrebinschi, T. (2013). "The Squared Circle of Participatory Democracy: Scaling Up Deliberation to the National Level," *Critical Policy Studies*, 7 (3): 219–41.

Pogrebinschi, T. and Samuels, D. (2012). "Can Participation Shape National Politics? An Empirical Answer for a Theoretical Question." Prepared for presentation at the 2012 Annual Meeting of the American Political Science Association, New Orleans, LA.

Pogrebinschi, T. and Samuels, D. (2014). "The Impact of Participatory Democracy: Evidence from Brazil's National Public Policy Conferences," *Comparative Politics*, 46 (3): 313–32.

Pokkharel, B., Branney, P., Nurse, M. and Malla, Y. B. (2007). "Community Forestry: Conserving Forests, Sustaining Livelihoods and Strengthening Democracy," *Journal of Forest and Livelihood*, 6 (2): 8–19.

Pollitt, C. (2008). *Time, Policy, Management: Governing with the Past*. Oxford: Oxford University Press.

Porritt, J. (1986). *Seeing Green*. Oxford: Blackwell.

Portney, K. E. (2003). *Taking Sustainable Cities Seriously: Economic Development, the Environment and Quality of Life in American Cities*. Cambridge: MIT Press.

Powell, G. (2013). "What is Environmental Citizenship?" http://authenticmeaning.wordpress.com/2013/10/17/environmental-citizenship

Price, A. (2012). *Recovering Bookchin: Social Ecology and the Crises of Our Time*. Porsgrunn, Norway: New Compass Press.

Przeworski, A. (2010). *Democracy and the Limits of Self-Government*. Cambridge: Cambridge University Press.

Purcell, M. (2008). *Recapturing Democracy: Neoliberalization and the Struggle for Alternative Urban Futures*. New York: Routledge.

Purdy, J. S. (2010). "The Politics of Nature: Climate Change, Environmental Law and Democracy," *Yale Law Review*, 119: 101–85.

Putnam, R. (1997). "Elite Transformations in Advanced Industrial Societies: An Empirical Assessment of the Theory of Technocracy," *Comparative Political Studies*, 10 (3): 385–7.

Putnam, R. (2000). *Bowling Alone: The Collapse and Revival of the American Community*. New York: Simon and Schuster.

Rahman, M. D. A. (1995). *People's Self-Development*. London: Zed Books.

Rancière, J. (1995). *On the Shores of Politics*. London: Verso.

Rancière, J. (2001). "Ten Theses on Politics," translated by D. Bowlby. *Theory and Event*, 5 (3): 1–16.

Rancière, J. (2015). *Dissensus: On Politics and Aesthetics*. London: Bloomsbury Publishing.

Randers, J. (2012). *2052*. White River Junction, VT: Chelsea Green.

Rao, V. and Mansuri, G. (2012). *When Do Participatory Development Projects Work*. World Bank Policy Research Report. Washington, DC: World Bank.

Raskin, P., Banuri, T., Gallopin, G., Gutman, P., Hammond, A., Kates, R. and Swart, R. (2002). *The Great Transition: The Promise and Lure of Times Ahead*. Stockholm: Stockholm Environment Institute.

Reason, P. (1994). "Three Approaches to Participatory Inquiry," in N. K. Denzin and Y. S. Lincoln (eds.), *Handbook of Qualitative Research*. London: Sage Publications, 324–39.

Reed, M. G. and Bruyneel, S. (2010). "Rescaling Environmental Governance, Rethinking the State: A Three-dimensional Review," *Progress in Human Geography*, 34 (5): 646–53. http://doi.org/10.1177/0309132509354836

Rees, C. (ed.) (2016). *The Sky is the Limit: Why the Paris Climate Goals Require a Managed Decline of Fossil Fuel Production*. Washington, DC: Oil Change International.

Rees, M. (2003). *Our Final Hour: A Scientist's Warning—How Terror, Error, and Environmental Disaster Threaten Humankind's Future in this Century—On Earth and Beyond*. New York: Basic Books.

Revkin, A. C. (2007). "Scientists Report Severe Retreat of Arctic Ice," *New York Times*, September 21. http://www.nytimes.com/2007/09/21/science/21arctic.html

Revkin, A. C. (2010). "The Idea That We're Going to Fix the Climate Change Problem or Solve Global Warming Has Always Been a Fantasy, Totally Wishful, From My Standpoint," *New York Times*, March 21. http://thinkprogress.org/climate/2010/03/21/205679/andrew-revkin-dotearth-global-warming/

Rhodes, R. A. W. (2012). "Waves of Governance," in D. Levi-Faur (ed.), *The Oxford Handbook of Governance*. Oxford: Oxford University Press, 33–48.

Rochloff, S. F. and Moore, S. A. (2006). "Assessing Representativeness at Different Scales of Decision-Making: Rethinking Local is Better," *Policy Studies Journal*, 34 (4): 649–70.

Rosenberg, S. W. (2003). *The Not so Common Sense: Differences in How People Judge Social and Political Life*. New Haven: Yale University Press.

Rothstein, H. (2013). "Domesticating Participation: Participation and the Institutional Rationalities of Science-Based Policy-Making in the UK Food Standards Agency," *Journal of Risk Research*, 16 (6): 771–90.

Rousseau, J. J. (1968). *Social Contract* (translated by Maurice Cranston). London: Penguin.

Rowe, G. and Frewer, L. J. (2004). "Evaluating Public Participation Exercises: A Research Agenda," *Science, Technology and Human Values*, 29 (2): 512–57.

Rowe, G. and Frewer, L. J. (2005). "A Typology of Public Engagement Mechanisms," *Science, Technology and Human Values*, 30: 251–90.

Royal Commission on Environmental Pollution (1998). *Setting Environmental Standards*. London: Royal Commission on Environmental Pollution.

Rutherford, P. (1999). "The Entry of Life Into History," in E. Darier (ed.), *Discourses of the Environment*. Malden, MA: Blackwell Publishers, 37–62.

Sabel, C. F. and Zeitlin, J. (2011). "Experimentalist Governance," in D. Levi-Faur (ed.), *The Oxford Handbook of Governance*, Oxford: Oxford University Press, 169–83.

Sale, K. (1980). *Human Scale*. New York: Coward, McCann & Geoghegan.

Sale, K. (2000). *Dweller in the Land: The Bioregional Vision*. Athens, GA: University of Georgia Press.

Salih, M. A. (2013). "Introduction," in M. A. Salih (ed.), *Local Climate Change and Society*. London: Routledge, 1–3.

Sanger, R. H. (1967). *Insurgent Era: New Patterns of Political, Economic, and Social Revolution*. Washington, DC: Potomac Books.

Sartori, G. (1987). *The Theory of Democracy Revisited: The Contemporary Debate*. Chatham, NJ: Chatham House.

Satterswaite, D., Saleemul, H., Reid, H., Pelling, M. and Lankao, P. (2009). "Adapting to Climate Change in Urban Areas: The Possibilities and Constraints in Low- and Middle-Income Nations," in A. Bicknell, D. Dodman, and D. Satterthwaite (eds.), *Adapting Cities to Climate Change: Understanding and Addressing the Development Challenges*. London: Earthscan, 3–47.

Scarse, I. and Smith, A. (2012). "The (Non-) Politics of Managing Low Carbon Socio-Technical Transitions," in H. Compston (ed.), *Climate Change and Political Strategy*. New York: Routledge, 49–68.

Schattschneider, E. E. (1960). *The Semi-Sovereign People: A Realist's View of Democracy in America*. New York: Holt, Rinehart and Winston.

Schlosberg, D. (1999). *Environmental Justice and the New Pluralism*. Oxford: Oxford University Press.

Schlosberg, D. and Coles, R. (2015). "The New Environmentalism of Everyday Life: Sustainability, Material Flows, and Movements," *Contemporary Political Theory*, 15. http://link.springer.com/article/10.1057%2Fcpt.2015.34.

Schumacher, E. F. (1973). *Small is Beautiful: A Study of Economics as if People Mattered*. London: Blond and Briggs.

Schwartz, P. and Randall, D. (2003). "An Abrupt Climate Change Scenario and Its Implications for United States National Security," October. http://www.oilcrisis.com/globalwarming/abruptclimatechange_schwartzrandall.htm

Schwartzmantel, J. (2012). "Insurgent Democracy." *Philosophers for Change*. https://philosophersforchange.org/2012/05/22/insurgent-democracy/

Sclove, R. (2010). *Reinventing Technology Assessment: A 21st Century Model*. Washington, DC: Wilson Center.

Scott, J. C. (1998). *Seeing Like a State: How Certain Schemes To Improve the Human Condition Have Failed*. New Haven: Yale University Press.

Security Council (2007). "Security Council Holds First-Ever Debate on Impact of Climate Change on Peace, Security, Hearing Over 50 Speakers," *Press Release*, April 17. New York: United Nations.

Shah, S. (2008). "Civil Society in Uncivil Places: Soft State and Regime Change in Nepal," *Policy Studies*, 48. Washington, DC: East-West Center.

Shaha, R. (2001). *Modern Nepal: A Political History*. Delhi: Manohar Publishers and Distributors.

Shearman, D. and Smith, J. W. (2007). *The Climate Change Challenge and the Failure of Democracy*. Westport, CT: Greenwood Publishing.

Shelton, D. (1991). "Human Rights, Environmental Rights and the Rights to Environment," *Stanford Journal of International Law*, 28 (39): 103–38.

Sheper-Hughes, N. and Courgois, P. (2004). "Introduction: Making Sense of Violence," in N. Sheper-Hughes and P. Bourgois (eds.), *Violence in War and Peace: An Anthology*. Malden, MA: Blackwell Publishing.

Sheridan, A. (1980). *Michel Foucault: The Will to Power*. London: Tavistock Publications.

Shiva, V. (2005). *Earth Democracy: Justice, Sustainability and Peace*. London: Zed Books.

Shove, E. and Walker, G. (2007). "CAUTION! Transitions Ahead: Politics, Practices and Sustainable Transition Management," *Environment and Planning A*, 39: 763–70.

Siller, P. (2010). *Demokratie und Klimawandel: Ökologen als Vordenker einer Expertokratie?* Heinrich Böll Stiftung. https://www.boell.de/de/demokratie/akademie-postdemokratie-expertokratie-8729.html

Skelcher, C. (2010). "Fishing in Muddy Waters: Principles, Agents, and Democratic Governance in Europe," *Journal of Public Administration Research*, 20 (1): 161–75.

Smith, A. and Kern, F. (2009). "The Transition Storyline in Dutch Environmental Policy," *Environmental Politics*, 18 (1): 78–98.

Smith, G. (2003). *Deliberative Democracy and the Environment*. London: Routledge.

Sorensen, E. and Torfing, J. (2007). *Theories of Democratic Network Governance*. Basingstoke: Palgrave Macmillan.

Sousa Santos, B. de (ed.) (2005). *Democratizing Democracy: Beyond the Liberal Canon*. London: Verso.

Sousa Santos, B. de (ed.) (2007). *Another Knowledge is Possible: Beyond Northern Epistemologies*. London: Verso.

Staeck, N., Malek, T. and Heinelt, H. (2001). "The Environmental Impact Assessment Directive," in H. Heinelt, T. Malek, R. Smith, and A. E. Töller (eds.), *European Union Environment Policy and New Norms of Governance*. Aldershot: Ashgate, 33–42.

Starn, R. (1971). "Historians and Crisis, Past & Present," *A Journal of Historical Studies*, 52: 3–22.

Stehr, N. (2016). "Exceptional Circumstances: Does Climate Change Trump Democracy," *Issues in Science and Technology*, Winter. https://www.academia.edu/20201986/Exceptional_circumstances_Does_climatechange_trump_democracy

Stein, T. (1998). *Demokratie und Verfassung an den Grenzen des Wachstums*. Bochum: Westdeutscher Verlag.

Stern, N. (2006). *The Stern Review on the Economics of Climate Change*. Cambridge: Cambridge University Press.

Stern, N. (2015). *Why Are We Waiting? The Logic, Urgency, and Promise of Tackling Climate Change*. Cambridge: MIT Press.

Stevenson, H. and Dryzek, J. S. (2014). *Democratizing Global Climate Governance*. Cambridge: Cambridge University Press.

Stewart, F. (1991). "Two Conceptions of Citizenship," *The British Journal of Sociology*, 46 (1): 63–78.

Stirling, A. (2009). "Participation, Precaution and Reflexive Governance for Sustainable Development," in W. N. Adgers and A. Jordan (eds.), *Governing Sustainability*. Cambridge: Cambridge University Press, 193–225.

Stirling, A. (2011). "Innovation, Sustainability, Development: A New Manifesto: Implications for New Practices in Research and Innovation Governance." Steps Centre, University of Sussex, UK. http://www.unesco.org/new/fileadmin/MULTIMEDIA/HQ/SC/pdf/SC_PSD_Innovation_Sustainability_Development_a_New_Manifesto_(STIGAP)_en.pdf

Stoker, G. (2011). "Was Local Governance Such a Good Idea? A Global Comparative Perspective," *Public Administration*, 89 (1): 15–31.

Stone, D. (2002). *Policy Paradox: The Art of Political Decision-Making*. New York: W.W. Norton.

Strassheim, H. (2015). "Politics and Policy Expertise: Towards a Political Epistemology," in F. Fischer, D. Torgerson, A. Durnova, and M. Orsini (eds.), *Handbook of Critical Policy Studies*. Cheltenham: Edward Elgar Publishing, 319–40.

Strassheim, H. and Ulbricht, T. (eds.) (2015). "Die Zeit der Politik–Ansätze einer Neuorientierung," in H. Strassheim and T. Ulbricht (eds.), *Zeit der Politik*. Baden-Baden: Nomos Verlagsgesellschaft mbH & Co. KG: Sonderband 30: 8–25.

Streeck, W. (1995). "Inclusion and Secession: Questions on the Boundaries of Associative Democracy," in E. O. Wright (ed.), *Associations and Democracy*. London: Verso, 184–92.

Streeck, W. (2014). *Buying Time: The Delayed Crisis of Democratic Capitalism*. London: Verso.

Stromberg, J. (2013). "What is the Antropocene and Are We In It?" *Smithsonian Magazine*, January. http://www.smithsonianmag.com/science-nature/what-is-the-anthropocene-and-are-we-in-it-164801414/?no-ist

Sugden, P. (2011). "Introduction: Mapping the Trajectory and Dynamics of the Nepal's 'Transition,'" *Nepal Journal of Social Science and Public Policy*, 1 (1): 1–16.

Sunam, R. Danjade, M. R., Paudel, N. S. and Kathri, D. B. (2010). "Can Bureaucratic Control Improve Community Forestry Governance? Analysis of Proposed Forest Act Amendment," *Discussion Paper Series* 10: 2. Nepal: ForestAction.

Sunstein, C. R. (2004). "Cost-Benefit Analysis and the Environment, *Law & Economics*," University of Chicago: Olin Working Paper No. 227. http://ssrn.com/abstract=604581 or http://dx.doi.org/10.2139/ssrn.604581

Sutherlin, J. W. and Willson, D. K. (2012). "Better Environmental Governance: A Function of Democracy or Authoritarianism?" *International Journal of Business, Humanities and Technology*, 2 (2): 173.

Swyngedouw, E. (2007). "Impossible 'Sustainability' and the Post-Political Condition," in M. Cerreta, G. Concilio, and V. Monno (eds.), *Making Strategies in Spatial Planning*. Heidelberg: Springer, 9: 185–205.

Swyngedouw, E. and Wilson, J. (eds.) (2014). *The Post-Political and its Discontents: Spaces of De-politicization, Specters of Re-Politicization*. Edinburgh: Edinburgh University Press.

Tageszeitung (2015). "Wie Geil War Das Denn?" December 15: 3.

Tandon, R. (1988). "Social Transformation and Participatory Research," *Convergence* 21: 5–18.

Taylor, B. P. (1996). "Democracy and Environmental Ethics," in W. M. Lafferty and J. Meadowcroft (eds.), *Democracy and the Environment: Problems and Issues*. Cheltenham: Edward Elgar, 86–107.

Taylor, P. and Buttel, F. H. (1992). "How Do We Know We Have Global Environmental Problems? Science and the Globalization of Environmental Discourse," *Geoforum*, 23 (3): 405–16.

Thapa, D. with Sijapati, B. (2003). *A Kingdom Under Siege: Nepal's Maoist Insurgency, 1996 to 2003*. Kathmandu: The Printhouse.

Thomas, W. (1995). *Scorched Earth: The Military's Assault on the Environment*. Gabriola Island, CA: New Society Publishers.

Thompson, M. R. and Ortmann, S. (2012). "China's Obsession with Singapore: Learning Authoritarian Modernity." Unpublished paper. Hong Kong University.

Thompson, P. (1993). *The Work of William Morris*. Oxford: Oxford University Press.

Thoreau, H. D. (1854). "Resistance to Civil Government." http://www.transcendentalists.com/slavery_in_ma.htm

Thorpe, C. and Jacobson, B. (2013). "Life Politics, Nature and the State: Giddens' Sociological Theory and the Politics of Climate Change," *The British Journal of Sociology*, 64 (1): 99–122.

Timsina, N. P., Luintel, H., Bhandari, K. and Thapaliyam, A. (2004). "Action and Learning: An Approach for Facilitating a Change in Knowledge and Power Relationships in Community Forestry," *Journal of Forest and Livelihood*, 4: 5–12.

Timsina, N. P., Ojha, H. and Paudel, K. P. (2004). "Deliberative Governance and Public Sphere: A Reflection on Nepal's Community Forestry 1998–2004," Fourth National Workshop on Community Forestry, Nepal, Kathmandu, Department of Forestry, Nepal. Vol. 6.

Torfing, J., Peters, G. B., Pierre, J. and Sorensen, E. (2012). *Interactive Governance: Advancing the Paradigm*. Oxford: Oxford University Press.

Torgerson, D. (1995). "The Uncertain Quest for Sustainability: Public Discourse and the Politics of Environmentalism," in F. Fischer and M. Black (eds.), *Greening Environmental Policy: Toward a Sustainable Future*. New York: St. Martins Press, 3–20.

Torgerson, D. (1999). *The Promise of Green Politics: Environmentalism and the Public Sphere*. Durham, NC: Duke University Press.
Torgerson, D. (2001). "Rethinking Politics for a Green Economy: A Political Approach to Radical Reform," *Social Policy and Administration*, 35 (5): 427–89.
Trainer, T. (2014). "Transition Townspeople, We Need To Think About Transition: Just Doing Stuff Is Far From Enough!" http://leipzig.degrowth.org/en/2014/08/transition-townspeople-we-need-to-think-about-transition-just-doing-stuff-is-far-from-enough/
Turner, T. (2010). *The Transition to a Sustainable and Just World*. Sydney: Envirobook.
Turner, T. (2010). "The Transition Towns Movement: Its Huge Significance and a Friendly Criticism." http://www.culturechange.org/cms/content/view/605/1/
United Nations Security Council (2007). "Security Council Holds First-Ever Debate on Impact of Climate." *Press Release*. http://www.un.org/press/en/2007/sc9000.doc.htm
United Nations University (2011). "Former National Leaders: Water a Global Security Issue." http://unu.edu/media-relations/releases/water-called-a-global-security-issue.html
United Nations World Water Assessment Programme (2009). *Climate and Water*. Special Report. France: UNESCO-WWAP 2009.
Urry, J. (2011). *Climate Change and Society*. London: Polity.
van Steenbergen, B. (1994). "Towards a Global Ecological Citizen," in B. van Steenbergen (ed.), *The Conditions of Citizenship*. London: Sage, 141–52.
Vaidyanathan, G. (2015). "Nuclear Power Must Make a Comeback for Climate's Sake," *Scientific American*. http://www.scientificamerican.com/article/nuclear-power-must-make-a-comeback-for-climate-s-sake/
Victor, D. G., Granger, M., Apt, J., Steinbruner, J. and Ricke, K (2009). "The Geoengineering Option: A Last Resort Against Global Warming," *Foreign Affairs*, March/April: 64–76.
Vinge, V. (1993). "The Coming Technological Singularity: How to Survive in the Post-Human Era," *Interdisciplinary Science and Engineering in the Era of Cyberspace*. NASA Publication CP-10129: 11–22.
Wagle, U. (2006). "Political Participation and Civic Engagement in Kathmandu: An Empirical Analysis with Structural Equations," *International Journal of Political Science*, 7 (3): 301–22.
Walker, G., Devine-Wright, P., Hunter, S., High, H. and Evans, B. (2010). "Trust and Community: Exploring the Meanings, Contexts and Dynamics of Community Renewable Energy," *Energy Policy*, 38 (6): 2655–63.
Walker, B., Holling, C. S., Carpenter, S. R. and Kinzig, A. (2004). "Resilience, Adaptability and Transformability in Social–Ecological Systems," *Ecology and Society*, 9 (2): 5.
Wall, C. (2008). "Ban Ki-moon Warns of the Coming Water Wars," *FP: The Magazine*, January 24. http://foreignpolicy.com/2008/01/24/ban-ki-moon-warns-of-the-coming-water-wars/
Wampler, B. (2009). *Participatory Budgeting in Brazil*. University Park: Penn State Press.
Warren, M. E. (2007). "Institutionalizing Deliberative Democracy," in S. Rosenberg (ed.), *Deliberation, Participation and Democracy: Can the People Govern?* New York: Palgrave MacMillan, 272–88.
Watts, M. and Peet, R. (eds.) (2004). *Liberation Ecologies: Environment, Development, and Social Movements*. New York: Routledge.

Weaver, P. J., Larson, L., van Grootveld, G., van Spiegel, E. and Vergragt, P. (2000). *Sustainable Technology Development*. Sheffield: Greenleaf Publishing.

Webb, D. (2007). "Thinking the Worst: The Pentagon Report," in D. Cromwell and M. Levene (eds.), *Surviving Climate Change: The Struggle to Avert Global Catastrophe*. London: Pluto Press, 59–81.

Weber, I. (2015). *Sozialismus in der DDR: Alternative Gesellschaftskonzepte von Robert Havermann und Rudolf Bahro*. Berlin: Christoph Links Verlag: 229–300.

Weingart, P. (2001). *Die Stunde der Wahrheit? Zum Verhältnis der Wissenschaft zu Politik, Wirtschaft und Medien in der Wissensgesellschaft*. Weilerswist: Velbrück Verlag.

Welzer, H. (2012). *Climate Wars: Why People Will Be Killed in the Twenty-First Century*. Cambridge: Polity.

Welzer, H. and Leggewie, C. (2009). *Das Ende der Welt, Wie Wir Sie Kannten: Klima, Zukunft und die Chancen der Demokratie*. Frankfurt am Main: S. Fischer Verlag.

Westall, A. (2011). "Time to Revisit Associative Democracy," *Open Democracy*. https://www.opendemocracy.net/andrea-westall/time-to-revisit-associative-democracy

Westra, L. (1993). "The Ethics of Holism and the Democratic State: Are They in Conflict?" *Environmental Values*, 2 (2): 125–36.

Westra, L. (1998). *Living in Integrity: A Global Ethic to Restore a Fragmented Earth*. Lanham, MD: Rowman and Littlefield.

Whelpton, J. (2007). *A History of Nepal*. Cambridge: Cambridge University Press.

White, D. F. (2003). "Hierarchy, Domination, Nature: Considering Bookchin's Social Theory," *Organization and Environment*, 16 (1): 34–65.

Williams, B. A. and Matheny, A. R. (1995). *Democracy, Dialogue and Environmental Disputes*. New Haven: Yale University Press.

Willis, J. (2015). *Locating Localization: Statecraft, Citizenship and Democracy*. Bristol: Policy Press.

Wilson, H. (2006). "Environmental Democracy and the Green State," *Polity*, 38 (2): 276–94.

Wolin, S. S. (1994). "Fugitive Democracy," *Constellations*, 1 (1): 11–25.

Wolin, S. S. (2008). *Democracy Inc*. Princeton: Princeton University Press.

Wood, E. J. (2003). *Insurgent Collective action and Civil War in El Salvador*. Cambridge: Cambridge University Press.

Woody, T. (2009). "Stewart Brand's Strange Trip: Whole Earth to Nuclear Power," *Environment 360*. http://e360.yale.edu/feature/stewart_brands_strange_trip_whole_earth_to_nuclear_power/2227/

Worker, J. and Ratté, S. (2014). "What Does Environmental Democracy Look Like?" World Resources Institute. http://www.wri.org/blog/2014/07/what-does-environmental-democracy-look

World Bank (1994). *The World Bank and Participation*, Operations Policy Department. World Bank. September.

World Bank (2008). "Promoting Active Citizenry—Advocacy and Participation Decision-Making," *Guidance Notes on Tools for Pollution Management*. http://siteresources.worldbank.org/intranetenvironment/Resources/GuidanceNoteonAdvocacyandDecisionMaking.pdf

World Commission on Environment and Development (1987). *Our Common Future.* http://www.un-documents.net/wced-ocf.htm

Wright, E. O. (1995). *Associations and Democracy.* London: Verso.

Young, S. (ed.) (2000). *The Emergence of Ecological Modernization: Integrating the Environment and the Economy?* London: Routledge.

Zahariadis, N. (2015). "Plato's Receptacle: Deadlines, Ambiguity, and Temporal Sorting in Public Policy," in H. Straßheim and T. Ulbricht (eds.), "Die Zeit der Politik: Democratisches Regieren in einer Beschlungten Welt," *Leviathan Sonderband,* 30: 113–31.

Zakaria, F. (2008). "A Bug to Save the Planet," *Newsweek,* June 16. https://www.highbeam.com/doc/1G1-179877307.html

INDEX

INDEX

ACKNOWLEDGEMENTS

This book is an outcome of a two-year worldwide collaboration of scholars and practitioners, experts in their fields and in their knowledge of the relevant aspects of the Syrian war. It is also a journey between disciplines, people, and perspectives that do not usually reside in a shared space and a partnership between Israelis and Syrians; lawyers and historians; academic scholars and opposition activists – all of whom deserve acknowledgement for their efforts and patience following a long process of editing and reviews.

We are extremely grateful to the scholars and activists who authored chapters in this book and whose insights provide enriching views of the conflict. We would also like to thank those who have commented on chapters throughout the editing process and guided us with modelling the book's overall framework: Professor Yuval Shany, Professor Uzi Rabbi, Dr. Brandon Friedman, and Professor Dalia Ofer. Special thanks are due to Ms. Liane Silver, for linguistic and content editing, and to Mr. Avraham Spraragen, for assisting in preparing the index.

Additional thanks are due to several institutions: The Moshe Dayan Center for Middle East and African Studies at Tel Aviv University, for academic and financial support, and the Academic College for Law and Science. We also thank Finola O'Sullivan and Marianne Nield from Cambridge University Press, who managed an efficient and seamless process that enabled this book to go to print.

Last but not least are our families, whom we would like to thank for their support, interest, and emotional involvement in the project. It is very much due to their encouragement throughout the years that this project finally sees the light of the day.

We conclude with a final remark. Although primarily a collaborative academic project, this work spreads beyond a purely academic framework. Headed by three individuals – a law professor in academia (Dr. Hilly Moodrick-Even Khen); a scholar and humanitarian activist who participated in the cross-border humanitarian work in Syria (Dr. Nir T. Boms);

and a scholar who is also an international criminal lawyer (Ms. Sareta Ashraph) – we hope that this project will leave a mark beyond the academic field. It is our wish that this academic research of the eight years of war in Syria will contribute not only to the learning and understanding of this conflict but also to a worldwide humanitarian effort to avoid and prevent the horrors of war.